Water Policy, Imagination Innovation

T0228159

This book explores creative interdisciplinary and potentially transformative solutions to the current stalemate in contemporary water policy design. A more open policy conversation about water than exists at present is proposed – one that provides a space for the role of the imagination and is inclusive – of the arts and humanities, relevant stakeholders, including landholders and Indigenous peoples, as well as science, law and economics.

Written for a wide audience, including practitioners and professional readers, as well as scholars and students, the book demonstrates the value of multiple disciplines, voices, perspectives, knowledges and different ways of relating to water. It provides a fresh and timely response to the urgent need for water policy that works to achieve sustainability, and may be better able to resolve complex environmental, social and cultural water issues. Utilising a broad range of evidentiary sources and case studies from Australia, New Zealand, Canada and elsewhere, the authors of this edited collection demonstrate how new ways of thinking and imagining water are not only possible but already practised, and growing in saliency and impact. The current dominance of narrower ways of conceptualising our relationship with water is critiqued, including market valuation and water privatisation, and more innovative alternatives are described, including those that recognise the importance of place-based stories and narratives, adopt traditional ecological knowledge and relational water appreciations, and apply cutting-edge behavioural and ecological systems science.

The book highlights how innovative approaches drawing on a wide range of views may counter prevailing policy myopia, enable reflexive governance and transform water policy towards addressing water security questions and the broader challenges posed by the Anthropocene and the UN Sustainable Development Goals.

Robyn Bartel is Associate Professor in Geography and Planning at the University of New England, Australia.

Louise Noble is Adjunct Senior Lecturer in English at the University of New England, Australia.

Jacqueline Williams is Senior Research Fellow at the Australian Centre for Agriculture and Law, University of New England, Australia.

Stephen Harris is Lecturer in English at the University of New England, Australia.

Earthscan Studies in Water Resource Management

For more information and to view forthcoming titles in this series, please visit the Routledge website: http://www.routledge.com/books/series/ECWRM/.

Water Policy, Imagination and Innovation

Interdisciplinary Approaches

Edited by Robyn Bartel, Louise Noble,
Jacqueline Williams and
Stephen Harris

LONDON AND NEW YORK

from Routledge

First published 2018 by Routledge

2 Park Square, Milton Park, Abingdon, Oxfordshire OX14 4RN

52 Vanderbilt Avenue, New York, NY 10017

Routledge is an imprint of the Taylor & Francis Group, an informa business

First issued in paperback 2019

British Library Cataloguing-in-Publication Data
A catalogue record for this book is available from the British Library

Library of Congress Cataloging-in-Publication Data
A catalog record for this book has been requested

ISBN: 978-1-138-72937-7 (hbk)
ISBN: 978-0-367-35227-1 (pbk)

Typeset in Goudy
by Swales & Willis Ltd, Exeter, Devon, UK

Contents

Figures

Tables

Contributors

Lorina L. Barker
University of New England (UNE)
Lorina Barker is a descendant of the Wangkumara and Muruwari people from northwest NSW, Adnyamathanha (Flinders Rangers SA), the Kooma and Kunja (southwest QLD) and the Kurnu-Baarkandji (northwest NSW). Lorina is an oral historian and filmmaker and teaches modern Australian history, Oral history and Local and Community Histories. Lorina uses multimedia as part of her community art-based projects to transfer knowledge, history, stories and culture to the next generations in mediums that they use and are familiar with, such as film, short stories, poetry and music. She wrote and directed the short film documentary *Tibooburra: My Grandmother's Country*.

Robyn Bartel
University of New England (UNE)
Associate Professor Robyn Bartel is a multi-award winning scholar with wide-ranging expertise in geography, law and education. Known internationally for her contribution to legal geography, her research encompasses regulation, regulatory agencies and the regulated, as well as the social, institutional and natural landscape in which all are situated. Her work has been influential in environmental policy development, heavily cited in the scholarly literature and handpicked for prestigious international collections and seminal texts in environmental law. Robyn is a founding member of AELERT, the Australasian Environmental Law Enforcement and Regulators network, as well as UNE's Water Research and Innovation Network (WRaIN).

Wendy Beck
University of New England (UNE)
Wendy Beck is an Adjunct Associate Professor in Archaeology at the University of New England. Her publications and research interests include plants in Archaeology, Place Studies in Archaeology and Teaching and Learning in Higher Education. She has recently retired from full time university teaching, but is an Australian Teaching and Learning Fellow. Much of her fieldwork has been carried out within the Murray-Darling hydrological Basin in eastern Australia,

and she is currently working in its upper reaches, researching the Aboriginal history of wetlands. She has a longstanding interest in transdisciplinary research projects.

Liz Charpleix
University of New England (UNE)
After growing up in Tasmania and Western Australia, Liz Charpleix worked in clerical, service and hospitality jobs around Australia, New Zealand and Europe. For 18 years until 2016 she has operated a Tasmania-based tax and public accounting business. Along the way, she gained tertiary qualifications in arts, fine arts and accounting. Drawing upon this experience and education, Liz is currently completing a PhD, researching how water is, and could be, valued, in economic and non-economic ways, with a particular focus on Indigenous understandings of water and place agency in Aotearoa/New Zealand and Australia.

Aaron B. Driver
Content Logic
Aaron B. Driver trained as journalist at Australia's oldest newspaper company, Fairfax Ltd, and then as an editor at the world's largest business publisher, Reed Elsevier. He has worked in politics, as a speechwriter and press secretary, and began his own consultancy in 1996. For the past 21 years he has advised on communications strategy for Australian governments, local councils, universities and dozens of blue chip companies. Aaron graduated from leading journalism school, CSU Mitchell, and completed his Master of Fine Arts in Creative Writing from the University of British Columbia. He is currently undertaking a PhD in climate change communications.

Michael Allen Fox
University of New England (UNE)/Queen's University, Canada
Michael Allen Fox received his PhD in Philosophy from the University of Toronto. He is Adjunct Professor, School of Humanities, University of New England, Australia, and Professor Emeritus of Philosophy, Queen's University, Canada. He has published six books, the latest of which is *Home: A Very Short Introduction* (Oxford University Press, 2016), and articles on a wide range of topics in nineteenth-century European philosophy, philosophy of peace, existentialism, environmental philosophy, and the theory of vegetarianism. These include essays in *Environmental Ethics*, *Ethics and the Environment*, *Organization and Environment*, *Ecosystem Health*, *International Journal of Applied Philosophy* and other journals and anthologies.

Stephen Harris
University of New England (UNE)
Dr Harris is a lecturer in the field of literary and cultural studies, with particular interests in American Literature and contemporary fiction. He has published books on the work of Gore Vidal and the historical novel in American culture, plus numerous articles and reviews. He also regularly collaborates on creative practice productions (music and theatre). His recent research focuses

on the relationship between literature and the environment, with a focus on eco-critical themes in Australian Literature. He is also a member of the inter-disciplinary research WRaIN (Water Research and Innovation Network) at the University of New England (UNE).

Donald W. Hine
University of New England (UNE)
Donald Hine is Professor of Psychology at the University of New England. He conducts research in the areas of environmental psychology and behaviourally effective communications. His work focuses on understanding situational and psychological factors that underlie environmental problems such as resource over-consumption, climate change, air pollution and the management of inva-sive species. Much of his work involves developing behaviour change strategies to encourage people to act in ways that benefit the common good.

Graham Marshall
University of New England (UNE)
Dr Marshall is an institutional economist at the University of New England, Australia, working primarily on problems of collective action in environmen-tal and natural resources management. He has authored the Earthscan book *Economics for Collaborative Environmental Management: Renegotiating the Commons* and a range of articles and chapters, including on governance of water resources in the Murray-Darling Basin. He is a Principal Research Fellow in the Institute for Rural Futures and a member of the Water Research and Innovation Network (WRaIN) at UNE.

Lynette McLeod
University of New England (UNE)
Lynette McLeod has just completed a PhD at University of New England investigating the links between human behaviour and participation in inva-sive and companion animal management. Previously she spent 20 years with NSW Department of Primary Industries working with primary producers and communities on invasive animal management projects and ecological research. Research interests include environmental psychology and philoso-phy, design of communication and behaviour change strategies and animal management.

Louise Noble
University of New England (UNE)/Queensland University of Technology (QUT)
Dr Noble is an Adjunct Senior Lecturer at the University of New England, Australia and Research Coordinator in the School of Justice at the Queensland University of Technology. Her recent publications consider the early mod-ern English ecological consciousness as manifested in literature and other works such as agricultural and scientific texts. Her current research takes the concept of the hydro-social cycle as a model for mapping the transference of the hydrological imagination across time and space, and how this imaginary

continues to influence water management practices and policies today. Louise is a member of WRaIN, UNE's Water Research and Innovation Network.

Melissa Parsons
University of New England (UNE)
Melissa Parsons is a river scientist with broad-ranging and interdisciplinary research interests in river and floodplain resilience, natural hazards, resilience assessment, river policy and management, river monitoring and assessment, large flood disturbances and river ecology. Melissa works at the interface between theoretical and applied science, examining the ways that concepts such as resilience can be applied to deliver management and policy outcomes. Melissa's current research is focused on the assessment of disaster resilience and river resilience.

Patricia Please
Griffith University/University of New England (UNE)
Patricia (Patty) Please migrated to Australia in 1988 and has spent most of the years since then working on various aspects of land and water issues in this country. During her career she has worked as an environmental psycho-social researcher, hydrogeologist, petroleum geologist, psychotherapist and policy analyst on water issues. Her more recent research position focused on the application of behaviour change practices and principles to promote community participation in pest animal management. She gets her greatest joy from work that involves exploring the subjective experience of a diverse range of individuals dealing with complex land and water issues – particularly the affective-emotional dimension of their experience.

Darren S. Ryder
University of New England (UNE)
Darren Ryder is an Associate Professor of Aquatic Ecology and Restoration in the School of Environmental and Rural Science at the University of New England, Armidale NSW. In this role he teaches units in freshwater ecology, management and restoration. When not teaching he reverts to life as a field ecologist with research interests in the effective management and restoration of freshwater ecosystems in rural landscapes. Darren's research focuses on community and ecosystem ecology, and has fostered an ambition to provide a scientific evidence-base for the management of water resources to sustain both ecological and rural communities.

Mark Shepheard
University of New England (UNE)
Dr Mark Shepheard has researched water law and natural resource management at the University of New England (UNE) 2014–17; Banting Fellow at McGill University 2012–13; Oxford University 2013; University of Lucerne 2011; and Lincoln University (NZ) 2010. Dr Shepheard is recipient (with Dr Bettina Lange, Oxford) of the Richard Macrory Prize for the best article in the Journal of Environmental Law during 2014. A focal point for Dr Shepheard's research is the stewardship responsibilities associated with property in natural resources.

This sits at the intersection of private rights and public interests in defining clearer obligations to achieve behaviour change.

Richard Stayner
University of New England (UNE)
Richard Stayner is Adjunct Research Fellow in the Institute for Rural Futures, University of New England. He is an agricultural and natural resource economist who has worked for over 30 years on issues concerning the changing economic and social condition of Australian primary industries and rural regions. His research has included several projects on the implications of water policy for the resilience of Australian rural industries and communities.

A. J. Walsh
University of New England (UNE)
A. J. Walsh is a Political Philosopher interested in a wide variety of normative issues concerning markets and environmental practice. He has written extensively on the ethics of marketisation and on the normative foundations of economic theory. He also works on questions of philosophical methodology and, in particular, on the use of thought experiments in ethical and political theory. He has published four books including *The Morality of Money* (2008) and *Ethics, Money and Sport* (2007). Most recently he was a co-editor of the collection entitled *The Ethical Underpinnings of Climate Economics* (2016). He is currently working on a book on Water and Distributive Justice.

Jacqueline Williams
University of New England (UNE)
Dr Jacqueline Williams is an Environmental Scientist with over 25 years applied experience in Natural Resource Management (NRM) in Australia having worked at local, regional, state and national levels encompassing sustainable agriculture, farm forestry, catchment management and community-based NRM. Her research work within the Australian Centre for Agriculture and Law at the University of New England has encompassed working extensively on water issues in the Northern Murray-Darling Basin, including the development of methods to analyze natural resource policy risks. More recently her research has focused on the next generation of rural landscape governance for Australia.

Chapter overviews

Chapter 1: Blue sky thinking in water governance: Understanding the role of the imagination in Australian water policy

Louise Noble, Stephen Harris and Graham Marshall

This chapter considers how the literary imagination has influenced cultural attitudes to water in the settler colonial context where an arid environment posed enormous imaginary challenges for those from regions with predictable rainfall patterns. By way of example, we focus on Australia, the driest inhabited continent on earth, and Dorothea Mackellar's 'My Country', to examine how understanding the relationship between literature and Australia's hydrological imagination might influence innovative, flexible water governance systems that are open to collaborative and creative strategies: in other words, to blue sky thinking. The European history of water management in Australia is primarily a history of the hydrological imagination. Understanding the shared imaginings or mental models that have set Australia down our current path of water governance enables us to interrogate and challenge our adherence to inherited ways of thinking that contribute to path dependence. Path dependence severely constrains the range of policy reform options available. We argue that how Australians imagine water is most coherently expressed in the poetry and fiction of the nation, and show how a poem such as 'My Country' both reflects and challenges outmoded ideas to prompt new ways of thinking, thus bringing fresh scrutiny to the broader practice of policy formulation.

Chapter 2: Aboriginal Rainmakers: A twentieth century phenomenon

Lorina L. Barker

Rainmaking ceremonies have been practiced for millennia in Australia, and these ceremonies and their associated rain objects are imbued with Aboriginal peoples' deep social, cultural and spiritual knowledge of the land and water. Australia is the driest inhabited continent in the world and subject to extreme periods of drought, and since colonisation people have had an overwhelming thirst for water. This aroused a frenzied desire to 'witness' rain ceremonies, and the outsiders went to tremendous lengths to arrange and transport the Rainmakers and rain objects across the country. This chapter focuses on the outsiders' fascination with Aboriginal customs and how they described and wrote about the Rainmakers, rain ceremonies and objects during the first half of the twentieth century. It examines both urban and regional Australian newspaper articles between the 1920s and 1950s, from the curious and often cynical outsider perspective to later writers interested in the Rainmakers' adventures in the 1960s, to those in the 1970s. Today, there are those who lament the passing of the Rainmakers and the perceived loss of water knowledge, which further emphasises the outsiders' uneasy relationship with water. As it was then, and still today, Aboriginal water knowledge and related ceremonies are important elements of Aboriginal culture and customs, and water is central to all life in Australia.

Chapter 3: 'Like manna from heaven?': Just water, history and the philosophical justification of water property rights

A. J. Walsh

Robert Nozick famously asserted in *Anarchy State and Utopia* (1974) that if goods fell 'like manna from heaven' then it would be acceptable to treat them as part of a collective pool to which some patterned principle of distributive justice might be applied. But, Nozick says, they do not and instead have a history, being formed by human hands, and, furthermore, that history must be central in determining what counts as a just set of holdings. Thus, any philosophical justification of property rights cannot disregard history. In this chapter, I explore *the role of history* in determining the legitimacy of property rights in water. Can there be a history of water, in the productive Nozickian sense, which underwrites ownership of particular parcels of water? Should a distinction be drawn between a history of production and a history of use and stewardship? Where might traditional Indigenous rights fit within this Nozickian schema? Conversely, what implications the case of water might have for historically based theories of distributive justice will also be considered. As well as addressing these philosophical claims, the paper lays out a number of fundamental conceptual features of theories of distributive justice that are essential when considering how water should be allocated.

Chapter 4: Progressing from experience-based to evidence-based water resource management: Exploring the use of 'best available science' to integrate science and policy

Darren S. Ryder

Current water resource management is dominated by an experience-based practice where management actions and policy development decisions are based on familiarity and institutional norms because time, personnel and scientific resources are limited. In contrast, evidence-based practice requires management actions and policy development based on the latest and objective information collected using a hypothesis-driven scientific method to remove the subjectivity inherent in many experience-based approaches. The move away from an experience-based mode of decision-making in water resources management requires the collaboration of scientists and policy decision-makers to provide an objective evidence-based approach that overcomes decisions that may be viewed as arbitrary, biased or politically motivated. A number of contentious resource management issues worldwide have led to the emergence of 'best available science' (BAS) as a concept to prioritise the use of evidence-based scientific information for resource management decisions. In most instances, the statutes of environmental law and policy requiring BAS have proceeded without a clear definition, often leaving legal interpretations of BAS to define what is, and what is not, best available science. In this chapter the principles that underpin BAS and its application are explored as an evidence base for water resource policy and management decisions. Environmental flows as a policy-science-management initiative implemented throughout the world is used as a case study to identify how BAS has contributed to contentious water resource decisions. The challenges and opportunities inherent in the BAS approach are identified to enhance the role of science in collaborative water resource management decisions.

Chapter 5: Accounting for water: From past practices to future possibilities

Liz Charpleix

Effective, equitable and sustainable management of the world's waters is becoming both increasingly necessary and difficult. Accurate accounting for the quantity and quality of the waters that exist on Earth is often considered an essential input to decision-making for the management of these waters. Determining how much water is available is one issue; another is how to value it for aspects beyond its quantity. Western capitalist approaches to valuing, recording and managing water may accentuate rather than resolve water management challenges such as climate change, Indigenous self-determination, expectations of constantly improving living standards and ever-increasing population. Resourcist paradigms package water as a commodity, to be valued only in financial terms, and position humans and the environment as consumers who must pay for access to water. Western capitalist approaches also adopt exclusionary and privileging perspectives, as well as practices informed by Eurocentrism and anthropocentrism, which together obstruct equitable water allocation and sustainability.

The aim of this chapter is to review and critique ontological conceptualisations and institutional arrangements that measure water as an economic commodity, as a relational substance and for its intrinsic qualities. A discussion about the ontology of water is the first step towards identifying its incommensurable values, which is necessary because primarily economic approaches do not attend adequately to the needs of natural features, processes and organisms such as wetlands and aquifers, animals, plants and the climate. We are living in what has been called the Anthropocene epoch, a term that refers to the impact on the environment of human activities. One among many responsibilities this places on humans is to ensure the sustainability of the Earth's water. In order to ensure the future of human and other life, the non-economic needs of the environment must be accommodated.

Chapter 6: Rethinking the value of water: Stewardship, sustainability and a better future

Michael Allen Fox

While awe-inspiring natural water features (such as waterfalls and wild lakes) are universally considered worthy of preserving and protecting, water as such is valued instrumentally and, in the absence of scarcity, generally taken for granted in everyday life and often considered by many kinds of users to be virtually free for the taking. From this anthropocentric and instrumental perspective, water is merely a resource that promotes human activities and aspirations. But a deeper understanding of both the properties of water and the value of nature directs attention toward the need for more intelligent and innovative decisions that will shape our common future in relation to this most precious life-supporting substance. Sustainability is often cited as a desirable governing principle for water use. This chapter makes a plea for viewing sustainability from within a reimagined philosophical perspective of environmental stewardship, and ends with some confronting thoughts about the future of water use and policy.

Chapter 7: Stewardship arrangements for water: An evaluation of reasonable use in sustainable catchment or watershed management systems

Mark Shepheard

Catchment or watershed management planning is adopted in many countries to help facilitate the implementation of sustainability into resource use practice. This chapter uses case studies from New Zealand, Switzerland, England and Quebec to evaluate whether catchment or watershed planning is effective in defining resource use rights and responsibilities for more sustainable performance.Stewardship accountability is described and further developed by the case studies as a model for motivating resource user appreciation of what is reasonable practice. The case studies demonstrate how existing arrangements for sustainable catchment or watershed management anticipate a reimagined standard of reasonable resource access and use for more sustainable catchment or watershed systems. Four dimensions of stewardship accountability are derived from the author's research on liability of natural resource users for environmental harm. These are: norms of practice, limits on exploitative freedoms, legitimacy and trust. Together these dimensions provide a basis from which to redefine underlying notions of what is reasonable natural resource use in the context of a sustainable catchment or watershed. These four dimensions are used to establish a link between strategic level statements about resource use, formal and informal resource use rules, landscape systems and the social systems in which natural resource users are accountable. Stewardship accountability is more likely to deliver resource use performance that meets the outcomes sought by collaborative catchment or watershed management plans, by redefining reasonable resource use in the context of sustainability.

Chapter 8: Water knowledge systems

Jacqueline Williams, Patricia Please and Lorina L. Barker

Current water governance systems place humans at the centre or 'above' nature, with water largely considered a commodity for human exploitation. This failure to recognise our dependence upon Earth as a source of life is reinforced by globalisation, which perpetuates unsustainable models of water management. In this chapter three researchers from very different disciplinary and cultural backgrounds explore Traditional Ecological Knowledge Systems (TEKS), a term used to describe Indigenous and other forms of traditional knowledge about sustainable water management. This chapter investigates the barriers to incorporating traditional knowledge systems into water governance models, as well as the benefits of empowering and recognising the importance of this knowledge. To achieve this, the researchers adopted a decolonising narrative approach through applying an Indigenous research framework to understanding traditional water knowledge. Incorporating an Indigenous research framework transformed the questions – instead of presuming that water is an economically defined commodity, the researchers asked: What is water knowledge and what forms of water knowledge exist? How do we go about finding out about them? What capacities do we need to develop within ourselves to engage with them? What are the social justice issues contained within water knowledge systems? How can traditional water keepers self-determine within colonised systems of water governance? Most importantly, how can traditional water knowledge be recognised and enabled within the current system to produce sustaining water governance models? With these questions at the centre of the inquiry, the authors are able to bring particular cultural and disciplinary approaches to these issues encompassing cross-cultural and interdisciplinary approaches, offering different insights into the barriers and opportunities presented by TEKS.

Chapter 9: Water policy for resilient agri-environmental landscapes: Lessons from the Australian experience

Richard Stayner and Melissa Parsons

The agricultural use of Australia's water resources following European settlement was driven by successive nation-building visions that imagined the natural environment as controllable and the future as a stable state. An inadequate understanding of a complex natural environment with high climatic variability resulted in mistakes that had high economic, social and ecological costs. Seeing water use in agriculture as a complex social-ecological system that needs to be managed on the basis of resilience principles provides lessons for water policy that might avoid repeating past mistakes. Key innovations in Australian water policy and management over the past 20 years have reflected some resilience principles, but remain an unfinished project. The perceived potential for further development of water resources in northern Australia offers an opportunity to apply resilience principles and so avoid the path dependence that constrains the implementation of recent policy innovation in the south. These principles continue to compete, however, with a frontier rhetoric and an inadequate recognition of inevitable climatic and economic shocks.

Chapter 10: Waterworks: Developing behaviourally effective policies to manage household water use

Donald W. Hine, Lynette McLeod and Aaron B. Driver

Australia is the driest inhabited continent and has made significant efforts to reduce water usage – but still has one of the highest per capita water consumption rates in the world. This chapter demonstrates how principles and methods from the social sciences can help formulate and deliver 'behaviourally effective' government policy to reduce household water use. We review six behavioural theories relevant to water conservation before introducing a framework for selecting high-impact target behaviours, evaluating drivers and barriers to behaviour change, and identifying the most appropriate behaviour-change interventions and policy delivery strategies. The framework provides policy makers and frontline practitioners with a common mental model for conceptualising and communicating about behaviour change, and an evidence-based foundation for launching more effective water conservation interventions and policies.

Chapter 11: Quixotic water policy and the prudence of place-based voices

Robyn Bartel, Louise Noble and Wendy Beck

This chapter charts several deficiencies in water policy identified through an examination of practices and paradigms of water management measured against the backdrop of environmental features and vernacular understandings of water. Frequently ill-matched, to both biophysical conditions and cultural appreciations emanating from place-based experiences, such management practices and paradigms include: raising water supply to meet growing demand, instead of matching demand to natural supply; governing water according to political rather than catchment boundaries; deploying government instead of governance; and the application of resourcist frames employing both commoditisation and commensuration. The trajectory of water policy in this Australian case study appears to be one of quixotic resilience – of resilient colonial hopes and dreams of abundance – which have been almost impervious to all evidence to the contrary. Irrigation schemes encouraged by government subsidies and an over-ambitious programme for 'making the desert bloom' provide one of the more potent examples of what may happen when material reality meets human hope and fallibility. Often absent from water policy are the views and knowledges of the peoples and communities who live in the landscape – this is a critical missing ingredient which must be more fully appreciated to ensure better outcomes.

Chapter 12: Heterotic water policy futures using place agency, vernacular knowledge, transformative learning and syncretic governance

Robyn Bartel, Louise Noble and Wendy Beck

The onset of the Anthropocene signals that the resilience of Earth systems and environments is becoming less forgiving of anthropocentrism and anthroparchy. In response, iterative reform of mainstream water policy has been attempted, albeit through simple and incrementalist learning cycles rather than the more fundamental and transformative change demanded by the circumstances. Transformation requires deeper questioning of assumptions, a process that may be assisted by the incorporation of alternative perspectives. Indigenous ontologies and knowledges recognise the agency of water, and local water stories generally demonstrate greater prudence and relationality than more dominant constructions. Similarly, envirosocial, hydrosocial and social-ecological system-based approaches may be used to re-connect water and humanity within place and the wider landscape. Transformative learning processes utilising different knowledge types and sources at multiple scales can be used to build syncretic governance systems combining a plurality of perspectives and approaches to build a heterotic water future. Syncretic governance would facilitate the incorporation of multiple ontologies, rather than relying on current monadic frames. A heterotic water policy future would exhibit a greater diversity and dynamism than in the past – more akin to the environment that it reflects and is intended to support.

Preface

Robyn Bartel, Louise Noble, Jacqueline Williams and Stephen Harris

The contributors to this work are members of a dedicated interdisciplinary research initiative, the Water Research and Innovation Network (WRaIN), at the University of New England in New South Wales, Australia.

The University of New England is located in a nation where water is of critical import – in the driest inhabited continent on Earth, which is frequently drought-stricken. The majority of authors live and work in the regional town of Armidale, situated on the drainage divide that bisects Eastern Australia from north to south – cleaving the continent unequally in two. To the east are the coastal lowlands, comparatively well watered and highly populated, with fiercely contested proposals for impoundments, water recycling and desalination plants. To the west lies the Murray-Darling Basin and the vast, relatively unpopulated tablelands, with numerous conflicts between agricultural, mining and environmental water users and contentious proposals to redistribute water from the east, or move agriculture to, or the water from, the tropical north of the continent.

Thus contestation and divisions over water issues are at the forefront of our minds, including those between academic disciplines and related ontological engagements with water. The aim of WRaIN was to bring all disciplines together in deep conversation about an alternative future for water policy.

This book is the result.

Introduction

Water policy and the Anthropocene

Stephen Harris, Robyn Bartel, Jacqueline Williams and Louise Noble

In observance of 'World Water Day' 2017, the World Water Council called on all governments to prioritise global water security (World Water Council, 2017). The World Economic Forum's 'Global Risk Report 2017' has declared water crises a 'societal risk' ranking within the top three in the high-impact category for the third consecutive year (World Economic Forum, 2017). Two billion people are affected by contaminated drinking water (World Health Organization, 2017) and the World Health Organization (2017) has identified that 'countries are not increasing spending fast enough to meet the water and sanitation targets under the Sustainable Development Goals (SDGs)'. Both surface and groundwater resources are in decline, with the United Nations predicting a shortfall by 2030, and there are concerns of 'water wars' in high risk countries resulting from the slowing of economic growth, food price spikes and increasing human migration (The National Geographic, 2016). Water conflicts are very much in the political psyche 'as much as oil shaped the global geopolitics of the 20th century, water has the power to reorder international relations in the current century' (Engelke & Sticklor, 2015). As we enter the 'International Decade of Action: Water for Sustainable Development 2018–2028' (United Nations, 2017), humanity is challenged by a critical juncture:

> What we do in the next 50 years will determine the outcome for the next 10,000 years. We are that generation right at that tipping point. We were alive at that exponential journey that took us here, we will probably be alive in the journey that will decide the outcome for the next 10,000 years (Rockström, 2017).

It is critical that future water policy achieves the maintenance of our environmental support systems and provides 'the safe operating space to be successful for a thriving humanity' (Rockström, 2017). We define policy broadly, after Schad (1991), as 'the total of all actions taken by individuals and entities in managing water resources' (p. 14). Schad (1991) was referring to national water policy, and several nations have taken bold and innovative steps. Israel has been identified as a water superpower, due to its development of ground-breaking technological approaches to water scarcity (Seigel, 2015). Other nations have explored institutional options, including according special legal status to rivers such as the Ganges River in India (The New Indian Express, 2017) and the Whanganui River in New Zealand (New Zealand Herald, 2017; see also Charpleix, this volume). Slovenia recently amended its constitution to include the right to water for all citizens, the first country in the European Union to do so (International Federation of Health and Human Rights Organizations, 2016). Yet, at the same time as such leadership is being demonstrated, so too is wholescale regression of environmental laws and protections in countries such as the USA and Australia (Riley, 2015; Percival, 2017).

With notable prescience Jonathan Ralston Saul (2005) described this space of uneven policy development as humanity struggling with the paradox of the 'chaotic vacuum' left due to the collapse of globalism and attempts 'to reinvent the world' within this vacuum (p. 58). We must now also face up to the challenges of the Anthropocene, an epoch defined by the twinned anthropogenic threats to the survival of the human species as well as the Earth's biomes: '"we", the human species, [have] unconsciously destroyed nature to the point of hijacking the Earth system into a new geological epoch' (Bonneuil & Fressoz, 2016, p. xii). The radical disruption of the hydrological cycle is one example:

> The modification of the continental water cycle is massive, with the draining of half the planet's wetlands and the construction of 45,000 dams with heights of more than fifteen metres, together retaining 6,500 cubic kilometres of water, some 15 per cent of the total flow of the world's rivers. These transformations have substantially modified the processes of erosion and sedimentation, without however freeing the greater part of humanity from water insecurity. (p. 8)

There is, therefore, growing urgency to the call to 'rethink our visions of the world and our ways of inhabiting the Earth', an approach which involves 'forging new narratives for the Anthropocene and thus new imaginaries' (Bonneuil & Fressoz, 2016, p. xiii). The anthropocentric celebration of humanity's capacity to adapt and innovate may not be commensurate to the gravity of the issues, and to accept such a challenge is necessarily to provoke a range of questions. Indeed, the title of this book anticipates the more immediate of such queries: how might the imagination work in relation to the ostensibly practical, and politically pragmatic, activity of creating water policy? What, given the connotative valency of the word 'imagination', is actually being denoted in this instance; and how, if at all, might such imaginative approaches be accommodated within bureaucratic systems and institutionally organized processes? The question as to what constitutes 'good' or sound policy is always one of scale and attenuating time frames: how does a given government contrive to fashion 'good' policy – in this case, water policy – while at once accommodating the broad needs and demands of diverse communities, observing the rules of best practice deemed appropriate for liberal democracies and incorporating the growing knowledge and understanding of the delicate dynamics that inhere in the earth system on which these same growing polities rely?

Approaches to the framing and implementation of water policy have become contorted, and increasingly reductive, through habituated reliance on practised interpretations of knowledge as motivated by competing political influences of an increasingly populist temperament. The problems afflicting policy formulation do not generally stem from a paucity of knowledge or information *in itself* – since, in the broad scientific understanding of the properties and bio-spherical dynamics of water, knowledge and understanding continues to develop. Rather, it is the manner in which such knowledge is interpreted, selectively used and 'strategically' applied in policy practice. While it is certainly true that there are recent attempts to address such shortcomings, the solutions have proven to be inadequate, or at least insufficient in relation to the forbiddingly complex task at hand. At large, this tendency describes the general narrowing of the creative lens through which such problems can be approached – a steady attenu-ation of the imaginative reflexes that serve best to encourage innovation and productive adaptation.

As this volume shows, water and related social and environmental issues are often complex and dynamic, and inter-disciplinarity proffers the much-needed combination and diversity of views required to comprehend, as well as ques-tion and critique, such features (Newell, 2001; and see also Karpouzoglou & Vij, 2017). Inter-disciplinarity can also provide the advantage of 'fresh eyes' – offering new and creative insights, and often also much-needed correctives – to narrow disciplinary gazes, as well as being able to fill the gaps that exist between the fields covered by traditional disciplines (Nissani, 1991).

Interdisciplinary approaches

It is being increasingly recognized that environmental problems, and per-haps particularly those involving fresh water – often referred to in popular discourse as 'the lifeblood of the planet' (Khadka, 2010) – require interdisci-plinary approaches (see for example Pahl-Wostl et al., 2013). As Krueger et al. (2016) have observed: 'An interdisciplinary approach to water research has long been considered indispensable for understanding the multifaceted issues surrounding water' (p. 370). Stember (1991) defines 'interdisciplinary' as the *integration* of one or more disciplines, 'multidisciplinary' as the *combination* of one or more disciplines and 'transdisciplinary' as *going beyond* the traditional disciplines.[1] Nissani (1991, p. 212) has observed that inter-disciplinarity may on occasion suffer from over-generalisation, which is where multi-disciplinarity approaches – that preserve the expertise of each antecedent discipline – may assist. By contrast, trans-disciplinarity is increasingly being viewed as neces-sary where current ontologies and epistemologies are proving inadequate for the task. Transdisciplinary approaches may go beyond knowledge generated by mainstream academic disciplines, and traditional methodologies for knowledge creation, to include the wider community and general public, and encompass the co-creation of new knowledges through collaborative processes (see for example Krueger et al., 2016).

In this volume, some of the contributions delve more deeply into trans-disciplinarity than do others – going beyond the disciplinary 'silos', and beyond integrating knowledge, towards creating something entirely new and potentially transformative for mainstream water policy (for example, Noble et al.; Williams et al.). Several have pursued this exercise through incorporating the perspectives of participants to co-create knowledge (as in the chapters by Bartel et al.), as well as alternative ontologies (see the chapters by Charpleix; Fox).

Several authors, galvanized by the challenges presented by water issues, have written chapters critical of their own disciplines and/or traditional modes of thinking. Through critiquing central tenets, and assumptions, they craft innovations in approaches to water problems as well as reimagining their disciplines and policy touchstones more fundamentally (for example, see the chapters by Noble et al.; Charpleix). The focus on policy and application is itself a disruptive endeavour for scholars and disciplines more traditionally focused on pure research (see for example the chapter by Walsh that combines philosophy and law, and by Ryder on science and policy). As author of Chapter 3, A. J. Walsh (pers. comm.) observes of the process: 'the primary challenge with which I have struggled in this project is a direct consequence of my disciplinary background. Philosophers (well at least analytic philosophers) have not discussed water policy in any detail at all'.

Other chapters incorporate cutting-edge research from disciplines that have emerged only relatively recently as relevant to law and policy, including ecology and psychology (see the chapters by Stayner and Parsons; Hine et al.). Stewardship (Chapin et al., 2011) and sustainability (Pezzey, 1992) are inherently interdisciplinary subjects. Fox, Shepheard and Williams et al. all tackle these topics from the applied perspective of generating better water policy. Fox brings a philosophical approach to an interrogation of the value of water, while Shepheard's comparative evaluation of catchment-based management techniques is more empirical. Williams et al. (this volume) have adopted an experimental dialogic style for their chapter (for a similar form, see Senge et al., 2004), which the authors refer to as a 'decolonising narrative', consciously adopted for its potential to disrupt traditional patterns of thinking and assist in generating knowledge in a more collaborative and inclusive, and ultimately more sustainable, fashion.

As a result of the diversity in disciplines, and range of inter-, multi- and trans-disciplinary endeavours pursued, the chapters demonstrate a breadth of epistemological positions and deploy a similarly extensive sweep of methodologies and a range of voices. Walsh and Fox provide largely analytical and conceptual examinations, while others draw on empirical research. Shepheard presents an international comparison of catchment management in Canada, Switzerland, England and New Zealand, and Ryder draws on examples from legislation and environmental flow approaches in Europe, the USA, Australia and South Africa. Charpleix focuses on the recent re-appreciation of the 'personhood' of the Whanganui River in New Zealand, an exceptional and potentially influential case that Charpleix uses as part of a much broader evaluation of water values and accounting.

Critically, several chapters use Australia as a case study, or examine Australian experiences (see the chapters by Noble et al.; Barker; Stayner & Parsons; Hine et al., this volume). The lessons from Australia have global relevance. Australia, the driest inhabited continent on Earth, has been described as the 'canary in the coal mine' for water issues and related social and environmental problems (Strang, 2009, p. 5). Australia is experiencing a precipitous deterioration in health of its river systems, the corresponding corruption of once-arable land through salinity, erosion and degradation, as well as increasingly hostile and dysfunctional political tournaments over water use and river management. In his chapter, Walsh helpfully qualifies the claim for Australian exceptionalism – for nearly all places and societies have water related issues, they are a global policy problem – and it is precisely for this reason that most other nations wish to learn from the Australian experience. In Chapter 11, the authors quote an observation made by one of their interview participants from a northern region of the Murray-Darling Basin (MDB) in Australia:

> …*the whole world's over-allocated their water and people are seeing how Australia's going to sort it out.* (Namoi 3)

The MDB is used as a case example in several chapters (see Stayner & Parsons; those by Bartel et al.), reflecting the worldwide fascination with the Basin in water policy discussions (World Economic Forum, 2015). As Bartel et al. (this volume) observe, this is in large part due to the unique features of this catchment – its capacity to support a diverse and rich agricultural industry responsible for producing much of Australia's exports in the sector, as well as goods for domestic consumption – in spite of a biophysically challenging environment combining unreliability of rainfall with poor soil fertility. Australia's climatic extremes present an ongoing challenge to policy-makers, while also featuring vividly in the cultural imagination of its citizens, inspiring prose (see for example Rolls, 1974) as well as poetry (see for example Pannell, 2012).

An exploration of the role of the imagination – an area that has been long overlooked in policy debates and in policy itself – forms the first part of the book. The second part presents innovations in traditional approaches to water policy, with a focus on the major contemporary mechanisms based on property, science and economics. The third part investigates the concepts of stewardship and sustainability and their potential for water policy innovation, from philosophical and legal angles. The fourth part examines innovation in different water policy contexts – rural and urban – drawing on insights from psychology. The fifth and final part explores the democratic and material turns in scholarship, and their potential for water policy innovation, especially place agency and collaborative governance, through processes of policy learning.

Imagination

In the opening chapter, Noble, Harris and Marshall outline how literary (and other creative modes of) expression currently exert influence on cultural imaginings of water, norms of behaviour and both formal and informal rules. Their examination reveals that this role is currently largely invisibilised and its power therefore unacknowledged and under-interrogated. For example, evocative expressions of colonial dominion in settler nations work both to inspire and legitimate water relationships that may be environmentally sub-optimal as well as socially inequitable, serving to exclude alternative renderings, including those of Australia's Indigenous (Aboriginal and Torres Strait Islander) Peoples. As Barker demonstrates in Chapter 2, the complexity and vitality of Aboriginal Rainmakers' knowledge and practices has been consistently misunderstood and misrepresented in European reportage.

When faced with issues concerning water quantity, quality and related ecosystem health issues, it seems that policy-makers rarely ask how people imagine or conceive of water, and how different relationships with water might have consequences for the environment, for water and for people. Instead, it appears to be assumed that people will primarily relate to water as a resource – instrumentally as an input – whether directly for consumption or indirectly for production, industry and recreation. The problem inheres not only in regard to management practices, as we have come to think of them, but the very assumptions informing the often-unchallenged concept of 'management' itself: 'to manage', intransitively speaking, reinforces the deep-seated assumption describing the relationship between 'user' and the 'resource', and, in turn, implicit beliefs about human behaviour. As Cameron Muir (2012) observes in regard to continuing efforts to protect one of Australia's most important wetlands – the Macquarie Marshes in central New South Wales – a succession of governments have blundered and dithered bureaucratically since the former state premier, William McKell, proposed and legislated for a range of conservation measures in the 1940s. The result at the governmental level is the institutionalisation of the 'split between production and protection', and, therefore, the continued political failure to organize integrated water policy: 'Almost sixty years after his [McKell's] promise to preserve the iconic river ecosystems like the Macquarie Marshes for all time, we're still talking about how to save rivers' (Muir, 2012, para. 16).

Water issues are often instead reduced to questions about supply and demand, and these issues are further reduced to problems of provision, to be addressed via technological and infrastructure solutions. Thus the water may be made invisible by the hardware. And increasingly, given the ascendancy of neoliberalism and current preference for market-based solutions, the role for government is also sometimes reduced, as they revert to 'steering' rather than 'rowing' (Berger, 2003; Clarke, 2004). In the process of bypassing government, people also may be passed over. Place stories and narratives in particular may serve as unique vehicles for accessing additional 'truths' about how water operates in our environment (Robertson et al., 2000; and see for example Cameron, 2014;

Humphreys, 2015), and provide 'another order of veracity' (Williamson, 2016) to science and economics. Through listening to people's stories, and in the inclusion of such narratives, policy processes as well as outcomes may be improved, not least through facilitating procedural justice. However, culture appears to be frequently overlooked in mainstream approaches to water policy. This silence is most surprising, given that it is culture that determines not only how much water we use and what we use it for, as well as how we dispose of it, but also, more fundamentally, how we view it, including as a resource, in the first place. Rarely is it acknowledged by policy-makers that the very choice of policy as a way of addressing water issues is itself influenced by cultural attitudes and behaviours, which are not immutable and rarely universal.

When faced with water shortages and water quality and waste issues, how may it assist policy-makers to ask questions about the way our culture influences our relationship with water? Firstly, culture accounts for many of the proximate causes of our relationships with water, including in many cases those that have become habitual and assumed, rather than themselves subject to enquiry. Secondly, it may also underpin the ultimate causes, and suggest that solutions may go beyond simply answering supply and demand questions and instead lie in more fundamental questions as to the nature of reality and how we conceive of it: is the natural world *out there* or *within?* Is water something we can manage, or even construct, or is it something to which we are inter-related and from which we are inseparable? Are water quality and availability issues a consequence of poor management or symptomatic of misguided human desires for control? The Aboriginal mythology that Barker describes may be contrasted with European-influenced mythologies relating to water as they have developed in, and been inflected through, Western cultural history. All mythologies carry an enduring as well as evolving set of associated beliefs and values, and in so doing, attest to the multitude of ways that water is, can and has been 'seen' in human history and culture.

In more recent times, many values may have been influenced by and over-written by the modern utilitarian attitudes that inform instrumentalist models of water use and governance. Cultural historians will refer to this as the cultural imaginary – the cumulative store of symbols, ideas and beliefs pervading cultural life, gathered and carried over time through the imaginative dimensions of language, rich in symbolic, allusive and figurative meanings. This deep cultural history of humanity's social 'creation' of water constitutes 'worlds of water' (Morgan & Smith, 2013, p. 106), which exist alongside, and submerged within (as it were), the dominant modes and scientific forms of knowledge of and about water in the modern age. To even consider incorporating this knowledge into contemporary, 'resourcist' approaches to water policy is to commence a necessary innovation. Indeed, to recognize the imaginative value of this 'world of water' marks a crucial first step in addressing both the problems attending the fashioning of current policy and, in turn, the limitations of Western(ised) water knowledge itself – at least, that knowledge awarded value for its use in an official and institutional capacity. It is not, that is to say, an 'alternative' view of water, but knowledge human beings collectively already inhabit, in the sense that we can be said to exist within this imaginative world. As Muir (2012) concludes, we must in the first instance 'understand the complex ways ecological metaphors, science and culture become entangled and change over time' (p. 5). We also need to see that the challenge at this level becomes one of fundamental worldviews – as these determine how humans 'work the world' – as Naomi Klein (2014) instructively highlights:

> Fundamentally, the task is to articulate not just an alternative policy proposal but an alternative worldview to rival the one at the heart of the ecological crisis – embedded in interdependence rather than hyperindividualism, reciprocity rather than dominance, and cooperation rather than hierarchy … because in the hot and stormy future we have already made inevitable through our past emissions, an unshakable belief in the equal rights of all people and a capacity for deep compassion will be the only things standing between civilization and barbarism. (p. 462)

Talk of 'imagination' and 'innovation' might induce scepticism in some – what 'good' is the use of the imagination in practical terms in the face of systemised practices and attitudes that have the power to actively resist change? That the public has grown increasingly wary and sceptical of the political process itself is an immediate symptom of the deeper malaise; that the rhetoric of 'contemporary innovation-speak' characterising our present age (Vinsel, 2016, p. 7) has so quickly been reduced to political shibboleth, debasing the very virtues of imagination and creativity that it aspires to promote and celebrate, further compounds the problem. The counterpoint is that it is *in* and *through* the powers of imagination – the capacity to conceive of alternatives; to re-form accepted ideas so as to induce new ways of 'seeing' and understanding – that formative change can be initiated:

> We need a mind shift to reconnect world development to planet Earth and we need to start recognizing that we are now in a dynamic, transformative, non-linear exponential phase where the key challenge is to start seeing the opportunities in transformations to global sustainability. (Rockström, 2017)

The ways in which humans 'see' water, and thus form their relationship to it over time and across cultural and geographic boundaries, is important in its own terms – even as contemporary developments in the handling of water influence and alter more deeply rooted ideas and beliefs. A deeper understanding and appreciation of both the culturally proximate and ultimate causes, and influences, of human relationships with water is required. Otherwise attempts to change the consequences of attitudes and behaviours that are currently putting water resources at risk of overuse and over-consumption, and thus also causing ecosystem collapse and threatening Earth support systems, may be unsuccessful. The relative power of different cultural constructs is important too. Institutional concepts of property and entitlement, crafted by the dominant legal system to protect the wealthy, continue to be deployed and assigned according to an economic, rather than ethical or environmental, basis to the highest price-giver – again preserving the powerful. This at least appears to be true of the colonial countries with which many of the authors are familiar, and although there are a great diversity of traditions, backgrounds and peoples within countries such as Australia, Canada and New Zealand, not all voices are equally enabled and entertained by policy conversations. Like-minds, like-views and like-ontologies are more likely to find a policy audience and a purchase. At the same time, alternative approaches are deployed informally and exert their influence on official framings – often unconsciously; hence they often have power that is under-acknowledged (see the chapters by Charpleix; Bartel et al.). However, the colonial and European-informed trajectory of water relationships in Australia, shared by similar colonial and settler nations, has privileged certain perspectives and prevented us from appreciating other ways of seeing and what they may have to offer.

In Australia, as in other colonial nations, some settlers adopted a 'cultural cringe' mentality towards the landscape (Phillips, 1950). This 'cringe' describes a preference for the colonising culture and environment of the 'Mother Country' to that of the new environment – and may mandate that the new and strange be 'tamed' and/or made to make way for something more familiar from 'back home'. In contemporary Australia such a mind-set may be considered antiquated, but vast areas of the continent, including the river systems, are now inhabited by introduced species – whether intentionally or accidentally, and whether productive or invasive – as well as foreign ideas, including in relation to water use and management.

Of course the new ideas were not uniformly disastrous nor were invaders all committed to transformation in the cringe sense, which is to say that they were not universally contemptuous, arrogant or ignorant (see for example Bonyhady, 1998). Various truths and beauties of the landscape were appreciated and some of the arrivals were keen of eye and mind, and more open and curious of heart. Dorothea Mackellar's 'My Country' (1908), one of Australia's most famous poems and the subject of the first chapter, is popularly considered a love letter to the 'real' dimensions of Australia – including its 'droughts and flooding rains' – in stark material and natural contrast to an imagined gentler and softer England. With a patriotic flourish, Mackellar claims that Australia is different. And significantly for the (then) new nation (Australia became a Federation in 1901), not only is it no less marvellous than the Mother Country, it is preferable. But what effect does such rendering have on a culture? As Noble, Harris and Marshall observe in the opening chapter, repeated mis-readings of the poem may risk 'creating, reinforcing and perpetuating path dependent attitudes and approaches to water governance'. Does a new nation, developing its own identity at the time the poem was written – and in part due to the poem having been written (for cultures co-create the country) – then configure an identity marked by love and deep reckoning, or is there still deep-seated ambivalence towards the features described and exulted by Mackellar, reflecting in part the ambivalence at its heart? As Barker (this volume) observes, the settlers were perceived to have had an uneasy relationship with the new country's 'rain, rivers and floods'. Are the droughts and flooding rains celebrated by the Europeans? Accepted? Attempted to be controlled and managed? Does the observation of almost perverse plenty and absence also suggest an engineering solution – to impound the bounty for the periods of dry? In rendering a larger canvas on which the forces of nature are seen to be at work, is there also a greater and therefore more attractive challenge for those who might wish to bring nature into line, to subdue and bind it to a progress-aligned will motivated to civilize the great untamed? The material may influence our imagination, but our imagined version of it may also be partial, moulded by our pre-existing ideas – and then these imaginings becoming real through our behaviours, including in policy, to make a landscape (and waterscape) in our own image: cultural legacies becoming 'locked-in' within policy. Present generations are thus held hostage to political legacies: all contemporary efforts to influence and transform water governance are bound by the enduring effects of previous decisions, both in regard to the lag in political and social processes and the actual ecological 'flow-on' effects.

Such 'lock-in' can affect appreciation of alternative perspectives in deleterious ways, as Barker reports in the second chapter. Barker analyses the stories that were written in the English language media about Aboriginal Rainmakers' knowledge and practices for European-Australian audiences. Mainstream media treatment of the Rainmaker's knowledge has been problematic: whether described as the work of frauds and subjected to ridicule, or viewed as curio and reified, it appears to be frequently misunderstood, objectified and reduced to novelty. There are resonances here with the first chapter and the mixed interpretations of the poem 'My Country'. Even the title of the poem can be considered troubling: it may communicate connection, but also, and perhaps more problematically, ownership, or at least entitlement for control. Property, long the basis for interaction with the natural world, and hence also at the root of many of our environmental problems (see for example Arnold, 2005; Graham, 2011), is interrogated in the chapter that opens the section on innovations in the traditional disciplines and mechanisms of water policy.

Innovation

The three chapters by Walsh, Ryder and Charpleix focus on property, science and economics respectively – the triumvirate of dominant techniques in the water policy domain. In terms of the legacy and path-dependency issues raised in the opening chapter, few social institutions cast shadows as long as these approaches. However, the authors of these chapters demonstrate that the ground on which they stand is shaky. Through supplying much-needed critique, they map the potential for more creative and effective policy approaches drawing on reformed and strengthened techniques.

Walsh questions entitlement to water and the origins of ownership in particular, and focuses his critique on the fundamental question of whether the privatisation of water can satisfy the original rules of property devised by its founding proponents and adherents. Walsh's analysis finds that water falls outside the definition of property, since the pre-requisite of entitlement through human input and labour cannot be satisfied, given that it is not produced but is instead 'manna from heaven'. This finding, derived through the application of first principles laid down by Nozick, is used by Walsh to call into question the capacity of property in water to deliver equity and fairness. The current system of water privatisation recognizes the (purchased) rights of water title-holders, while the interests of Indigenous peoples, whose ties to water are deeply historic and cultural, but may not be proprietary, are often not only de-prioritised but de-legitimised and ignored. The interests of the environment itself are almost completely excluded. Walsh concludes that it is the property-based claims that are illegitimate. When relationships are sought to be recognized, or reconfigured – for example where water titles have had to be resumed or extinguished, as in Australia for environmental purposes – the perceived impact on property rights often sparks debates about liberty and justice, even though the conversation is narrowly centred on property, rather than on interests in water. These perverse, and inevitably inequitable, consequences are a result of the institution of property being extended to cases, like water, that it was never intended to reach: it is an inappropriate fit for the institution – stretching it to breaking point. Walsh critiques how the policy response to such questions has become constrained by the pre-eminence of property, entitling owners not only to their so-called right to water, but the right for their grievances to be heard, for their interests to be valued and considered, for their access to justice and fairness to be assured, while excluding others perhaps more deserving, not least of all water itself. And perhaps even more concerningly, the very construction of the problem as a contest between competing rights-holders is problematic, as, through the unquestioned acceptance of relations based on competition over access and use, the dominant framing perpetuates not only conflict but also consumption and therefore resourcist approaches.

Undesirable and inappropriate framings are also central to the focus of the next chapter by Ryder, particularly the ways in which scientific knowledge is framed by and within policy. Ryder's chapter concerns itself with the question of how science may be undervalued and side-lined within policy and by policy processes. He argues that while scientific knowledge may be perceived as the obvious 'go-to' source of knowledge for policy-makers, that 'truly' science-based policy is often the exception rather than the rule. The use of 'best available science' is prescribed by legislation and policies in Australia, Europe and the United States, but these requirements may actually fail to effect the inclusion of scientific findings, and may mask a lack of appreciation of science, in terms of standards and quality, and the processes by which it may be derived, incorporated into policy and applied in practice. Such provisions may be a way of 'window-dressing' otherwise *status quo* (or worse, corrupt) approaches, neglecting not only science but also excluding valuable but technically 'unscientific' knowledge, such as traditional ecological knowledge. Thus, policies may attempt to cloak themselves in the respectability of science, while the broader institutional apparatus neglects the production of science: it is simply too expensive, time-consuming and risky. The best science available may not be necessarily the best science possible, and even the best science possible will be contingent and in many cases science of the requisite standard may be simply impossible to obtain: policy-makers may thus misunderstand the fundamentals of the scientific method and complexity science (see for example Ravetz, 2006). Science is also often contested, conflicting and self-correcting, but this inherent diversity and fallibility may be uncomfortable for policy-makers. Adaptive management principles similarly require a commitment to iterative improvement as more information comes to hand, is applied and is evaluated for its outcomes. However, risk-aversion in policy circles is very high (see for example Bartel, 2016) and policy certainty is preferable, particularly for its impact on economic decisions (see also Houck, 2003). The market purports to thrive on certainty, which science is not only unable to provide, but with which it is philosophically at odds.

Furthermore, given the wide range of scientific research that may be available, policy-makers may, through normative isomorphism, cognitive bias and path dependency, prefer to choose the results that they find more familiar and palatable. This could include the repeated use of out-dated information or mis-application of knowledge from one area to another. Old ideological preferences and habitual (and habituated) rules-of-thumb may not only remain untested and un-interrogated, but continue to form the basis of policy decisions – rather than reliance on knowledge – of any kind. Water policy may remain 'informed' by what Ryder refers to as experiential knowledge – a term that is contrasted with traditional ecological knowledge – and which Ryder identifies as equally applicable, if not in the same category as science. Perhaps worse, science may be used as a political weapon to justify a pre-existing decision, rather than, ideally, presented as informing a range of potential policy options.

Ryder identifies several major pitfalls to avoid, including overcoming the many inaccurate views and treatments of science in the environmental flow domain, as well as the growing influence of market-based ideology informing the direction of water policy. This may outweigh the influence of scientific and other knowledge. In the preceding chapter, Walsh also identifies economic thinking as increasingly prominent in policy circles. Charpleix explores economic values in further detail in her assessment of accounting models for the evaluation and measurement of water quantity and quality. Charpleix articulates how many of these models are predicated on the objectification of water and predisposed towards its commodification and commensuration. Attempts to extend such models to encompass intangible values and ecosystem benefits are critiqued as insufficient to cure inherent anthropocentrism and abstraction of water. The landmark recognition of the Whanganui River in New Zealand is used as an example of relational ontological approaches which present an alternative to current resourcist orientations, and which may be able to address deficiencies in dominant characterisations: recognizing both the agency of water, as well as cultural connections. Charpleix draws on the concept of 'hybridity' to explore the possibility of dialogue between colonial and Maori knowledge systems in order to transform historic and current conflicts, and also to chart a co-creative path forward incorporating, and deeply respectful of, multiple perspectives.

In the following chapter exploring stewardship and sustainability, Fox agrees that economic valuations are unable to capture all the values of water, nor achieve its protection. He argues that sustainability principles, and strong rather than weak sustainability, offer a pathway out of the current anthropocentric mindset and predominance of utilitarian and instrumental approaches to valuing water. Fox concludes that policies based on notions of environmental stewardship and a duty of care, that embrace both the intrinsic and instrumental values of nature, rather than management and arrogant 'dominionism', would be more beneficial and equitable. Reimagining the scope and role of stewardship is also central to Shepheard's chapter, which examines the experience of legislated stewardship duties for catchment management and sustainable development in four case studies. The chapter reveals both the good and the bad in catchment management in New Zealand, Canada, Switzerland and England. Shepheard argues for better stewardship and accountability through a form of social contract that puts the onus on cooperative individual behaviour to ensure sustainable catchments and the amelioration of watershed damage.

Williams, Please and Barker (Chapter 8) argue that water management must be decolonised in order to be sustainable. They demonstrate how Traditional Ecological Knowledge Systems may be used to transform Anthropocentrism and Anthroparchism, as well as counter the commodification of water that is reinforced through neo-colonialism, neo-liberalism and globalisation. Their chapter innovatively presents each author's experiences, interactions and insights, drawing on stewardship principles, eco-psychology and Indigenous research methodologies, in a collaborative and generative conversation.

The next two chapters also incorporate lessons drawn from psychology, although in very different ways. Stayner and Parsons (Chapter 9) extend the idea of resilience to water management in agriculture while Hine, McLeod and Driver (Chapter 10) use cutting-edge behavioural science to examine household water use. Walker and others (2004) have defined resilience as 'the capacity of a system to absorb disturbance and reorganise while undergoing change so as to still retain essentially the same function, structure, identity and feedbacks' (online). The term is borrowed from ecology – its original meaning referred to the capacity of species and ecosystems to be adaptive to changing conditions (Holling, 1973), as well as psychology – where resilient behaviour was first described in research in the early 1970s (Werner, 1971). Stayner and Parsons also apply Biggs, Schluter and Schoon's (2015) principles, and argue that adaptive management, appreciation of socio-environmental linkages, dynamism and complexity would require a significant overhaul of existing institutional and policy arrangements. They describe simplistic visions of nation-building as having undermined resilience, leaving a legacy of not only circumscribed imaginings, as Noble, Harris and Marshall critique in their opening chapter, but also physical infrastructure, which curbs the ability of future communities to adapt to change. Path dependence is assured because the dams have been built, and the channels dug, but also because of changes to our climatic and ecological systems. These biophysical changes may prove insurmountable, given that they may be irreversible. The prospects for dam decommissioning in Australia, as is happening elsewhere in the world, particularly the USA (Grant, 2001), seem slim, especially as major political parties continue to promote such options, and are therefore destined to repeat past errors. This vastly altered environment may require a resilience which goes beyond 'bouncing back' from perturbations, but one that is able to 'bounce into' a new state (although preferably not the one that is presently being created). The former type of resilience describes a response to a specific threat, i.e. the ability to 'return to normal' after a particular type of trauma, as well as a generalised capacity to respond well to shocks of any kind (Folke et al., 2010). Resilience has thus been used to describe coping strategies in response to adversity and may also be described as being *positive* and proactive, rather than reactive. In this conception of resilience, changeability is a *prerequisite* rather than a consequence: change can occur out of *choice* in order to *enable* resilience (Folke et al., 2010; and see also Smit and Wandel, 2006). What is described as 'transformational' change may be adopted if the current state is untenable (Folke et al., 2010).

Transformational change is also required at the level of the individual, and Hine et al. (this volume) address this aspect in terms of household water use, which in Australia exceeds that of most other countries. Water restrictions are commonly applied during drought periods but these are often viewed as temporary, to be shelved when the inevitable floods return (see Noble et al., this volume). They induce few lasting changes – instead both policy proscriptions and behaviours revert to the unsustainable *status quo*. More sophisticated interventions are proposed by the authors, drawing upon the Behaviour Change Wheel (Michie et al., 2014) and community-based social marketing. Such interventions are empirically based and, importantly, rather than designed to be universal, are intended to be tailored to the specific intended operating environment. There is no one-size-fits-all model for human behavioural change, just as there is no one-size-fits-all water policy for the environment. Legislation and regulation remain one of the few interventions available which draw on, or which may be able to utilise, all of the functions of the intervention spectrum (see Hine et al., this volume). The effect of legislation and policy trajectories over time are explored further in the final part of the book.

The past and future of water policy

In the final two chapters, Bartel, Noble and Beck traverse the history of water management in a catchment of the MDB, utilising legal and policy analysis, as well as environmental and interview data, to map a path for future policy. They conclude that water policy has been maligned by a modernist and resourcist orientation largely divorced from biophysical reality. Through their empirical work the authors find that people living in the Basin appear to possess knowledge with greater place-congruence than that reflected in the objectives of disconnected, aspatial policy. One of the assertions informing several chapters in this collection is that such 'folk' and place-based knowledge matters a great deal in any serious discussions of and deliberations over water, most particularly concerning the design of innovative water policy. This is especially true for Australia. In discussing the relation between nation building projects and pioneering figures of 'Big Development' schemes in Australia (central among which was J. J. Bradfield's proposal to 'turn the rivers inland and "save" the dry interior'), Libby Robin (2007) notes that such individuals, in seeing nature as a resource, independent of people, thus assumed that people were infinitely mobile: 'Big Development can build national pride, but does not build a sense of "home" anywhere in particular' (p. 198).

As antidote, Bartel et al. recommend the incorporation of *genius loci*, via local, emplaced, vernacular knowledge, through processes of collaborative governance and more sophisticated policy learning cycles than have been hitherto adopted. The authors also marshal the relational and democratic (also known as deliberative) 'turns' in the social sciences, as well as new materialism (see Bartel, forthcoming), to craft the neologisms of syncretic (drawing on multiple ontologies) governance using transformational (triple-loop) learning cycles for a heterotic (more diverse and dynamic) water policy future. They identify that the voices of place speak of the particularities of local water and its requirements – a counter to the universal and anthropocentric frames that have dominated historic approaches to water policy.

The editors of this volume agree that the dominant approaches to water policy are limited, and thus a limiting approach to policy development and practice. Indeed all of the authors assembled here are as one in submitting that water issues require more than just *status quo* and more-of-the-same policy solutions. In particular, the measurement and management agenda is not only narrow, but narrowing, and both propagates and perpetuates a view of water that is little more than commodification (see for example Weir, 2009 and the chapter by Charpleix, this volume). Similar to memes, the prevailing cultural and policy orthodoxies – perpetuated and compounded by path dependence (see Noble et al.; Stayner & Parsons, this volume) – of commoditisation and commensuration, centralisation of economic and cultural power, runs counter to the insights here of deep connection, dynamism, heterogeneity and the need for tailored solutions profoundly empathetic to the nuances of water, place and people. To continue on the current path is to guarantee that policy innovation will be stifled and more sustainable outcomes hindered. It would also be to deny the full reality of experience and humans as feeling and imaginative beings co-creating the world with the world. To recognize this greater reality, and in so doing furnish a better future, we need new ways of thinking about water – new water-ways – new ways of acting, thinking, being and governing water.

Final introductory remarks

The 'legacy of extremes of instrumentalism' (Smith, 2016) – the deeper and abiding disruptions to complex ecosystems – cannot simply be annulled through the revision of policy; and where activists such as Klein (above) affirm the imperative of direct political action, the chapters in this volume complement more overt forms of protest by seeking to influence such 'worldviews'. For the guiding objective here is to encourage a concerted reassessment of the value and place of water in human social and cultural experience in order to influence water policy. To do so involves confronting the deep-seated assumptions that are reinforced not only in broadscale management and political approaches to water governance, but also the much wider social assumption that water is, in the first and irrefutable instance, a utility to be supplied and a commercialised resource to be 'used' – a material good that exists *for* humanity's convenience and need. Directly confronting such beliefs is necessary in order to encourage a fundamental innovation – in approach and process of understanding – as the basis for promoting more nuanced approaches to water policy, which, in turn, and ideally, will engender the more considered understanding that water constitutes our place in this world – it represents the irreducible interconnection between human beings and the non-human natural and material worlds.

As humanity becomes more conscious of the Anthropocene, the plight of water and our very survival, the opportunity of an awakened consciousness towards the planet and our place in it is emerging. Goswami (1995) claims that consciousness is the ground of all being and the exponential growth of human innovation that Rockström (2017) illustrates as already underway, offers inspiration, hope and guidance. And new ways already exist: local people are practising them, the sciences, the humanities and the arts are all brimming with ideas, and the chapters herein demonstrate examples from around the world. Ecological approaches, brought famously to bear on the question of economically motivated weed and pest control (Carson, 1962), have demonstrated the myth of narrowness and linearity in appreciating natural systems and today these are described in the scientific literature as social-ecological systems (Berkes et al., 2003) – comprised of not only food chains and ecosystems but also the societies and institutions embedded in them. The humanities and social sciences, with remarkable perspicacity and synchronicity, have arrived at similar notions, including socionatures (White, 2006), naturecultures (Haraway, 2008; White, 2006), envirosocial approaches (Bartel et al., 2014) and the hydrosocial cycle (see Linton, 2010; Swyngedouw et al., 2002), that recognize the inter-related nature of nature and its co-production by both human and non-human agents (and not to diminish the uniqueness of nature, but to recognize its agency, cf. Plumwood, 2006). These all may be described as 'new' water-ways of thinking, but they have roots in ancient thought as well as in contemporary practices of relating to and understanding the non-human world (see for example Linton, 2010).

In Aqua sanitas: 'in water there is health'. So concludes the Latin proverb extolling the revelatory effects of wine (*in vino veritas*). To consider, then, the utility of such homiletic wisdom, and the deeper well of cultural knowledge to which it gestures, is but one way of pressing into view the perennial question concerning knowledge and its uses – that is, what we understand and accept as knowledge at any given time and how and why such knowledge is 'put to use', to invoke the pervasive utilitarian language of the present day. For example, we 'know' unarguably that water is essential to life: 'it is the substance . . . and giver of life . . . no known life exists without it' (Jha, 2015, pp. 2–3). Yet, rightly, statements pronouncing incontestable truths should engender scepticism, if for no other reason in this instance than the fact that this statement reflects a predictably 'hydrocentric view coming from a water-based life form' (p. 309). Indeed, by way of completing his recently published 'life-story' of water, Jha observes that 'we have come to accept water as the medium of life for sensible reasons . . . but if there are any lessons in the human story of water . . . it is surely that, in our privileged position as the products of water, we should be cautious in assuming that anything, water's singular role included, is as simple or definite as it might initially seem' (pp. 309/318–9). There are, as Jha (2015) reports, hypothetical alternatives to water as life-supporting substance; yet the great majority of humans, along with the non-human life forms with whom we share the vast, intricate, life-sustaining biosphere we know and think of as our home, must continue to live at the ground level of daily subsistence where water is indispensible.

It is those commonplace words 'share' and 'home' that point towards the prevailing challenges at present; challenges that knot together deep and powerful notions of belonging – of what it means as humans to inhabit and flourish; to assume responsibly our place in the encompassing natural world – and the multiplying complications of humanly engineered, ever-concentrated modes of life that inspire, and distortedly magnify, Promethean notions of unassailable entitlement over the natural world and its processes. The seventh Sustainable Development Goal (United Nations, 2015) is 'to ensure the availability and sustainable management of water and sanitation for all'. Related targets include water use efficiency, improved water quality and environmental health, restoration of aquatic ecosystems, as well as equitable access to drinking water and sanitary conditions, with processes including integrated transboundary governance, international cooperation and community participation. It is the latter which appears crucial for the achievement (and transformation) of all of the preceding objectives. Hopefully this book will demonstrate the value of a plurality of perspectives – drawing instructive lessons from a wide range of disciplines as well as inter-, multi- and trans-disciplinary engagements – to learn from and to heed the contributions of many disciplines, many peoples, many cultures, many places and many voices, including that of water itself.

We are returned, then, to the importance of understanding the wider narrative reach of 'water stories' in human culture – and the broader cultural imaginary constituting a valid form of knowledge in itself, and to understand this in relation to the production of policy. In turn, such a proposition reinforces the growing acknowledgement, across the broad enterprise of government and the socially oriented functions of associated offices, of the practical value of the interdisciplinary approaches we refer to in this collection's title – approaches that, in defying long-standing 'rules' of intellectual engagement based on tightly defined paths of specialist knowledge and expertise, constitute an important, and innovative, step in the positive uses of the imagination.

Note

1 Related terms include cross-disciplinary/arity and co-disciplinarity; see for example Hsy (2014). See further discussion in Chapter 12 (this volume).

References

Arnold, C. A. (2005). Introduction. Integrating water controls and land use controls: New ideas and old obstacles. In C. A. Arnold (Ed.), *Wet growth: Should water law control land use* (pp. 1–56). Washington, D. C.: Environmental Law Institute.

Bartel, R. (2016). Legal geography, eography and the research-policy nexus. *Geographical Research*, 54(3), 233–244.

Bartel, R. (Forthcoming). Place-thinking: The hidden geography of environmental law. In A. Philippopoulos-Mihalopoulos and V. Brooks (Eds.), *Handbook of research methods in environmental law*. Cheltenham, UK: Edward Elgar Publishing.

Bartel, R., McFarland, P., & Hearfield, C. (2014). Taking a de-binarized envirosocial approach to reconciling the environment vs economy debate: Lessons from climate change litigation for planning in NSW, Australia. *Town Planning Review*, 85(1), 67–96.

Berger, G. (2003). Reflections on governance: Power relations and policy making in regional sustainable development. *Journal of Environmental Policy and Planning*, 5(3), 219–234.

Berkes, F., Colding, J., & Folke, C. (2003). *Navigating social–ecological systems: Building resilience for complexity and change.* Cambridge: Cambridge University Press.

Biggs, R., Schluter, M., & Schoon, M. L. (Editors). (2015). *Principles for building resilience: Sustaining ecosystem services in social-ecological systems.* Cambridge, UK: Cambridge University Press.

Bonneuil, C. & Fressoz, J.-B. (2016). *The shock of the Anthropocene.* London and New York: Verso.

Bonyhady, T. (1998). *The colonial earth.* Melbourne University Press.

Cameron, A. (2014). *The last pulse.* Sydney and Melbourne: Penguin.

Carson, R. (1962). *Silent spring.* Boston: Houghton Mifflin Company.

Chapin III, F. S., Pickett, S. T. A., Power, M. E., Jackson, R. B., Carter, D. M., & Duke, C. (2011). Earth stewardship: A strategy for social-ecological transformation to reverse planetary degradation. *Journal of Environmental Studies and Sciences*, 1, 44–53.

Clarke, J. (2004). Dissolving the public realm? The logics and limits of neo-liberalism. *Journal of Social Policy*, 33(1), 27–48.

Engelke, P. & Sticklor, R. (2015). Water wars: The next great driver of global conflict? *The National Interest.* September 15, 2015. Retrieved at http://nationalinterest.org/feature/water-wars-the-next-great-driver-global-conflict-13842

Folke, C., Carpenter, S. R., Walker, B., Scheffer, M., Chapin, T., & Röckstrom, J. (2010). Resilience thinking: Integrating resilience, adaptability and transformability. *Ecology and Society*, 15(4), 20.

Goswami, A. (1995). *The self aware universe: How consciousness creates the material world.* New York: Tarcher.

Graham, N. (2011). *Lawscape: Property, environment, law.* London: Routledge.

Grant, G. (2001). Dam removal: Panacea or pandora for rivers? *Hydrological Processes*, 15, 1531–1532.

Haraway, D. (2008). *When species meet.* Minneapolis: University of Minnesota Press.

Holling, C. S. (1973). Resilience and stability of ecological systems, *Annual Review of Ecology and Systematics*, 4, 1–23.

Houck, O. (2003). Tales from a troubled marriage: *Science* and law in environmental policy. *Science*, 302(5652), 1926–1929.

Hsy, J. (2014). Co-disciplinarity. In E. Emery and R. Utz (Eds.), *Medievalism: Key critical terms* (pp. 43–51). Cambridge: D.S. Brewer.

Humphreys, H. (2015). *The river*. Toronto: ECW Press.

International Federation of Health and Human Rights Organizations (2016). *Slovenia: The right to clean drinking water constitutionalized*. Retrieved at http://www.ifhhro.org/news-a-events/617-slovenia-the-right-to-clean-drinking-water-constitutionalized

Jha, A. (2015). *The water book*. London: Headline Publishing Group.

Karpouzoglou, T. & Vij, S. (2017). Waterscape: A perspective for understanding the contested geography of water. *WIREs Water*, doi: 10.1002/wat2.1210

Khadka, A. J. (2010). The emergence of water as a 'human right' on the world stage: Challenges and opportunities. *International Journal of Water Resources Development*, 26(1), 37–49.

Klein, N. (2014). *This changes everything: Capitalism vs. the climate*. London: Allen Lane.

Krueger, T., Maynard, C., Carr, G., Bruns, A., Mueller, E. N., & Lane, S. (2016). A transdisciplinary account of water research. *WIREs Water*, 3, 369–389.

Linton, J. (2010). *What is water? The history of a modern abstraction*. Vancouver and Toronto: UBC Press.

Michie, S., Atkins, L., & West, R. (2014). *The behaviour change wheel: A guide to designing interventions*. Great Britain: Silverback.

Morgan, R. A. & Smith, J. L. (2013). Pre-modern streams of thought in twenty-first-century water management. *Radical History Review*, 116, 105–129.

Muir, C. (2012). Preserved for the people for all time. *Inside Story*. Retrieved at http://inside.org.au/preserved-for-the-people-for-all-time

The National Geographic (2016). *What you need to know about the world's water wars*. Retrieved at http://news.nationalgeographic.com/2016/07/world-aquifers-water-wars/

Newell, W. H. (2001). A theory of interdisciplinary studies. *Issues in Integrative Studies*, 19, 1–25.

The New Indian Express (2017). *Uttarakhand High Court recognises river Ganga as living entity*. Retrieved at http://www.newindianexpress.com/nation/2017/mar/20/uttarakhand-high-court-recognises-river-ganga-as-living-entity-1583592.html

New Zealand Herald (2017). *Whanganui River given legal status of a person under unique Treaty of Waitangi settlement*. Retrieved at http://www.nzherald.co.nz/nz/news/article.cfm?c_id=1&objectid=11818858

Nissani, M. (1991). Ten cheers for interdisciplinarity: The case for interdisciplinary knowledge and research. *The Social Science Journal*, 34(2), 201–216.

Pahl-Wostl, C., Vörösmarty, C. J., Bhaduri, A., Bogardi, J., Röckstrom, J., & Alcamo, J. (2013). Towards a sustainable water future: Shaping the next decade of global water research. *Current Opinion in Environmental Sustainability*, 5, 708–714.

Pannell, D. J. (2012). The Murray-Darling basin plan: Economic and community perspectives. In J. Quiggan, T. Mallawaarachchi, and S. Chambers (Eds.), *Water Policy Reform* (pp. 198–199). Cheltenham, UK: Edward Elgar Publishing.

Percival, R.V. (2017). Environmental law in the Trump administration (February 24, 2017). *Emory Corporate Governance and Accountability Review*, 4, 225–235.

Pezzey, J. (1992). Sustainability: An interdisciplinary guide. *Environmental Values*, 1, 321–362.

Phillips, A. (1950). The cultural cringe. *Meanjin*, 9(4), 299–302.

Plumwood, V. (2006). The concept of a cultural landscape: Nature, culture and the agency of land. *Ethics and the Environment*, 11(2), 115–150.

Ravetz, J. R. (2006). Post-normal science and the complexity of transitions towards sustainability. *Ecological Complexity*, 3, 275–284.

Riley, S. (2015). Country reports: Australia. *IUCN Academy of Environmental Law Journal*, 6, 117–126.

Robertson, M., Nichols, P., Horwitz, P., Bradby, K., & MacKintosh, D. (2000). Environmental narratives and the need for multiple perspectives to restore degraded landscapes in Australia. *Ecosystem Health*, 6(2), 119–133.

Robin, L. (2007). *How a continent created a nation*. Sydney: UNSW Press.

Rockström, J. (2017, January). *Beyond the anthropocene*. World Economic Forum, Davos-Klosters, Switzerland. Retrieved at http://www.stockholmresilience.org/research/research-news/2017-02-16-wef-2017-beyond-the-anthropocene.html

Rolls, E. (1974). *The River*. Cremorne, Sydney: Angus and Robertson Publishers.

Saul, J. R. (2005). *The collapse of globalism: And the reinvention of the world*. Toronto: Viking Canada.

Schad, T. M. (1991). Do we have a national water policy? *Journal of Soil and Water Conservation*, 46(1), 14–16.

Senge, P. M., Scharmer, C. O., Jaworski, J., & Flowers, B. S. (2004). *Presence: Human purpose and the field of the future*. New York: Society for Organizational Learning, Crown Business.

Siegel, S. M. (2015). *Let there be water: Israel's solution for a water-starved world*. New York: St Martin's Press.

Smit, B. & Wandel, J. (2006). Adaptation, adaptive capacity and vulnerability. *Global Environmental Change*, 16, 282–292.

Smith, J. L. (2016). I, River: New materialism, riparian non-human agency and the scale of democratic reform. *AsiaPacific Viewpoint*, Nov. 9, n.p. Retrieved at http://onlinelibrary.wiley.com/doi/10.1111/apv.12140/abstract

Stember, M. (1991). Presidential address: Advancing the social sciences through the interdisciplinary enterprise. *The Social Science Journal*, 28(1), 1–14.

Strang, V. (2009). *Gardening the world: Agency, identity and the ownership of water*. New York: Berghahn Books.

Swyngedouw, E., Kaika M., & Castro, E. (2002). Urban water: A political-ecology perspective. *Built Environment*, 28, 124–137.

United Nations (2015). *Transforming our world: The 2030 Agenda for Sustainable Development*. A/RES/70/1

United Nations (2017). *Resolution adopted by the General Assembly on 21 December 2016, 71/222. International Decade for Action, 'Water for Sustainable Development', 2018–2028*. Retrieved at http://www.un.org/en/ga/search/view_doc.asp?symbol=A/RES/71/222

Vinsel, L. (2016). Silicon folly: The pitfalls of innovation policy. *The Saturday Paper*, August 27, Issue 123.

Walker, B. H., Holling, C. S., Carpenter, S. R., & Kinzig, A. (2004). Resilience, adaptability and transformability in socio-ecological systems. *Ecology and Society*, 9(2), 5.

Weir, J. (2009). *Murray River Country: An ecological dialogue with traditional owners*. Canberra: Aboriginal Studies Press.

Werner, E. E. (1971). *The children of Kauai: A longitudinal study from the prenatal period to age ten*. Honolulu: University of Hawaii Press.

White, D. (2006). A political sociology of socionatures: Revisionist manoeuvres in environmental sociology. *Environmental Politics*, 15, 59–77.

Williamson, G. (2016). Australian literature crisis as government funding cuts kick in. *The Australian*, 2 July 2016.

World Economic Forum (WEF) (2015). *Global Risks*, 10th Edition, Insight Report, Geneva Switzerland.

World Economic Forum (WEF) (2017). *The Global Risks Report*, 12th Edition. Retrieved at http://www3.weforum.org/docs/GRR17_Report_web.pdf

World Health Organization (2017). *Radical increase in water and sanitation investment required to meet development targets*. Retrieved at http://www.who.int/mediacentre/news/releases/2017/water-sanitation-investment/en/

World Water Council (2017). *World Water Council calls on all governments to prioritize global water security*. Retrieved at http://www.worldwatercouncil.org/news/news-single/article/world-water-council-calls-on-all-governments-to-prioritize-global-water-security/

1 Blue sky thinking in water governance

Understanding the role of the imagination in Australian water policy

Louise Noble, Stephen Harris and Graham Marshall

'My Country'
Dorothea Mackellar

The love of field and coppice,
Of green and shaded lanes.
Of ordered woods and gardens
Is running in your veins,
Strong love of grey-blue distance
Brown streams and soft dim skies
I know but cannot share it,
My love is otherwise.

I love a sunburnt country,
A land of sweeping plains,
Of ragged mountain ranges,
Of droughts and flooding rains.
I love her far horizons,
I love her jewel-sea,
Her beauty and her terror –
The wide brown land for me!

A stark white ring-barked forest
All tragic to the moon,
The sapphire-misted mountains,
The hot gold hush of noon.
Green tangle of the brushes,
Where lithe lianas coil,
And orchids deck the tree-tops
And ferns the warm dark soil.

Core of my heart, my country!
Her pitiless blue sky,
When sick at heart, around us,
We see the cattle die –
But then the grey clouds gather,
And we can bless again
The drumming of an army,
The steady, soaking rain.

Core of my heart, my country!
Land of the Rainbow Gold,
For flood and fire and famine,
She pays us back threefold –
Over the thirsty paddocks,
Watch, after many days,
The filmy veil of greenness
That thickens as we gaze.

An opal-hearted country,
A wilful, lavish land –
All you who have not loved her,
You will not understand –
Though earth holds many splendours,
Wherever I may die,
I know to what brown country
My homing thoughts will fly.
                Published by Arrangement with
    the Licensor, The Dorothea Mackellar
Estate, c/- Curtis Brown (Aust) Pty. Ltd.

Introduction

To show how literature has influenced the hydrological imagination of arid settler colonial countries other than Australia requires deeper culturally and historically specific analyses than can be accommodated here. However, this type of analysis, which shows how a popular literary representation of water can become deeply ingrained in the cultural subconscious of a nation and have the power to influence major water management decisions, has important future benefits beyond Australia. In Dorothea Mackellar's 'My Country', the stark metaphorics of aridity rescued occasionally by rain bear witness to the difficult reality of water in the Australian landscape. It is no surprise then that approaches to addressing the manifold issues relating to water policy and governance in Australia are complex in hydrological, ecological, social, cultural, political and economic terms, as well as contested and conflicted. Inspired ways forward are limited by the legacies of past decisions and prevailing attitudes to policy formulation, which are themselves subject to an intersecting range of influences involving knowledge and the imagination. Einstein's (1929) much quoted statement, 'Imagination is more important than knowledge' offers a suggestive starting point for thinking about the function of imagination and knowledge in our decision making processes. If we consider the imagination as facilitating a path to knowledge through the creation of new possibilities, then we can imagine our way to a better future.

One approach is to examine how the imagination has helped to create and perpetuate the often-uneasy relationship Australians have with the land and water. This chapter looks at the extent to which the message of 'My Country' has, either consciously or unconsciously, been narrowly interpreted and recruited into accustomed ways of seeing, and thus knowing, that frequently favour certain cultural, environmental, scientific and political interests. Only recently the Australian Prime Minister, Malcolm Turnbull, drew imagery from the poem in an attempt to explain excessive flooding in Tasmania in his declaration that 'we live in the land of droughts and flooding rains' (Hunter & Lee, 2016). In some ways, the poem can be considered part of a convenient historical national narrative of water relationships since early European settlement, which excluded Indigenous Australians, and established and legitimised water management practices that are environmentally detrimental. As Cathcart (2009) points out, 'The colonists had the values, aesthetics and aspirations of a culture that was steeped in water'. Theirs was a waterscape totally at odds with the arid land in which they found themselves (p. 47).[1] However, the poem also offers Australians alternative ways of envisioning their relationship with the water and the land that nourishes them and on which they depend and, in doing so, challenges habituated colonial worldviews by inspiring new, imaginative possibilities.

It is therefore important to recognize the role the imagination has played historically in producing and encouraging certain modes of thought, which lock us into patterns of behaviour that often hinder creative and adaptive ways forward. One way to think about the constraints such legacies impose on the innovative development of water governance is in terms of path dependence, a concept drawn from institutional economics and related disciplines that offers a useful lens for identifying and understanding behavioural impediments to new ways of approaching water governance. Wheeler (2014) identifies this as an important lesson in water policy reform: 'Water reforms are driven by a complex interaction of multi-layer, path-dependent influences, with reforms building upon many previous waves of institutional reform' (p. 62).[2] We argue that the concept of path dependence can be more widely applied to help us understand the role of the imagination in decision-making processes – imagination as it works in established practice but also as it might work in more nuanced, collaborative and 'open' ways. Understanding the influence of the imagination on paths chosen in water policy and governance requires, in the first instance, a rethinking of the kinds of evidence that traditionally supports water governance policies and decisions. To our minds, weighing the imaginative history of water as a form of knowledge alongside other more familiar types of knowledge (for example, scientific, engineering) not only promises to enrich current understandings of our complex relationship with water, but also to challenge the current water policy paradigm with new ways of thinking. After all, while materially essential to all life on earth, water is in many ways also an abstraction, a product of both natural and social processes and their representations. As Linton (2010) writes, 'Water is what we make of it . . . every instance of water that has significance for us is saturated with the ideas, meanings, values and potentials that we have conferred upon it' (pp. 3–5). In these terms, then, we produce the kind of water we want.

Current understandings of water have an imaginative history that influences how Australians understand water in the landscape, and by corollary, how these imaginative habits can act to constrain the design of creative water policy. However, as Sardar (2009) points out,

> Given that our imagination is embedded and limited to our own culture, we will have to unleash a broad spectrum of imaginations from the rich diversity of human cultures and multiple ways of imagining alternatives to conventional, orthodox ways of being and doing. (pp. 443–444)

Acknowledging the influence of literature on cultural attitudes to water and landscape – a cumulative influence expressed through the collective imagination – paves a way for thinking about the drivers and consequences of institutional path dependence in this domain from a perspective that the sciences (social and physical) cannot always confidently address. For this reason, a closer 'interdisciplinary' partnership between literary studies and the sciences can enable greater progress in charting hitherto little explored ways in which water policy and governance processes in the Anthropocene might be made more sustainable.

The question of how Australians have imagined water over time is rich with collaborative possibilities. If, as Cathcart (2009) points out,

> we are to learn from our successes and our failures, we must first come to grips with the diverse ways in which Australians have struggled to understand the country. We have to articulate the values, myths, aspirations and anxieties that have shaped our cultural geographies. (p. 2)

To this end, we bring together two unlikely disciplinary bedfellows, economics and literary studies, to explore the role of the imagination in its potential to influence present efforts to shift water policy and institutions onto more sustainable paths. Where the notion of path dependence is a fundamental problem that stymies creative and collaborative change in water governance, we encourage the consideration of the ways that literary representations of water at once serve to reinforce 'locked in' ways of thinking, but also to offer the means of reshaping knowledge for the future. In using Mackellar's poem as a case study, we aim to contribute to a better understanding of how the powerful images created by literature – images, readily popularised, that inform mental models of the land's character – can offer new possibilities for water policy reforms, even as such images have the potential to reinforce prevailing collective understandings that impede such reforms. In acknowledging the nuanced dynamic between the literary and collective imaginations – to see this as important and useful knowledge in itself – we can appreciate how the Australian cultural imaginings of water might more directly inform approaches to solving water issues of escalating seriousness and contention. In turn, this provides three main lessons:

Lesson One	The limitations that path-dependent processes place on policy and governance reform;
Lesson Two	The complex role played by the imagination, in particular in path-dependent processes;
Lesson Three	The ways in which the social and physical sciences could work more effectively with literature and cultural studies in building the kinds of knowledge urgently needed in coming decades to overcome such limitations.

Path dependence

The process of reforming water policies and institutions is fundamentally path-dependent – past governance and policy choices in this domain limit current options, just as present choices will constrain the set of future reform options (Challen, 2000; Garrick, 2015; North, 1990). Policy choices in the contemporary Australian context are in turn increasingly compromised in imaginative scope and character by a 'confluence of effects', notably the erosion of the public service as the result of a rapidly changing political culture, the pervasive influence of managerialist practices on the bureaucracy, the narrowing obsession with pragmatic 'outcomes'-based methods of application and accountability and, of particular relevance in this instance, the increasingly impoverished nature of public discourse as concerns the dissemination of ideas and range of debate. Kashoor (2016) argues the point in this way:

> Our policy approaches have long fled from engagement with questions of values and ethics – of which policy settings aid individuals to engage in family or communal life, to participate economically, to be politically active and engaged … Principles about right action, flourishing lives, and the highest ideals or our politics are at stake here – not just the efficient use of our scarce resources … [P]olicy-makers [must] embrace their roles as protagonists in the drama of public ideas … understand their creative task in respect of the very hardest of those ideas … [and] engage with ideas as a way to reimagine, and ultimately reshape, our social and political worlds. (p. 69; see also Tingle, 2015)

With water policy facing escalating uncertainty as we proceed deeper into the Anthropocene, and the growing pressure for robust governance to contend with both unforeseen events and the turbulent dynamics of contemporary societies, there is an urgent need to better understand path dependence in its broadest complexities. To do so will help mitigate its negative consequences for institutional adaptability and transformability, and thus promote robustness. Characterising robustness as the ability of institutions to perform well despite uncertainty, Schindler et al. (2015) propose in this vein that 'the best we can likely do is enable our abilities to change course as new limits . . . or new opportunities are discovered' (p. 954).

However, with its tendency to normalise decision-making behaviour, path dependence works against the adoption of alternative approaches because it typically leads to the 'lock-in' of institutional choices, in the sense that the choices become more costly to reverse than to implement in the first place. One way that lock-in occurs is through the incentives that an institutional choice (for example, to publicly fund an irrigation scheme) creates for subsequent behavioural decisions by individuals and organisations. These decisions affect learning-by-doing and, in turn, 'the perceptions people have about the world around them and hence the way they rationalize, explain and justify that world' (North, 1990, p. 76). An institutional choice thereby comes to influence the beliefs, mental models and worldviews (Marshall, 2005) through which individuals and organisations perceive where their interests lie in any given situation. These mental models and worldviews tend to be reflected inter alia in literature which, through its ability to represent our world back to us in language (and also images, tropes and symbols), has a powerful influence on how we imagine and understand our world and thus the choices we make at the individual and institutional levels. Such generalised worldviews, plus perceptions of vested interests, influence and adapt to the institutional choice and contribute towards 'locking it in'. As observed by Bromley (2012) 'the habituated mind comes to see current practices, current choices, and current actions as normal, right, and correct' (p. 5). Such habituation is in turn reinforced by the shifting dynamics of the political and social context – the pragmatic imperatives to 'win' and 'succeed', for example, which demonstrably compromise more imaginative approaches to policy formulation.

Constraints on institutional reform from vested interests are not immutable, despite mainstream economists typically assuming the contrary. Critiquing this mainstream economic stance, Rodrik (2014) argues that

> new ideas about what can be done – innovative policies – can unlock what otherwise might seem like the iron grip of vested interests ... Political agents design their own strategy space. The available instruments are up for grabs and limited only by their political imagination. (p. 194–195)

Given that individuals and organisations 'often have only a limited and preconceived idea of where their interests lie' in making political choices, imagination, as it represents the creative capacity to open us to entertaining more possibilities, can enable them to reconceive their interests, or else devise innovative strategies for realising their existing perceived interests (Rodrick, 2014, p. 206). Either way, institutional lock-in can be weakened and institutional adaptability strengthened through the exercise of political imagination. A framework for understanding path dependence thus needs to account for the role of imagination in ameliorating lock-in. Of particular interest is the role that literature can play in this respect by contributing to the larger imaginary that is the nation – in fact this applies to all countries with a national literature – and, with growing emphasis in the Anthropocene, its relationship to geographical place.

How water has been imagined over time, particularly in representational forms such as literature, is deeply implicated in the path dependence that determines Australian water policy. The example of Mackellar's poem is especially revealing in this respect because at one level it supports its popular interpretation as a sentimental 'song' celebrating the modern and rightful enterprise of land management in Australia, suggested by the natural world conforming to human needs and economic demands – 'she pays us back threefold' (line 36). Later we show that reading this poem against the political grain (as it were) provides a sharper understanding of how imaginative conceptions of place are readily influenced in ways that prove particularly resistant to change. Such understanding can foster the kind of creative thinking, innovation or 're-imagining', that leads to a more sustainable path of institutional and policy development in the hydrological sphere. Here we echo Homer-Dixon's (2006) more general insight on sustainability that 'we can't hope to preserve at least some of what we hold dear . . . unless we're open to radically new ways of thinking about the world . . . We need to exercise our imaginations so we can challenge the unchallengeable and conceive the inconceivable' (p. 282). These are compelling words, but we can go further. Just as we need strategies that will inspire 'the public imagination with ideas based on the best science' (Marshall, 2010, p. 55), so we need to foster the means of imagining the 'best science' and, in turn, inspire different ways of imagining new approaches to water policy and governance. Furthermore, Scheffer et al. (2015) suggest that

> artists have a way of extracting meaningful aspects of the complex world around us that is quite complementary to what scientists tend to do, and thus may help to map some of the 'sea of ignorance' when it comes to finding interesting input for our hypothesis testing machinery. (p. 3)

It is promising in this respect that scientists are coming to recognize the value of the arts as a partner in catalysing novel ideas.

Of equal importance is an understanding of the complex dynamics informing the manner in which we imagine and represent the landscapes and waterscapes of nature and of the imagination that form an enduring mythography of water, which in turn organises a particular worldview. Here we use the term 'mythography' in its most fluid sense to mean the assembling and interpreting of a disparate set of texts and points of view that tell the story of water in the cultural imagination over time and as it shifts and alters under changing circumstances. As one instance of imagining, literature comes together with other instances, such as science and engineering, to form a mythography of water that transfers spatially and historically. An obvious example of this can be seen in the myths of the early British settlers in Australia, and by extension in other arid contexts, who brought to this landscape entrenched eighteenth-century expectations and processes of managing what Yi-Fu Tuan (1968) calls a 'well-watered earth [as] an unexamined article of faith' (p. 144). This imaginary of water was underpinned by little understanding of, even contempt for, aridity (Linton, 2010, p. 123). The newcomers imagined this dry land in a green English guise and believed themselves entitled to unlimited access to fresh water. Such instances of imagining have shaped and continue to shape attitudes to water in Australia and have stubbornly endured despite contrary evidence.

Acknowledging the historically significant role the imagination has played in bringing us to our current state of 'locked in' patterns of thinking is a critical first step in this process. So too is thinking about the role of the imagination in moving us out of what Sardar (2009) calls our 'current impasse.' In fact, Sardar identifies the imagination as 'the only tool, which takes us from simple reasoned analysis to higher synthesis' (pp. 443–444). Currently, considerations of the power of the imagination to influence, positively and negatively, innovation and change are largely absent in path dependence analysis. This is a critical omission in our view, given that our ideas about, and economic and emotional investment in, water have been shaped and sustained by cultural beliefs, values and myths of landscape carried over since European settlement.

It has become increasingly recognized, including by economists (see for example Crase et al., 2009b; Marshall, 2013; Marshall et al., 2016; McCann et al., 2014), that cultural barriers to water reforms – formed in part through just such beliefs and myths – need to be taken seriously because they are adding, sometimes prohibitively, to the costs of reform efforts. For instance, economists lament how the costs of recovering 'environmental water' (largely from the irrigation sector) for Australia's Murray-Darling Basin (MDB) have been increased considerably by political opposition from the irrigation sector and irrigation-dependent communities to water 'buybacks' (i.e., government purchases of water rights from willing sellers) while preferring the higher-cost infrastructure approach of recovering environmental water – behaviour that historically has been synonymous with attempts to realise imaginings of a well-watered MDB. Australia's Productivity Commission (2010, p. xxxv) found that policy-makers' acquiescence to relying on this system for most of its environmental water recovery 'can be seen as the price the Australian Government was prepared to pay to make progress on important reforms'. This price is reflected in the over $A20 billion that Loch et al. (2013) estimated has been spent on environmental water recovery in recent decades. Even though a large proportion of the existing MDB water recovery target has already been achieved (Murray-Darling Basin Authority, 2015), questions remain regarding the adequacy of this target (see for example Pittock et al., 2011). It is likely, moreover, that additional major water recovery efforts will be needed in coming decades for reasons including longer-term climate change impacts not having been accounted for in calculating the target (Alexandra, 2016)[3] and a likelihood that water savings from infrastructure projects have been over-estimated (Crase et al., 2009a).

Such factors emphasise the importance, from economic and many other perspectives, of understanding path dependence in water policy development in order to progress future required reforms in this domain with less friction from institutional lock-in. While Horn (1995) remarked that achieving institutional reform is unlikely unless the political (and thus underlying cultural) calculus that sustains existing institutions is accounted for, Wilson (2014, p. 2) proposed that understanding how the history of institutional practice has 'locked in' present water institutions 'can suggest both how they might come apart and how their permanence will shape water policy options and limit the politically possible.' Drawing on their imagination and creativity, political entrepreneurs capitalise on this knowledge to 'notice and exploit those loose spots in the structure of ideas, institutions, and incentives' towards more innovative approaches (Leighton et al., 2013, p. 134). This highlights how a beneficial approach to the manifold problems surrounding the present management of water in Australia and else-where would be to encourage the recognition that there are more nuanced ways of thinking about water than those that currently exist.

Imagining water and the role of literature

Literature in its various forms has had an important influence on the cultures of stakeholders in water reform efforts. Throughout the written and oral cultures of the world, story-telling stands as a rich repository of the beliefs, assumptions, fears and desires that influence and mould cultural imaginations; thus, gaining a clearer understanding of this deeper level of influence is important for identifying some of the obstacles that path dependence places in the way of reforming policy prac-tice. In this way, imagination is recognized as a valid form of knowledge in itself and, as such, can be included in deliberations over water policy and governance. Philosopher of the imagination, Bachelard (1983), calls this a 'a water mind-set': the manifestation of images of water in human life, dreams and creative forms such as poetry (p. 6). Stripping such cultural concepts as landscape, environment and water of their imaginative histories makes them unrecognizable. Picture, for exam-ple, a lake, river or stream that has significance for you. Now attempt to describe that water without recruiting your imagination or deploying metaphorical lan-guage. The task is all but impossible because the act of picturing is in itself an act of the imagination, which is to say that we do not in any simple way 'see' a body of water as an objective material entity – as noted above, we 'see' through a set of associated ideas, images and influences, whether we acknowledge this or not. How we, individually and collectively, have imagined and represented water over time makes it recognizable in the present and into the future: it is how we *know* water.

Furthermore, if we think of literature as a barometer of the world in which it is produced, then we can understand literary representations, in this instance of water, as taking on the characteristics – social, emotional, intellectual and historical – of that world and in turn participating in the production of cul-tural meaning, in the same way as do ideologies, customs, rituals and beliefs. Yet, literature also has the ability to make us see the world anew, to challenge and replace old assumptions and beliefs with new insights, and to open up our minds and imaginations, thus allowing new possibilities, perspectives and ideas to

Figure 1.1 A man in front of a sign pointing to water near the Murray River.
National Archives of Australia: A1200, L7328.

take hold. In this way, literature can create an imaginative space for us to think about water in different ways. Acknowledging this has the potential to inform a more nuanced and creative approach to water policy and governance than currently exists, because literature in all its forms influences the knowledges of the various stakeholders in this domain, and thereby contributes to shaping their identities and the emotions they attach to water in the landscape, to particular water 'objects' (for example lakes, wetlands, streams) and to the various kinds of water use, livelihoods and lifestyles associated with them. At the same time, this imaginative potential can work to reinforce undesirable attitudes and behaviour: literature in the past has contributed towards some of the cultural obstacles to water reform efforts experienced currently, the most obvious and enduring example being the biblical injunction for humans to exercise dominion over the natural world. The important point is that a greater understanding of the subtle and cumulative manner in which historical myths inform and create cultural obstacles offers a positive way forward. One recent instructive Australian example of how this process is made visible is dramatised in Cameron's (2014) satirical novel, *The Last Pulse*, in which the history of Australia's relationship to water – centring on the protagonist's farcical 'war' against the exploitation of the rivers of the Murray Darling Basin by farmers and graziers upstream in Queensland – is told in the form of an absurd comedy. Yet, literature also has the capacity to inspire creative leaps in the public and political imagination that will enable such resistance to be overcome.

Literature has a long historical relationship with the water in the landscape, a testament to the imaginative pull of both the reality and idea of water – its myth-inspiring power in fact. This is reinforced by Herendeen (1986), who points out that '[a]ny "place", whether real or imaginary, becomes the product of the creative mind' (p. 15), and Schama (1995), who notes that '[l]andscapes are culture before they are nature; constructs of the imagination projected onto wood and water and rock' (p. 61). As a product of the imagination, literature becomes a repository for these projections. Further, Cathcart pays particular attention to the role of the literary imagination in generating ideas and powerful misconceptions about water in Australia. From early settlement, commentators, lost for words when compelled to describe an unfamiliar land that eluded recognition, turned to the language of nation-building epic poetics of earlier English writers such as the seventeenth-century works of the poet John Milton (Cathcart, 2009, p. 80). Here we see the resilience of inherited mental models transferred across time and space to continue to influence the cultural imagination. In this way, myths of place and subconscious attitudes to water form and proliferate, contributing to path dependence by reinforcing entrenched ways of thinking about water and other natural resources. Importantly, myths are not necessarily static: they can change to shape our 'sense of place', and to influence stewardship of water and other natural resources in new ways. This is because an intimate and caring '[s]ense of place . . . increases the likelihood of prioritizing long-term solutions over short-term benefits' (Chapin III et al., 2012, p. 16). It seems obvious, then, that developing an empathetic connection goes a long way towards ensuring smart, creative and sensitive water management decision-making. With its ability to stir the imagination, give reflective voice to ideas and touch our emotional chords in surprising ways, literature has the power to influence this process.

'My Country' and the resilient myths of landscape

Known to most Australians for generations, taught widely in schools and re-imagined in song, film and art, the imagery of 'My Country' is deeply embedded in Australia's cultural memory. Referring to the poem, the New South Wales State Library's (2017) website states that 'Dorothea Mackellar's iconic verse is now regarded by many Australians as the universal statement of our nation's connection to the land', thus perpetuating the notion that Mackellar's poem inspires a shared attachment to the natural landscape. The resilience of

the poem and its imagery is also ensured in such creations as the memorialisa-
tion of Mackellar in a life-sized bronze statue at Gunnedah where a 'talking rock'
features the poet's voice reciting the poem, the adaptation of the poem into a
song, the conversion of the poem into parody and the production of a limited
edition commemorative coin set entitled 'Sunburnt Country' by the Perth Mint
in Western Australia. This speaks not only to 'My Country' as a cultural asset
in which Australians are heavily invested, but also to the cultural currency and
durability of its landscape imagery.

But what landscape does the poem represent? The poem has been interpreted
as representing, for example, a single landscape stereotype, or Australian nation-
alism, or a nostalgic yearning for England, or as reinforcing a colonial worldview.
For example, the image of '[a] stark white ring-barked forest / All tragic to the
moon' (lines 17–18) represents just one of the many painful scars inflicted on
the landscape by settler colonialism.[4] However, there is much more going on in
this poem, for it can also, and perhaps most convincingly, be read as a poignant
description of the deep interrelationship between water, landscape and culture.
Thought to have been written around 1904, the poem follows the breaking of the
Federation Drought (1895–1902), which had devastated livestock numbers, and
comes midway between the first major irrigation schemes of the 1890s and the
signing of the River Murray Waters Agreement by the Australian government
and the governments of Victoria, New South Wales and South Australia in 1915:
activities that set Australia on a path of ambitious water resource development
that has seen little change since.[5] The poem responds to this historical moment
by offering a timely reminder of the unpredictability and vulnerability of water in
the Australian landscape and what is at stake if we fail to understand and value
its ecological complexity.

In literary terms, 'My Country' is essentially a poignant love poem, an elegy
in the classical sense of the form, and herein lays its persuasive power. The first
stanza's concluding line, 'My love is otherwise', introduces the metaphorics of
love for the Australian landscape that the poem performs, which is reinforced
throughout with expressions of intense emotional attachment for this 'sunburnt
country' (line 9), 'her far horizons' (line 13) and 'her jewel-sea' (line 14). The
country is felt emotionally and physically; it is '[c]ore of my heart' (line 25) in
a way those 'who have not loved her / . . . cannot understand' (lines 43–44). At
times the land is personified as a changeable female lover – '[a] wilful, lavish
land' (line 42), she is at once punishing with '[h]er pitiless blue sky' (line 26)
and benevolent in her 'opal-hearted[ness]' (line 41). Furthermore, this land is
celebrated in its own uniqueness. The fact that it stands in opposition to that
landscape of settler nostalgia, of 'field and coppice, / Of green and shaded lanes. /
Of ordered woods and gardens / . . . of grey-blue distance / Brown streams and
soft dim skies' (lines 1–3), has captured the speaker's heart and is brought home
powerfully in the poem. Of equal importance, however, is that such imagery can
be considered as a particular kind of knowledge that circulates culturally and
politically to challenge the values, assumptions, beliefs and behaviours we currently bring to water policy design.

This is heady stuff. What we can identify here is recognition, conveyed with a deeply felt passion, of the uniqueness and complexity of the land. In many ways, this represented at the time a way of looking at the colonised Australian landscape through fresh eyes. However, the ability of this imagery to instil a new way of being in relation to the land has been compromised by the tendency to abbreviate the poem to the few lines that engage us at an emotional rather than intellectual level. Although 'My Country' actually consists of six stanzas (not one or two as often assumed), Australians are most familiar with the oft repeated first four lines of the second stanza: 'I love a sunburnt country, / A land of sweeping plains, / Of ragged mountain ranges, / Of droughts and flooding rains' (lines 9–12). Unfortunately, this abbreviation sacrifices the important nuances and tone of the poem. Moreover, what has become a nationalised declaration of love for the land has not resulted in a duty of care. The idea of 'loving' the land in this glib manner elides the more important task of acting out that love in a process of respect and caring of the kind that Alexandra advocates when he urges us to 'become intimate with our places, learn about, love and accept the "nature" of "a land of droughts and flooding rains" and get to know, understand and respond to its rhythms and patterns' (Alexandra, 2016, p. 2). Reducing the many layers of meaning in the poem to a few lines privileges a shallow, sentimentalised relationship with place over a deeper and more challenging connection that inspires commitment and action to ensure the wellbeing of the land. The former, by failing to prompt any kind of more serious re-imagining, risks creating, reinforcing and perpetuating path-dependent attitudes and approaches to water governance. Such cultural barriers to change need to be taken seriously because they are adding significantly to the cost and time involved in addressing issues concerning the regulation and governance of water supplies.

It should be emphasised that such an effect is not a matter of ascribing intentions to the author, or of arguing that there is a 'right' or 'wrong' way to interpret the poem's evocative imagery; rather, this effect occurs through the much subtler process of what might be called a form of translation, whereby 'misreadings' of 'My Country', repeated over generations, are woven into the collective mindset and recruited into a persistent belief in the possibility of water supply predictability. Selective images from the poem have been used to support a colonialist ontology that reinforces a particular water world-view. This is an instance of path dependence, wherein mental modes inspired by colonial imaginings of a well-watered land tend to filter out interpretations of the poem that are inconsistent with these imaginings. Accordingly, as stated above, the poem is reduced to a few simple ideas useful for recruitment into water, environment and climate policy and debate, as well as a whole slew of creative enterprises. For example, each silver coin in the Perth Mint's tribute to 'My Country' takes as its heading not the actual title of the poem but the familiar first line of the popularised second stanza, 'I love a sunburnt country', thereby reducing the poem to this single, memorable line, and confirming its mythic status in the national narrative. Further, two notable recent examples can be seen in the titles of, respectively, the formal report on the inquiry into the release of the Murray-Darling Basin

Plan in 2010, *Of Drought and Flooding Rains: Inquiry into the Impact of the Guide to the Murray-Darling Basin Plan* (House of Representatives Standing Committee: Commonwealth of Australia, 2011)[6] and the title of the book *Land of Sweeping Plains: Managing and Restoring the Native Grasslands of South-eastern Australia* (Williams et al., 2015). The frequent use of the image of 'droughts and flooding rains' in Australian vernacular accords with the kind of wishful thinking that underpins water policy based on modeling for predictable rainfall. The tendency to focus on this image from the poem fails to acknowledge the poem's reminder that the default position of this land is 'brown country' (line 47). In terms of the popular understanding, a selective reading of Mackellar's poem has influenced the way in which Australians envision their relationship with land and country, earth and ecology.

As mentioned, cultural myths, occurring through accretion and invariably entwined with associated images, ideas and beliefs, reinforce broad assumptions that often qualify as common knowledge, even accepted truths. In turn, such assumptions contribute directly to path dependence. Consequently, just as the collective 'misreading' of the poem fosters misunderstandings of country, so too are dreams of reliable water flows resilient to evidence of its scarcity (see Chapter 11, this volume). Representations of the land in 'My Country' suggest that notions of Australia's water abundance are wishful thinking. In fact, dreams of achieving such abundance through attempts to 'drought proof Australia', and thus tame its nature to conform to the idealised English landscape of the coloniser's past, have long been strong drivers of Australian water policy and governance arrangements (Williams, 2003, p. 40–42). The poem addresses this idealised image with the familiar trope of the ideal green mother country (England) of myth and nostalgia; however, by turning her back on the attachment the English have for this landscape, which she 'know[s] but cannot share' (line 7), the poem's speaker makes a strong statement about yearning for this green world as misplaced regret. Instead the poem embraces a landscape that stands in direct opposition to England: one that has been repeatedly compromised and defined in terms of this contrast since European settlement. In the harsh environment of the poem, the tragedy of drought in this 'wide brown land' of 'flood and fire and famine', '[w]hen sick at heart, around us / We see the cattle die', is rescued by '[t]he steady, soaking rain' (line 32) that produces a 'filmy veil of greenness . . . Over the thirsty paddocks' (lines 37–39). However, such verdancy is fleeting and unreliable. Mackellar's poem, then, does not imagine a predictable, hopeful land with promises of recurring water abundance, as is the case with England; rather, scarcity is its natural mode. Thus, the poem makes an appeal for respecting and embracing – and by corollary, planning for – hydrological uncertainty in the Australian landscape. This uncertainty is, however, at odds with a history of water governance decisions of over-allocation (see Chapter 11, this volume) based on flawed models of certainty, naturalised (as it were) in the national imaginary and memorialised for a range of purposes.

A greater attendance to literary representations of water such as offered by Mackellar's poem and its on-going public reception promotes a richer comprehension of the subtle and various human relationships to water. Such insights have immediate epistemological value and are readily translatable in practical terms, if not always apparent to those customarily engaged in formulating policy or contending with practical problems of supply and access. Consequently, the immediate relevance to policy formulation of such an understanding of the poem is to highlight the importance of being open, not only to alternative approaches and forms of knowledge, but also to doubt and uncertainty as intrinsic and necessary to the constructive process of water governance. Lawrence (2016) expresses the latter point well when she asks rhetorically,

> What's wrong with admitting that this [the Coalition's direct-action climate change policy] is a policy area where much is yet to be learned, that there are many proposed solutions, that most of the serious commentators are critical of his [the Minister for Environment's] policy, or at least, that while the jury is still out on many of the policy options, and new evidence will help shape future policy? (p. 29)[7]

Such a willingness to accept uncertainty indicates maturity in the exercise of political power and the deliberations over policy decisions; it also denotes the complementary acknowledgement that the past provides important answers to questions concerning human motivation and behaviour – questions customarily ignored in the contemporary situation: 'Even a cursory look at the language and images embedded in politics and public policy shows that asking serious questions about what the past can tell us about the likely effectiveness of proposed policies is rare' (Lawrence, 2016, p. 24).

Moreover, in encouraging a greater receptivity to the role of the literary imagination in the formulation of water policy and governance, we are promoting the fundamental rethinking – effectively, the re-imagining – of the means and very bases of what we presume and accept as knowledge regarding the most essential of life-supporting substances: water. 'Water,' as J. M. Powell (1997) observes in charting the origins of modern water management practice in Australia, 'never makes a mistake', but it is now alarmingly apparent that humans have, will and do make mistakes concerning this most vital of elements (pp. 102–121). Acknowledging the propensity to repeat those mistakes is an important step towards a richer understanding; by including literature in our deliberations, we create an opportunity to free up the imagination and thus mitigate future lock-in.

The fact that path dependence crucially influences worldviews and political approaches bearing on water policy and governance, and that culturally instilled ideas and beliefs – aspects of what might be called 'common knowledge' – play a central role in reinforcing these 'paths' supports our central claim that literature makes an important contribution to how Australians imagine water in the landscape and their relationship to their water environment. This raises the question as to why the abbreviated form of Mackellar's poem has enjoyed such enduring popularity. On the one hand the poem is worthy of popularity for its own sake: it is an expression of deep attachment to and respect for the beauty and ecological uniqueness of the Australian landscape. Read this way the poem has the capacity to instruct and inspire Australians to reassess their relationship with and experience of the land and water. On the other hand, as we have been arguing, selective imagery from the poem has been subverted, popularised and recruited into a convenient ideology not only of love for the land without the associated requirements of respect and care, but also as reinforcing the myth of predictable seasons. This is a mis-reading that has suited a range of cultural and political purposes and practices to do with water, thus encouraging instrumentalism, resourcism and a sense of entitlement. Such selective readings of the poem have worked to reinforce misconceptions about water in the landscape in the national psyche. 'My Country' is far from the only example of a literary work that has appealed to Australia's cultural imagination and, in doing so, influenced to some degree the national understanding of, and ideas about, land and country. Yet, equally, it has, in its bowdlerised form, come to encapsulate a powerful notion of postcolonial Australians' relationship with the country of the poem's title – a curiously sentimental and personalised relationship that acquires greater resonance in its continuing status as the nation's unofficial 'song'.

Conclusion

We argue in this chapter that processes of path dependence have in the past and continue to severely limit the scope of politically feasible water policies and governance options, and that by understanding these processes and being open to new, shared ways of thinking and different forms of knowledge and evidence, we can work towards the creative, innovative and collaborative approaches that are urgently needed. We consider the imagination as influenced by and influencing path dependence in Australia in complex ways and take the literary imagination, using Dorothea Mackellar's iconic poem, 'My Country', as an example of how this occurs. Further, we believe that the approach taken here provides a useful model for similar analyses beyond Australia. We propose that an enlightened way forward would be for the social and physical sciences to collaborate with literary studies to seek alternative ways of thinking and building the kind of knowledge urgently needed for robust water policy now and into the future.

Bringing the social and physical sciences together with literary studies is not as big a step as it seems. As this chapter has demonstrated, interdisciplinary research of this kind produces not only unique and valuable perspectives, but also encourages a deeper understanding of and respect for each other's discipline. The rewards are immense. The management of water is at once an environmental and a human problem. If we are to do what is best for both we need to fully understand the human story of water – literature is, after all, about what it means to be human. The human capacity to imagine as expressed in literature can be seen as a specific domain of knowledge with the power to influence how humans think about themselves in relation to the natural world. Because literature is both a product of, and has an influence on, the individual and cultural imagination as it relates to water, literature and literary studies has much to offer the on-going deliberations over water policy and governance. The description of the ecological uniqueness of the Australian landscape and waterscape in 'My Country', with its drought and dryness and unpredictable rainfall, offers an important lesson for policy makers.

We realize that policy makers are becoming more accepting of the need to incorporate different forms of knowledge, for example local, community and vernacular, into their deliberations, and encouraging progress has already been made towards meaningful reform in water governance. This suggests that there is, in some areas, a willingness to do things differently; however, the entrenched nature of path dependent processes makes this a challenging task. If the spirit for change is genuine, it seems not such a big next step for less tangible forms of knowledge such as the imagination and literature, which are deeply implicated in local and national knowledge systems, to be included in policy deliberations. Furthermore, as we note above, government committees and scientific researchers have appropriated inspirational catch phrases from Mackellar's poem. Indeed, if key players in the policy ambit can selectively utilise literature in this way, then it seems reasonable that they might look more deeply and consider the effect of such imagery on the cultural psyche and what else they might learn from the poem, from literature and from our wider imaginary engagements.

Notes

1 Gibbs (2009) observes that 'Australia's colonists brought with them ideas for ordering nature and tools for transforming landscapes that led to inappropriate and destructive water governance, and to the silencing and exclusion of indigenous voices and local knowledge systems' (p. 2964).
2 In discussing the impacts and challenges faced by climate change, Alexandra (2017) also draws attention to the need for 'adaptive water governance' and 'innovative approaches' (p. 1).
3 The first Basin Plan has a 10-year horizon. The Murray-Darling Basin Authority (MDBA) found that it was not possible over this period to distinguish the effects of climate change from the Basin's natural climate variability. The Minister for the Environment has suggested to the MDBA that further consideration of climate change risks occur when the Basin Plan is next reviewed.
4 Ring-barking refers to the removal of a deeply cut strip of bark from around the circumference of a tree, resulting in the death of the tree. European settlers used the practice extensively as an early method of land clearing.
5 The poem was first published in 1908 with the title 'Core of My Heart' and republished several times. In 1911 it appeared as 'My Country' in Mackellar's first book of poetry, *The Closed Door, and Other Verses*. Beverly Kingston, *Australian Dictionary of Biography*. <adb.anu.edu.au/biography/Mackellar-isobel-marion-dorothea-7383> 9 July, 2015.
6 The line in the poem actually reads 'droughts'.
7 The 'Coalition' is a shorthand reference to the political alliance of the Liberal and National Parties in Australia; the Liberal-National Coalition is one of the two dominant political parties in Australia, the other being the Australian Labor Party (or The Labor Party).

References

Alexandra, J. (2016). Water reform in Australia – 'experiments' in adaptive governance? Unpublished article (used by permission).

Alexandra, J. (2017). Risks, uncertainty and climate confusion in the Murray-Darling Basin Reforms. *Water Economics and Policy* 3(3), 1–21.

Bachelard, G. (1983). *Water and Dreams: An Essay on the Imagination of Matter*. Dallas: The Pegasus Foundation.

Bromley, D. W. (2012). Environmental governance as stochastic belief updating: Crafting rules to live by. *Ecology and Society 17*(3), 14. doi.org/10.5751/ES-04774-170314

Cameron, A. (2014). *The Last Pulse*. North Sydney: Vintage Books.

Cathcart, M. (2009). *The Water Dreamers: The Remarkable History of Our Dry Continent*. Melbourne: Text Publishing.

Challen, R. (2000). *Institutions, Transaction Costs and Environmental Policy: Institutional Reform for Water Resources*. Cheltenham: Edward Elgar.

Chapin III, F. S., Mark, A. F., Mitchell, R. A., & K. J. M. Dickinson. (2012). Design principles for social-ecological transformation toward sustainability: Lessons from New Zealand sense of place. *Ecosphere 3*(5), 1–21.

Crase, L., & O'Keefe, S. M. (2009a). The paradox of water savings: A critique of 'Water for the Future'. *Agenda 16*(1), 45–60.

Crase, L., O'Keefe, S. M., & Dollery, B. E. (2009b). The fluctuating political appeal of water engineering in Australia. *Water Alternatives 2*(3), 441–447.

Einstein, A. (1929). What Life Means to Einstein: An Interview by George Sylvester Viereck. *Saturday Evening Post*. 117. Retrieved from http://www.saturdayeveningpost.com/wp-content/uploads/satevepost/what_life_means_to_einstein.pdf

Garrick, D. E. (2015). *Water Allocation in Rivers under Pressure: Water Trading, Transaction Costs and Transboundary Governance in the Western US and Australia*. Cheltenham: Edward Elgar Publishing.

Gibbs, L. (2009). Just add water: Colonisation, water governance, and the Australian inland. *Environment and Planning 41*(12), 2964–2983.

Herendeen, W. H. (1986). *From Landscape to Literature: The River and the Myth of Geography*. Pittsburgh: Duquesne University Press.

Homer-Dixon, T. (2006). *The Upside of Down: Catastrophe, Creativity and the Renewal of Civilization*. Melbourne: Text Publishing.

Horn, M. J. (1995). *The Political Economy of Public Administration: Institutional Choice in the Public Sector*. Cambridge: Cambridge University Press.

Hunter, F., & Lee, J. (2016). Election 2016: Malcolm Turnbull's warning on climate change. *The Sydney Morning Herald*. 9 June. Retrieved from http://www.smh.com.au/federal-politics/federal-election-2016/election-2016-malcolm-turnbulls-warning-on-climate-change-disasters-20160609-gpf65k.html

Kashoor, V. (2016). Avoiding the simplicity trap: The predicament of public policy. In J. Schultz & A. Tierman (Eds.), *Griffith Review 51: Fixing the System* (pp. 54–73). Melbourne: Text Publishing.

Lawrence, C. (2016). The memory ladder: Learning from the past, living with doubt. In J. Schultz & A. Tierman (Eds.), *Griffith Review 51: Fixing the System* (pp. 24–32). Melbourne: Text Publishing.

Leighton, W., & López, E. (2013). *Madmen, Intellectuals, and Academic Scribblers: The Economic Engine of Political Change*. Stanford, USA: Stanford University Press.

Linton, J. (2010). *What Is Water? The History of a Modern Abstraction*. Vancouver: University of British Columbia Press.

Loch, A., & McIver, R. (2013). The Murray-Darling Basin Plan and public policy failure: A transaction cost analysis approach. In H. Bjornlund, C. Brebbia & S. Wheeler (Eds.), *Sustainable Irrigation and Drainage IV* (pp. 481–494). Southampton, UK: WIT Press.

Marshall, G. R. (2005). *Economics for Collaborative Environmental Management: Renegotiating the Commons*. London: Earthscan.

Marshall, G. R. (2010). Governance for a surprising world. In S. Cork (Ed.), *Resilience and Transformation: Preparing Australia for Uncertain Futures* (pp. 49–56). Melbourne: CSIRO Publishing.

Marshall, G. R. (2013). Transaction costs, collective action and adaptation in managing complex social-ecological systems. *Ecological Economics* 88, 185–194.

Marshall, G. R., & Alexandra, J. (2016). Institutional path dependence and environmental water recovery in Australia's Murray-Darling Basin. *Water Alternatives* 9(2), 203–221.

McCann, L., & Garrick, D. (2014). Transaction costs and policy design for water markets. In K. W. Easter & Q. Huang (Eds.), *Water Markets for the 21st Century: What Have We Learned?* (pp. 11–34). Dordrecht: Springer.

Murray-Darling Basin Authority. (2015). Towards a healthy, working Murray-Darling Basin: Basin annual report 2013–14. Canberra: MDBA.

North, D. C. (1990). *Institutions, Institutional Change and Economic Performance.* Cambridge: Cambridge University Press.

Pittock, J., & Finlayson, C. M. (2011). Freshwater ecosystem conservation: Principles versus policy. In D. Connell & R. Q. Grafton. (Eds.), *Basin Futures: Water Reform in the Murray-Darling Basin* (pp. 39–58). Canberra: ANU E Press.

Powell, J. M. (1997). Enterprise and dependency: Water management in Australia. In T. Griffiths and L. Robin (Eds.), *Ecology and Empire: Environmental History of Settler Societies* (pp. 102–124). Washington: University of Washington Press.

Productivity Commission. (2010). Market mechanisms for recovering water in the Murray-Darling Basin. Canberra.

Rodrik, D. (2014). When ideas trump interests: Preferences, worldviews, and policy innovations. *Journal of Economic Perspectives* 28(1), 189–208.

Sardar, Z. (2009). Welcome to postnormal times. *Futures* 42(5), 435–444.

Schama, S. (1995). *Landscape and Memory.* London: Harper Collins Publishers.

Scheffer, M., Bascompte, J., Bjordam, T. K., Carpenter, S. R., Clarke, L. B., Folke, C., Marquet, P., Mazzeo, N., Meerhoff, M., Sala, O., & Westley, F. R. (2015). Dual thinking for scientists. *Ecology and Society* 20(2), 3.

Schindler, D. E., & Hilborn, R. (2015). Prediction, precaution, and policy and global change. *Science* 347(6225), 953–954.

State Library New South Wales. (2017). *My Country: Dorothea Mackellar.* Retrieved from http://www.sl.nsw.gov.au/stories/my-country-dorothea-mackellar

Tingle, L. (2015). Political amnesia: How we forgot how to govern. *Quarterly Essay*, 60.

Tuan, Y.-F. (1969). *The Hydrologic Cycle and the Wisdom of God: A Theme in Geoteleology.* Toronto: University of Toronto Press.

Wheeler, S. A. (2014). Insights, lessons and benefits from improved regional water security and integration in Australia. *Water Resources and Economics* 8, 57–78.

Williams, J. (2003). Can we myth-proof Australia? *Australasian Science* 24(1), 40–42.

Williams, N., Marshall. A., & Morgan, J. (2015). *Land of Sweeping Plains: Managing and Restoring the Native Grasslands of South-eastern Australia.* Clayton: CSIRO Publishing.

Wilson, P. I. (2014). The politics of concrete: Institutions, infrastructure, and water policy. *Society and Natural Resources* 28(1), 109–115.

2 Aboriginal Rainmakers

A twentieth century phenomenon[1]

Lorina L. Barker

Introduction

In Australia, the importance of water knowledge reverberates through the work of the Aboriginal Rainmakers, the highly revered Elders renowned for their knowledge of water business (Rose, 2007), and is evident in the high esteem in which they are held. The knowledge of how to care for Country (land and water) through dance, song and ceremony is passed down by Aboriginal Elders from one generation to the next. It is the men and women of this fraternity who are commonly referred to as Rainmakers. Throughout most of the twentieth century, there was a fascination with the mystical abilities of Aboriginal Rainmakers. It was an interest that intensified during extreme drought conditions. People travelled to the 'outback' and 'red centre' to witness an 'authentic' Aboriginal ceremony, and some expressed their curiosity and scepticism in newspaper columns.

It is not the intention of this chapter to provide intricate details about rain ceremonies or their associated rain objects, as some are sacred and only accessible to the men of the Rainmaking fraternity. This chapter does, however, explore and unpack the way in which the Rainmakers are portrayed and how rain ceremonies were described and interpreted by outsiders in newspaper clippings during the first half of the twentieth century. It also looks at the interest in Rainmakers during this period and why they were in such demand. What does this tell us about Aboriginal people's water knowledge and the perceived uneasy relationship of the settlers with rain, rivers and floods? While it is acknowledged that the language used in these newspaper articles is set in the historical context of its time, the language used and the assignment of 'fake' names like 'Pepeorn' (Shaw, 1946, p. 4), 'King Billy Willy' (Bowyang, 1930b), 'Mick' and 'Jim' (Plenty of rain soon, 1934, p. 10) to Rainmakers continues to affect outsiders' beliefs about and attitudes toward Aboriginal people, our cultural beliefs, stories, songs and ceremonies. These perceptions were influenced by early nineteenth and twentieth century literature. Poetry and prose, as Shoemaker (1992) points out, 'characterised Aborigines according to damaging and degrading stereotypes' and made our ceremonies novelty 'objects' to despise or praise (p. 80).

The following is an analogy that provides a context for the focus on public rainmaking ceremonies, which men, women and outsiders are permitted to attend. The documentary *Bush Plum: The contemporary art of Angelina Pwerle* features Angelina, her two sisters, Polly and Kathleen Ngala, and friend Gladdy Kemarre from Utopia in the Northern Territory, as they travel through and paint Country (Hylands and Hylands, 2011). The artists explain the multi-layered stories embedded in their artworks, including the storyboard version, which is the officially accepted desensitized version for the public and the secret/sacred knowledge meant for community members, which is never disclosed to outsiders (Hylands and Hylands, 2011). This deliberate form of storytelling was most likely what the Rainmakers employed when performing rain ceremonies, choosing a more general version for outsiders, while the secret/sacred dances, songs and stories were only performed in secret among the men of the Rainmaking fraternity. And if a particular aspect is shared with an outsider, as Langton (2006) explains, the telling is done in a cryptic manner so as not to disclose the sacred nature of the story, place or item (p. 145). Therefore, this chapter focuses on the public Rainmaking ceremonies that men, women and outsiders are permitted to attend.

To understand this phenomenon from the perspective of the outsider I analysed 43 newspaper clippings from regional and urban Australian newspapers. The articles range from the negative and dismissive sceptic to the supportive believer espousing idealistic beliefs and attitudes, while other articles reiterate the mantra of a supposedly 'dying race' whose culture needs preserving. As Shoemaker (1992) points out, 'when viewed as objects of praise this could easily develop into idealisation, most frequently of traditional Aboriginal ceremonies and beliefs' (p. 79). There are a few articles that describe the Rainmakers as crafty negotiators who would threaten to cause a flood or refuse to 'persuade a drop to fall' unless their requests for payment were met in either cash or rations (Bedford, 1924, p. 15). Several articles report how the Rainmakers' power and competence were challenged (Drain, 1958, p. 10). Among them are articles that are tinged with ridicule and derogatory comments about the Rainmakers and Aboriginal peoples' beliefs and practices (Melon, 1936, p. 2). One article quotes the sermon of a superintendent of Forrest River Mission, near Wyndham in Western Australia, who attempted to show 'the link between the native superstitions and the religions teaching' – fire 'is the element that dispels evil spirits' – and the Rainmakers are God-like 'who control all terrestrial supplies of water' (Tribal myths, 1933, p. 6).

Rain dreaming

Rain ceremonies have been performed for millennia, fashioned by the ancestral beings during their epic journeys and exploits through Country. These spiritually imbued places are 'embodied in songs', which are performed by the Rainmakers (Tonkinson, 1972, p. 21). The documentary film *Putuparri and the*

Rainmakers: A remarkable story of national significance (2015) is evidence of this continuous practice of 'Water Knowledge' (see Chapter 8, this volume). The deep knowledge and reverence for water of the Martu people of the western deserts in Western Australia is expressed in the virtual reality film *Collisions* (2016) and through rain ceremonies (see Chapter 8, this volume). The Martu at Jigalong would annually perform the *Da:Wajil* Rainmaking ceremony where they sang about and re-enacted the exploits of *Wirnpa* and 'other ancestral rainmaking beings' (Tonkinson, 1972, p. 88). After his epic journey, *Wirnpa* turned into a snake and entered the *jila* (waterhole) that bears his name. Many western desert language groups are connected through the songlines of *Wirnpa* that 'travel underground, imbuing the Country with power' (MIRA, 2017). Today, the people who moved away from Country travel back to reconnect with *Wirnpa Jila*, also introducing *Wirnpa* and passing the responsibility of the songs onto the next generation of young people (MIRA, 2017). The Muruwari in northwest NSW also sang and performed many songs to *Bida-Ngulu*, the Great Creator ancestor. According to Muruwari custom, as my great grand-father Jimmie Barker explained, certain rain songs were sung only by cultural Elders, both men and women: 'the younger ones [could listen and learn them] but were not allowed to sing [them]' (Mathews, 1977, p. 32). There are, for example, the songs Jimmie often heard sung by Muruwari Elders, Old Maria and Hippai:

Give us rain	*Purtu nguwa ngana*
The ground is like dust	*Mayi ngara thulukala*
We have done you no wrong	*Wala yural thanu ngana*
Give us rain	*Purtu nguwa ngana*
Give us bread	*Manu nguwa ngana*[2]

This synchronized performance connects the Rainmakers to the ancestral beings and to all living things and in so doing evokes the Dreaming and the traditional law.

An outsider's view inside the ceremonies

The performance of these rain songs and dances attracted people from rural towns, cities and outback stations, and among them were researchers, writers, reporters, station owners, local town folk and tourists. These curious observers shared their experiences by publishing their accounts in local and regional newspapers. They were supposedly writing for an interested audience awaiting the next instalment of stories about the 'outback' and Aboriginal Rainmakers' 'power to produce rain' (Bowyang, 1930a, p. 8). Some used it as an opportunity to ridicule and express their own derogatory beliefs, while concealing their identity in a pseudonym like Paddy Melon, who wrote about 'Bogey, the demon rain-maker and "bone-pointing" mischief maker' (Melon, 1936, p. 2). Along with W. R. R. and James Devaney who used demeaning terms to describe the Rainmakers as 'devil-man', 'wizard' and 'charlatan' (Devaney, 1932, p. 21; W. R. R., 1932, p. 9), there was the movie enthusiast who wrote an article about *Kangaroo*, a Hollywood Western set in outback South Australia that featured a Rainmaking ceremony (Rain-making ritual shown in 'Kangaroo', 1951, p. 61).

The rain stories collated from newspaper clippings were mainly authored by men. If women wrote publically about their experience and knowledge of Rainmakers most remained anonymous, like the writer who used the pseudonym 'Crystal' (1912). The exceptions are Dorothy Drain (1958), a writer and columnist with *The Australian Women's Weekly*, Eleanor Barbour (1952), a columnist with the Adelaide *Chronicle* who featured 'eyewitness' accounts about the Rainmakers, and later Monica Heary (2002) a writer for *The Daily Telegraph*. Daisy Bates (1929) published her ethnographic and controversial findings of Aboriginal customs and ceremonies and her account of the Ooldea Rainmakers in 1929 (Bates, 1929, p. 6). While some of the male authors used pseudonyms, others simply used their initials followed by a last name, and a few used their full names, but this was generally the professionals and researchers, such as anthropologist Charles Percy Mountford. Some authors provided a retrospective account, such as the person who went by the name 'Bill Bowyang' (1926) who published stories in the *Sunday Mail* in December 1926, claiming to have witnessed rain dances and songs '[m]any years back when travelling from Winton to the Gulf' (Bowyang, 1926, p. 8). An amended version of these rainmaking stories was reproduced in the regional newspapers the *Narromine News and Trangie Advocate* in January 1930, and it was also circulated in northwest New South Wales in *The Western Champion* in February 1930 (Bowyang, 1930a, p. 16; 1930b, p. 8). When the article 'Rainmakers as profession' was published in *The Sydney Morning Herald* in November 1945, the author Leo C.F Law used his full name, but when the story was republished with a slight change of title 'Rainmakers as a profession' in *The Advertiser* in February 1946, there was a misspelling of his last name to 'Shaw' (Law, 1945, p. 6 and Shaw, 1946, p. 4). The article was also embellished with further details and circulated later that year in the *North Eastern Advertiser* under the initials L. C. F. L. (L. C. F. L., 1946, p. 4).

Who speaks for whom?

Most of the authors in these newspaper clippings used Pidgin English to represent the 'authentic' voice of Rainmakers. This linguistic feature is reflective of wider societal attitudes and beliefs, and the authors, as Jacquelyn Kilpatrick (1999) points out, 'cannot help but invest in their work their own preconceptions and attitudes' (p. 1). This selection also explains the continuing influence of nineteenth century Australian literature, which portrayed Aboriginal peoples' cultural beliefs and practices in a demeaning manner (Brantlinger, 2011, p. 127). As Patrick Brantlinger points out, nineteenth century poetry and novels depicted the 'unimproved savage' of a 'dying race' and bands of 'marauding natives' terrorising the settlers (p. 129–30). In these newspaper clippings the dialogue between the outsider and the Aboriginal Rainmaker is in Pidgin English; for example, on a western district property in Queensland, the station owner provided inducements for a ceremony to be conducted. Agreeing to the terms, the Rainmaker replied, 'All right, me make him plenty of rain come' (Brantlinger, 2011, p. 129–30). When asked about their impressions of the bright lights and big city of Melbourne, the Rainmaking troupe from Central Australia replied in unison, 'Good-oh' (Plenty of rain soon, 1934, p. 10). Sydney Kidman told stories of his travels through southwest Queensland and how he 'discovered a rainmaker named Coongy Bill', who was known to bring floods to which delighted community members would respond, '[p]lenty feed come along, and great big bullock jump up fat' (How the floods come, 1903, p. 4). Using this form of speech implies that Pidgin English is the only means of communicating with Aboriginal people and also emphasizes the Rainmakers' 'poor use' of English (Brantlinger, 2011, p. 130), further suggesting that their English skills are 'in some way linguistically underdeveloped or lacking in grammatical competence' (Meek, 2006, p. 100). In North America, the use of Pidgin English in early colonial literature and later in Hollywood films helped to create the imaginary 'generic Indian image' where the character speaks Tonto-talk – 'How! Me Injun' – reinforcing the stereotype of what Barbara Meek refers to as Hollywood Injun English: 'Me smoke-um peacepipe and How! Accompanied by a raised hand' (Meek, p. 96).

The good and the bad magic

Rainmakers are learned men who know Country and understand the changing atmospheric and environmental conditions. This was considered by outsiders as an act of 'gaining knowledge in a scientific way from nature study' (L. C. F. L., 1946, p. 4). When the Rainmakers performed for outsiders at public events, they were described favourably as the 'good' and obliging Aborigines from a 'noble race' who were invited to entertain international dignitaries, such as the Queen during a visit to Toowoomba in 1954 (Rain-making in the north, 1954, p. 2). When it rained, the spiritual practice of Rainmaking was considered to be 'powerful' and 'mystical'. The Rainmakers were considered to be as good as the 'white' meteorologists, whose expertise in weather patterns were thought to be 'not vastly different from the method of observation of the primitive black (sic)' (L. C. F. L., 1946, p. 4). However, when no rain came, attitudes toward the Rainmakers changed, and this was reflected in the negative textual representations of the Rainmakers and their spiritual beliefs, making it easy to dismiss them as mere forecasters of rain.

The outsider perspective as published in some newspapers encouraged readers to believe that the Rainmakers' status and reputations were dependent on their 'power to produce rain at will' (Bowyang, 1930b, p. 16). For example, in the Kimberley district of Western Australia, the *Examiner 1937* reports how a local Rainmaker's reputation is 'suddenly enhanced by a fall of rain'. The initial ceremony brought no rain and after a week, 'the rainmaker asserted that an evil spirit had taken away the rain'. But following the second performance, it rained in what is said to have been a 'phenomenal occurrence at this time of year' (Rain ceremony succeeds, 1937, p. 8). Being praised by a meteorologist and fellow Rainmaker further increased one's reputation, as Billy Bowyang professed in *The Western Champion*: 'Clement Wragge, one time Government Meteorologist of Queensland, in his lectures often referred to King Billy Willy, a famous [A]boriginal rainmaker, of the Kalkadoon tribe' (Bowyang, 1930b, p. 16). However, within their own communities Rainmakers are highly revered and respected Senior men and cultural leaders, known throughout the district and interstate for their rainmaking abilities. The need to impart status and prestige on a Rainmaker is perhaps a way for the outsider observer to maintain their scepticism or, in some cases, justify their scathing reports of the ceremony in question if there was no rain, thus making it easier to distance themselves from the Rainmaker and ceremony and disregard it as 'hocus pocus' (How much was faked, 1950, p. 6). However, like their nineteenth century forebears, the Rainmakers' decisions to bring rain, dust or a storm or chase it away were perhaps a form of tactical resistance to further encroachment on their lands (Smith & Reynolds, 2008). Henry Reynolds points out that these ceremonies and the use of 'magic' in secret also prevented violent retaliations on the frontier (Smith & Reynolds, 2008). Threatening to cause a natural disaster was also a way for senior cultural men during the twentieth century to acquire essential necessities such as rations. For example, great grandfather Jimmie Barker explained how Muruwari Elder Peter Flood was considered by white station owners in northwest NSW as 'an Aborigine of some importance' and how, if he did not receive extra rations he would 'threaten to stop the rain, bring a flood or produce some other dire calamity' (Mathews, 1977, p. 19–20). Irrespective of whether the granting of additional rations was out of fear or pity, it was an acknowledgement by the outsiders, the station owners, of the Rainmakers' or 'clever men's' powerful 'magic' (Smith & Reynolds, 2008).

Dancing up a storm

The performance of rain songs and dances attracted large regional and city audiences. For example, on returning from a trip of the Mann Ranges in 1940, Charles Percy Mountford, an anthropologist with the South Australian Museum, explained how he had witnessed Rainmakers 'break a months-old drought with impressive speed in four days, to be exact' (They almost make drought breaking look easy, 1940, p. 2). Similarly, a local from the Mt. Lawley region in Western Australia described an Arunda rainmaking ceremony that was said to have 'continued throughout the night', and 'two days later it rained' (Rainyerri, 1938, p. 44). Another observer, Dr J. G. Woods (1944) wrote of his 1944 trip to Innamincka in the northeast corner of South Australia with the New South Wales Royal Flying Doctor Service: 'I was very interested in the "rainmaking" which was going on at the time among the blacks (sic) on Innamincka Station'. He sceptically added:

[f]ailure during the first few days was accounted for by the old boy as being due to another rainmaker at Cordillo Downs, 100 miles to north, who was working against him. Later he claimed he was waiting for the wind to change ... Since the wind had been in the south for days, and all agree that a northerly wind precedes all rains in that district. (p. 2)

However, despite such scepticism, to 'break the drought' desperate station owners and companies went to great lengths to employ Aboriginal Rainmakers or their magically infused instruments. For example, in 1934 a group of Rainmakers from Central Australia performed at the Town Hall in Melbourne and, as honoured guests, they toured the city, which included a trip to 'the big waterhole – Port Phillip's Bay' (Plenty of rain soon, 1934, p. 10). With a slight change of title, the story of the Rainmakers performance in Melbourne was also circulated in other regional newspapers in Queensland in the *Evening News* (Native Rainmakers from the interior, 1934) and the *Maryborough Chronicle, Wide Bar and Burnett Advertiser* (Rainmaking in Melbourne, 1934) and was featured in the Brisbane *Sunday Mail* (Rainmakers from the desert, 1934). During the 1952 drought, the station manager of Etoduna Station near Birdsville in Queensland requested that rain stones be flown up from Copely in South Australia. It is said that the request was made on the advice of the local Birdsville Rainmaker who informed the station manager that 'Birdsville stones were not suitable and Copley stones "more better"' (Black's request fly rain stones, 1952, p. 3 and Rain is 'stone certain', 1952, p. 1). The manager of Trans-Australia Airlines who chartered the flight also mentioned how the prized rain stones 'enhanced' the weekly flight 'from the south to the parched north' (Rain stones for Darwin, 1952, p. 5).

The sharing of rain songs and artefacts is culturally practiced throughout Australia. For Aboriginal groups it was a way of maintaining connections to the ancestral creators and the places through which they travelled on their epic journey through Country and other neighbouring groups – what anthropologist Robert Tonkinson (1972) calls 'intercommunity transmission' of knowledge (p. 53). The neighbouring groups in the western deserts of Western Australia shared rain objects; for example the Elders at Jigalong gave two men rain stones to take south to Laverton (p. 53). Wangkumara Elder *Kaku* Jack O'Lantern performed rain songs from the Northern Territory given to him by his grandfather. *Kaku* performed these songs as part of the Weilmoringle Rainmaking ceremonies and later recorded them with Janet Mathews and linguist Lynette Oates.[3] Kaku was a part of a four-men troupe known as the Weilmoringle Rainmakers and included Muruwari song men Bertie 'Murdi-Murdi' Powell, Robin 'Quart-pot' Campbell and Cyril 'Shillin' Jackson. These men spoke and understood each other's languages and they came together to share and maintain rain songs, stories and dances.

The Weilmoringle Rainmakers, like their counterparts in northern, western and southern Australia, performed rain ceremonies on request for station owners and the general public and, in doing so, showcased as well as preserved their cultural beliefs and practices. For example, throughout the drought of 1965–1968, the Weilmoringle Rainmakers were invited to Sydney and Melbourne to perform 'ceremony for rain and for country, for people and for life' (Rose, 2007, p. 10). The invitations were probably triggered by their 1965 public performance during the Culgoa Sports carnival at Weilmoringle. *The Sun Herald* reported:

> They danced in the dust before a campfire and clicked boomerangs and twigs and chanted strange incantations. They wore emu and fowl feathers which they stuck to their bodies with their own blood. (Gill, 1996, p. 10)

It is said, as news of the performance spread, that some people even chartered airplanes to Weilmoringle and it is estimated that up to 500 people watched the ceremony on the day (Heary, 2002, p. 104). That same year, the Weilmoringle Rainmakers were invited by a metropolitan newspaper to perform in Sydney and, like their Central Desert counterparts, toured the city (Black outlook, 1965, p. 4; Gill, 1996, p. 10; Heary, 2002, p. 104). Later, in 1971, two travelling journalists, Gary Steer and John Davis, on a 'bush Christmas' experience, interviewed Robin Campbell and Jack *Kaku* O'Lantern at Weilmoringle about their rainmaking adventures to the 'big smoke'. When asked if the ceremony had brought rain Robin replied, 'Yes. We danced in Sydney and it rained in Melbourne' (Steer and Davis, 1971, p. 60–1). The Weilmoringle Rainmakers' 1965 trip to Melbourne was funded by the plastic raincoat manufacturing firm Plastalong, as part of their advertising campaign. The local *Brewarrina News* reported:

> [Plastalong] hired a plane to bring the men from Weilmoringle. They gave two performances, interspersed by models displaying raincoats, in the packed auditorium of Walton's store. (Gill, 1996, p. 10)

The continuous practice and exchange of cultural knowledge between Aboriginal groups was modified by the Rainmakers to suit engagements with the settlers, thus becoming a reciprocal cross-cultural transaction with perceived benefits for both Aboriginal and settler groups throughout the first half of the twentieth century.

While they were praised for their charismatic and mystical 'drought-breaking' abilities, the Weilmoringle Rainmakers, like their counterpart, the Booroloola Rainmaker Bill Hooker, became scapegoats when there was 'too much rain' and the threat of possible flooding. The *Centralian Advocate* in 1954 reported that after Bill Hooker had performed a rain ceremony, it 'rained and rained and kept on raining' until the Booroloola people 'got sick of it'; they 'told Bill Hooker a thing or two about his "art", grabbed the rainmaking stones and pelted them into the McArthur River' (Much too wet, 1954, p. 1). Similarly, Bertie, Robin, Jack and Cyril danced for days in 1956, and when the rain finally came down it inundated the Weilmoringle community for months and the Rainmakers blamed themselves. It is also said that the Springwood community blamed the Weilmoringle Rainmakers for the torrential rains that hammered the Blue Mountains in 1965, and the Docker River Aboriginal community held the Weilmoringle Rainmakers responsible for the 1974 flood (Heary, 2002, p. 104).

Robert Tonkinson (1972) was told by Aboriginal Rainmakers at Jigalong and other Western Desert groups that 'prolonged flooding rain is caused by the actions of unfriendly Aborigines in other areas' (p. 69).

Rain birds

While visiting Innaminka, Dr. D. G. Woods of the Royal Flying Doctor Service consulted a local from the Diamentina River area who explained how seeing or hearing the flood bird or rain bird signalled the 'come down' of water into the Cooper's Creek and Diamentina River (Woods, 1944, p. 2). Woods dismissed this explanation, preferring the scientific reason that 'the Cooper's Creek comes down any time, but regularly each third year in the case of the Diamentina'. His scepticism is further confirmed by news of pending floods via radio that 'an approaching "Cooper's" is common knowledge a month in advance, depending as it does on copious rain hundreds of miles away' (Woods, 1944, p. 2). Woods, like other outside observers, was ignorant of Aboriginal peoples' engagement with and knowledge of water. Similarly, when particular birds are sighted they signal flood or rain: 'birds are directly associated with Aboriginal Rainmaking beliefs and activities' (Tonkinson, 1972, p. 74).

Rain songs and ceremonies

While some Aboriginal groups continue to pass on water knowledge and songlines of Country, others mourn the loss of the Rainmakers who have joined our ancestors. For example, the Martu, who were Rainmakers and firemakers, are unable to complete some of their seasonal rituals, which means 'less rain and the reduced productivity of Country' (Walsh, 2008, p. 245). The intergenerational trauma of frontier violence, the loss of water knowledge and songs are evident in the oral histories of the Yanyuwa in the Gulf of Carpentaria. Elders describe how Yanyuwa senior cultural leaders and Rainmakers were attacked and killed by a group of white-men on horseback while performing a ceremony (Roberts, 2009, p. 5). The loss of the Rainmakers and the rain songs echoes today, as 'certain songs were only known by the old men who were killed, and so were lost forever' (Roberts, 2005). According to the Yanyuwa, this horrific event and the loss of most of their Rainmakers in the Kawurrungkuma (McPherson Creek) massacre resulted in the 'bad droughts in the 1890s' (Roberts, 2009, p. 5). Today, outsiders are still interested and fascinated by Aboriginal songs and ceremonies, especially during droughts when people are reminded of the human and divine intervention of the Aboriginal Rainmakers. As Merri Gill (1996) laments, '[i]n the recent long drought [of the 1990s], Weilmoringle could no longer call upon its Rainmakers for spiritual and practical help. We all felt part of that loss' (p. 10). So too did journalist Monica Heary (2002), who in 2002 interviewed Elaine Thompson at Caringle Station, about twenty kilometres southeast of Weilmoringle. Like Merri Gill, Thompson is also reminded of the Rainmakers' 'magic' during extreme drought conditions. She says, 'the emus are dying – they are the first to go. It's conceivable the Weilmoringle rainmakers of old could have helped' (Heary, 2002, p. 104).

Conclusion

The newspaper articles discussed in this chapter provide details of Aboriginal people's Rainmaking abilities; they also demonstrate the outsiders' contemporary beliefs, scepticism and their uneasy relationship with extreme weather – being inundated by floodwater one season and struck by very long periods of drought in another. This is perhaps one of the reasons why the Rainmakers were in so much demand during the first half of the twentieth century. The Rainmaker phenomenon highlights what is perhaps a fascinating and important episode of cultural interaction and negotiation between Aboriginal people and outsiders, particularly those individuals who managed remote properties and those who travelled from urban centres to the outback in search of water knowledge – water songs and ceremonies.

Notes

1 Sections of this chapter are included in Chapter 3, 'Ngarntu' of my Phd thesis, L. L. Barker, 'Ngarraka Yarn: A Murdi History of Weilmoringle', University of New England, Armidale, 2014, pp. 84–121.
2 These recordings form part of the Muruwari collection held at the AIATSIS. The above rain song was sung by Jimmie Barker and recorded in the 1970s for the researcher Janet Mathews, and later by Elder Robin Campbell in recordings with linguist Lynette Oates. Although there are two versions it appears to be the same song, although Jimmie's version is longer, while the above version includes the Muruwari language. See L. F. Oates, *The Muruwari Language*, Pacific Linguistics, ANU Printing Service, Canberra, 1988 and J. Mathews, *The Two Worlds of Jimmie Barker*, Canberra, Australian Institute of Aboriginal Studies, 1977.
3 These recordings form part of the Muruwari and Wangkumara collection held at the Australian Institute of Aboriginal and Torres Strait Islander studies, Canberra.

References

Barbour, E. (1952, April 3). Rain-Makers work in vain. *Chronicle*, p. 29.

Bates, D. M. (1929, December 7). Aboriginal Rainmakers. *The Australasian*, p. 6.

Bedford, R. (1924, July 19). West of the Warrego: Dry cleanliness. *The Brisbane*, p. 15.

Black outlook (for porcupines). (1965, September 21). *The Daily Telegraph*, p. 4.

Black's request fly rain stones. (1952, March 28). *The Courier Mail*, p. 3.

Bowyang, B. (1926, December 12). The Rainmakers: Aboriginal methods. *Sunday Mail*, p. 8.

Bowyang, B. (1930a, January 24). Aboriginal Rainmakers: Breaking droughts by magic. *Narromine News and Trangie Advocate*, p. 8.

Bowyang, B. (1930b, February 8). Aboriginal Rainmakers: Breaking droughts by magic. *The Western Champion*, p. 16.

Brantlinger, P. (2011). Eating tongues: Australian colonial literature and 'the Great Silence'. *The Yearbook of English Studies*, 41(2), 125–39.

Crystal. (1912, December 4). Aboriginal Rain-Makers. *Sydney Mail*, p. 27.

Devaney, J. (1932, June 11). Aboriginal lore. *The Brisbane Courier*, p. 21.

Drain, D. (1958, May 28). It seems to me. *The Australian Women's Weekly*, p. 10.

Gill, M. (1996). *Weilmoringle: A Bi-cultural Community*. Dubbo: Development and Advisory Publications of Australia.

Heary, M. (2002, October 25). Raining champions. *The Daily Telegraph*, p. 104.

How much was faked? Odd native happenings: Aboriginal 'magic'. (1950, February 9). *Geraldton Guardian*, p. 6.

How the floods come. (1903, July 14). *The Registrar*, p. 4.

Hylands, A. (Director) & Hylands, P. (Director) (2011). *Bush Plum: The Contemporary Art of Angelina Pwerle*. South Yarra Victoria: Creative Cowboy Films.

Kilpatrick, J. (1999). *Celluloid Indians: Native Americans and Film*. Lincoln, Nebraska: University of Nebraska Press.

Langton, M. (2006). Earth, wind, fire and water: The social and spiritual construction of water in Aboriginal societies. In B. David, B. Barker and Ian J. McNiven (Eds.), *The Social Archaeology of Australian Indigenous Societies* (pp. 139–69). Canberra, Australia: Aboriginal Studies Press.

Law, L. C. F. (1945, November 17). *The Sydney Morning Herald*, p. 6.

L. C. F. L. (1946, July 23). Aboriginal Rainmakers: Observers of nature's magical ritual. *North Eastern Advertiser*, p. 4.

Mathews, J. (1977). *The Two Worlds of Jimmie Barker*. Canberra, Australia: Australian Institute of Aboriginal Studies.

Meek, B. A. (2006). And the Injun goes 'How!': Representations of American Indian English in white public space. *Language in Society*, 35(1), 93–128.

Melon, P. (1936, February 20). Demon vs. demon. *The Queenslander*, p. 2.

MIRA. (2017). *Canning Stock Route Project Archive*. Retrieved from http://mira.canning stockrouteproject.com/node/2548

Much too wet. (1954, February 19). *Centralian Advocate*, p. 1.

Native Rainmakers from the interior: To demonstrate powers in Melbourne. (1934, September 24). *Evening News*, p. 9.

Oates, L. F. (1988). *The Muruwari Language*. Canberra, Australia: Pacific Linguistics, ANU Printing Service.

Plenty of rain soon? Aborigines and 'black magic'. (1934, September 24). *Advocate*, p. 10.

Rain ceremony succeeds: Native enhances reputation. (1937, June 12). *Examiner*, p. 8.

Rain is 'stone certain'. (1952, March 28). *The Argus*, p. 1.

Rainmakers from the desert: Melbourne tests black magic. (1934, September 23). *Sunday Mail*, p. 2.

Rain-making in the north. (1954, March 22). *Sydney Morning Herald*, p. 2.

Rainmaking in Melbourne. (1934, September 24). *Maryborough Chronicle, Wide Bay and Burnett Advertiser*, p. 9.

Rain-making ritual shown in 'Kangaroo'. (1951, March 24). *The Australian Women's Weekly*, p. 61.

Rain stones for Darwin. (1952, March 29). *Barrier Miner*, p. 5.

Rainyerri. (1938, September 15). Rain in the desert. *Western Mail*, p. 44.

Roberts, T. (2005). *Frontier Justice: A History of the Gulf Country to 1900*. St. Lucia, Brisbane: Queensland University Press.

Roberts, T. (2009). The brutal truth: What happened in the Gulf Country. *Monthly: Australian, Politics, Society and Culture* (The Monthly Essays), November, 1–30. Retrieved from https://www.themonthly.com.au/issue/2009/november/1330478364/ tony-roberts/brutal-truth

Rose, D. (2007). Justice and longing. In E. Potter, A. Mackinnon, S. McKenzie and J. McKay (Eds.), *Fresh Water: New Perspectives on Water in Australia* (pp. 8–20). Carlton Victoria: Melbourne University Press.

Shaw, L. C. F. (1946, February 16). Rainmaking as a profession. *The Advertiser*, p. 4.

Shoemaker, A. (1992). *Black Words White Page: Aboriginal Literature, 1929–1988.* St Lucia, Brisbane: University of Queensland Press.

Smith, M. & Reynolds, H. (2008). *Aboriginal Sorcery with Henry Reynolds.* Interview by Matt Smith with Henry Reynolds. Retrieved from http://www.latrobe.edu.au/news/articles/2008/podcasts/aboriginal-sorcery-with-professor-henry-reynolds/transcript

Steer G. & Davis, J. (1971, December 8). Midsummer in the Outback. *The Australian Women's Weekly*, pp. 60–1.

They almost make drought breaking look easy. (1940, October 31). *The Argus*, p. 2.

Tonkinson, R. (1972). *Da:Walji: A Western Desert Aboriginal Rainmaking Ritual.* Vancouver, Canada: PhD thesis, University of British Columbia.

Tribal myths: Spiritual beliefs of Aborigines. (1933, December 11). *The Daily News*, p. 6.

Walsh, F.J. (2008). *To hunt and to hold: Martu Aboriginal people's uses and knowledge of their country, with implications for co-management in Karlamilyi (Rudall River) National Park and the Great Sandy Desert, Western Australia.* Perth: PhD thesis, University of Western Australia.

Woods, J. G. (1944, October 10). Rain making and the flood bird. *Barrier Miner*, p. 2.

W. R. R. (1932, July 2). Aboriginal lore: Rain-making magic. *The Sydney Morning Herald*, p. 9.

3 'Like manna from heaven?'
Just water, history and the philosophical justification of water property rights

A. J. Walsh

Introduction: The role of historical claims in theories of just water

> If things fell from heaven like manna, and no one had any special entitlement to any portion of it . . . there might be a more compelling reason to search for a pattern. But since things come into existence already held (or with agreements already made about how they are to be held), there is no need to search for some pattern for unheld holdings to fit.
>
> (Nozick, 1974, *Anarchy, State and Utopia*, pp. 198, 219)

How should water be distributed? What principles of justice should be employed when making decisions about the fair or just allocation of water between competing uses? These normative questions regarding the distribution of water resources are undoubtedly of great social significance and yet, somewhat surprisingly, philosophers have had less to say about them than one might well have expected, given the scarcity of water in many parts of the world (Goudie, 2000, pp. 203–260). This lack of philosophical engagement is odd when one considers the extent of water shortage across the world. Not only is water scarce but, as a consequence of competing demands from industrial, agricultural and environmental groups, as well as from urban populations who require water for domestic use, there are increasing pressures on the water resources we have at our disposal. These demands create conflicts that will, in all likelihood, be exacerbated by climate change. Indeed, such pressures on water resources are already evident right across the globe, especially as a consequence of population growth. Is it possible, then, to develop a theory of Just Water? Furthermore, to what extent must the distribution of a resource like water be determined – as Nozick suggests above – by its history of production and development? Is history an important element of distributive justice?

Clearly, how water resources should best be distributed will be a key issue for public policy makers at both a national and an international level. At the present time in the Western world there is an increasing reliance on markets to solve problems of allocation and this is likely to be true well into the foreseeable future (see Chapter 6, this volume). In order to facilitate market-based solutions, markets have been created via the *privatization* and *commodification* of many water resources. Thus, water resources – which had previously been publicly owned or part of a national or global commons – have been *privatized*, and individuals or commercial groups have been assigned property rights over those resources. They have also been *commodified* in the sense that the water resources themselves and the rights to those resources can be bought and sold on open markets (see Chapter 5, this volume).

The justifications for these processes associated with commodification are worthy of closer consideration. In particular, it might be questioned why, given the distributive implications, previously commonly owned goods should be subject to appropriation by private individuals or corporations. One obvious response to this question is that privatization is a necessary precursor to the commodification that is central to market-based public policy; in order for resources to be bought and sold on a market and thus to be subjected to the efficiencies that such markets allegedly bring, then those goods presumably need to be owned privately. The pressure for such privatization, at the level of public policy, then, has often come from those who believe markets bring efficiency and thus the *justification* for privatizing provided by such groups is typically utilitarian and, hence, extrinsic. But equally, the call for the privatization dovetails neatly with the claims of those who regard privatization as intrinsically justifiable because private property rights are deserved in terms of the history of how 'goods' have come into existence. The history of the goods provides *fundamental rights* of ownership and control over them. The concern of this chapter then is with those historical claims. Justificatory claims based on utilitarian considerations will need to be dealt with elsewhere in another forum.

Private property claims – historical or otherwise – in general have distributive implications. If the pattern of access to and use of water resources across a society is questioned, and if there are legitimate property rights associated with them, then the presumption is that those water resources should be held by persons with such rights and that this will have distributive implications. In this scenario, could *history* provide grounds for the legitimate acquisition of property rights in water? How might claims of history do so? These are fundamental questions for those who wish to marketize water on the grounds that property rights generally involve respect for a human right to own goods.

With these considerations laid out, it is now time to sharpen the analytic focus; in particular this chapter is concerned with the following problems:

1   The role that historical considerations should play in the allocation of water.
2   Whether there are circumstances in which historical considerations are entirely irrelevant.

These are important considerations for *any* theory of Just Water. The evaluation of these questions will begin with a brief consideration of the essential nature of the problem of distributive justice itself.

Distributive justice and water: Some conceptual preliminaries

The issues raised in this chapter are part of a set of larger abstract issues concerning the just distribution of water (or the theory of Just Water). Some salient features of the problem of distributive justice will be explored, in particular as those problems relate to water. The concern in this chapter is with *historical approaches to the problem of the just distribution of water*. But what is distributive justice? What exactly does this term mean?

Distributive justice concerns itself with the fair or just allocation of *benefits* and *burdens*; that is, it concerns decisions about the distribution of *both*: (i) the goods that society procures or produces and (ii) the harmful activities that are necessary to secure those goods and to maintain flourishing human lives. Thus, when social actors argue in the public domain about the fair way of ensuring that all have clean drinking water, they are debating an issue of distributive justice.

In addition to the idea of benefits, it is important to include *burdens* – or, as David Miller calls them, 'disadvantages other than punishment' – in our definition of distributive justice (Miller, 1976, p. 22; Walzer, 1983, p. 165). Miller takes burdens simply to be negative benefits. It is appropriate (and essential) that burdens be included in our discussions of distributive justice, for when societies develop systems for the allocation of resources – whether they involve principles enacted by the state or market distributions – they also allocate burdens. These distributive burdens involve activities that are required for the production of benefits or for the prevention of harms. Consider the following issues. Who lives near the rubbish tip? Who has to fight in a war? Who has to undertake the unpleasant work that is necessary for the production of social benefits or for the prevention of further harms? How should the sick and elderly be looked after? These are all questions that raise vexing problems of distributive justice. Turning to water, there are important distributive problems concerning burdens that arise, for example, in relationship to flooding, wastewater and sanitation. Where should wastewater go? Should towns be allowed to pump wastewater into streams and into the ocean? Who has responsibility for the costs associated with wastewater? It is important to remember, then, that water is both a benefit and a burden. Theories of 'distributively just' water must take into account both of these considerations. How a society should fairly or justly deal with the burdens associated with water will also be a significant consideration for any theory of Just Water.

Some of the issues that will be raised in what follows about distributive justice will be the following:

- What it is that makes a distribution just.
- What principles, if any, should govern the just distribution of social resources.
- What mechanisms should be employed to distribute benefits and burdens.

In relation to water itself there are a series of significant issues, including:

- How the allocation of water can be be fairly determined.
- What the normative principles that underpin the allocation of water should be.
- How past injustices in relationship to the allocation of water might be rectified.
- Whether there are *specific* features of water itself that will influence the shape of the distributive principles that are appropriate to water.

These are the central issues for any theory exploring the just distribution of water.

An important distinction to keep in mind when discussing theoretical issues about distributive justice is that between the *concept* and *conceptions* of justice. Originally drawn by the philosopher John Rawls, this distinction highlights the difference between the territory on which debates take place and the content of any such debates themselves (Rawls, 1972, p. 5). In Rawls's usage, a concept presents a problem, while a conception presents a solution. A conception is, it might be said, a specification of a concept. The concept of justice involves questions concerning the nature of the problem of justice. A conception of justice, on the other hand, involves concrete proposals about how goods should be allocated. Thus, if it is the question of how problems of distributive justice arise that is being discussed, it is the *concept* of justice that is under consideration. If the question at hand concerns whether it is fair that certain goods be allocated equally, then it is the legitimacy of a specific *conception* of justice that is of concern. Again, the concept of justice involves the notions of arbitrariness (or lack thereof) and of striking a proper balance between things; a conception of justice will offer interpretations of arbitrariness and propriety. It is vital when discussing questions of distributive justice to be aware of the *level of the analysis*, for all too often debates in this area become confused when discussants move between these two levels. It is useful, therefore, to set out important conceptual features before considering the role that historical considerations should play in developing a conception of Just Water.

In what immediately follows, a number of *conceptual* points about distributive justice and especially about distributive justice in the context of water policy will be made. In re-imagining water policy it is important to be mindful of the very nature of the problem of distributive justice and various constraints and limitations upon its implementation. Thus, it will be argued that:

(i)   Scarcity is a necessary condition for the existence of the problem of distributive justice.

(ii) Human beings have both an identity and a conflict of interests, and recognition of this must be central to our understanding of the problem of distributive justice.

(iii) It is important to be mindful of the fact that questions of justice are so central that our capacities for unbiased judgments are much lower than in many other areas of inquiry.

(iv) Loss on the part of any individual in some form of redistribution does not necessarily entail moral harm or moral wrongdoing.

(v)   It is important to also be vigilant that the allocation of burdens as a central part of the problem of distributive justice is not ignored.

First, how problems of distributive justice arise should be considered. The primary reason such issues surface is because of the *scarcity* of social goods. Imagine, by way of contrast, a world of *absolute* abundance. In such a world – given appropriate means of access for all members of the world – there would be no need to choose between different uses of resources, since all possible interests could be satisfied. In such a world there would be no competition for resources. However, in our world, the situation is such that choices must be made between competing users and competing uses. This is a fundamental condition of the world as it is currently constituted and thus forms part of the circumstances of justice around which principles of justice must be framed. It should also be clear that water policy is very much shaped by the fact of scarcity. Droughts are common events in many parts of the world today, and with this comes a dearth of clean water. This is likely to increase with climate change. These are all central issues for water policy.

In fact, scarcity is at the heart of water policy and can be usefully divided into two kinds: namely *moderate* scarcity and *hard* scarcity (Walsh, 1997, p. 119). In conditions of moderate scarcity it is possible to distribute goods universally. However, it is characteristic of hard scarcity that resources cannot be distributed universally. In conditions of hard scarcity, supplies of goods reach such low levels that if they were to be distributed universally they would no longer be the goods that they are. For example, if a case in which there are only two kidney machines in a remote village and fifty people who need to use that resource on a 24-hour basis is considered, then this involves hard scarcity. Lifeboat problems, where one has not enough seats to save all who require saving, also involve hard scarcity. Despite the difference noted here, *both* hard and moderate scarcity give rise to questions of distributive justice – the problems are simply more severe in cases of hard scarcity.

A second point concerns the nature of human interests. John Rawls suggests in *A Theory of Justice* that society is a cooperative venture with both an identity and conflict of interests (Rawls, 1972, p. 4). There exists an *identity of interests* since social cooperation makes possible a better life for all than they would have if each were to live solely by his or her own efforts. Additionally, there exists a *conflict of interests* since persons and collectives are not indifferent as to how the greater benefits produced by their collaboration are distributed.

Third, it is important to be mindful of the fact that distributive justice is not a topic to which any of us are wholly neutral (and in this sense it is perhaps very different to many other areas of philosophical speculation). Further, it is often the case that the ordinary views expressed about the fair allocation of goods in any particular case reflect positional interests. This might usefully be referred to as *positional bias*. Think here of a media magnate who stridently asserts that progressive taxation is unfair and harms the economy. Equally, think of those poorer members of society who argue that the wealthy should be taxed at higher rates, and that these taxes be used to provide better social services for everyone and to ensure that all have an equal opportunity to develop their talents. This is also evident in debates over water allocations. So irrigators, for instance, present very different claims of justice from those who wish to use water for industrial processes, and, in both instances, those claims are intentionally developed to undergird their own rightful claims to the resources in question. This looks rather like what might be termed 'positional bias'.

Faced with such bias, some readers might be tempted towards a form of general skepticism about the *very idea* of distributive justice. If claims of injustice simply reflect the positional bias of competing persons and groups, then why not regard the whole intellectual enterprise of arguing about distributive justice as intellectually bankrupt? This is a good question and raises serious problems for the justice theorist. Indeed, it was a concern with such bias (although he did not describe it directly using this terminology), whilst maintaining a commitment to distributive justice as a plausible and coherent intellectual pursuit, that led Rawls to develop the device of the so-called 'Original Position'. The Original Position is a thought experiment that involves asking what determinate principles of justice agents would agree to if they did not know inter alia: (i) our position in society, (ii) their levels of wealth and (iii) the talents that they possess (Rawls, 1972, p. 19). According to Rawls, agents should adopt as just or fair those principles to which they would agree in the Original Position. In this way, Rawls provides one means of avoiding such skepticism. Regardless of whether one agrees with Rawls's solution or whether one thinks of justice as objective more generally, one should be mindful of the fact that our views in this area are often biased towards our interests, and if an agent is concerned with determining what *really* is fair, he or she should reflect closely upon his or her own judgments and be careful that those judgements do not simply reflect what benefits him or herself or their own group.

A fourth point to note is that loss on the part of any individual in some form of redistribution does *not necessarily entail moral harm*. In many discussions of justice and distribution there exists a tendency to assume that if there is a loss on the part of an individual, then this implies or entails some wrongdoing on the part of some other person or institution. Indeed, it is a common tendency within everyday social discourse to regard *any loss – or the fact* that one is a loser in some changed set of circumstances – as *irrefutable evidence* of injustice. But this is fundamentally mistaken. If the preceding discussion regarding the necessity of sacrifice given the fact of scarcity is recalled, then it should be clear exactly why this is mistaken. Equally, rectifying past injustice will typically involve some loss for parties other than those who have been injured or are viewed as the victims of some injustice. If distributive justice involves the allocation of scarce resources, then the fact that some person has little or no access to a set of goods or, alternatively, that changed arrangements mean that they have less than they had before, is not *in and of itself* evidence of injustice. There will always be some who do not receive all that they need or require. Of course, equally, that is not in any way to suggest that when people lose access there is never any injustice. In many instances there will be considerable injustice perpetrated against people as a result of decisions about how social goods should be allocated – indeed rectifying and avoiding injustice is a large part of why discussions of distributive justice matter. The point here is simply the conceptual one that *mere loss is not by itself* a sign of moral harm or moral wrongdoing on the part of those responsible for the new distributive arrangements.

A fifth conceptual point concerns distributive burdens. The great bulk of philosophical discussion of distributive justice focuses on social benefits. However, concern with social allocation also encompasses questions of social harms or socially generated evils. In thinking of water as a problem of distributive justice, it is important to ensure that the allocation of burdens as also an important element of the problem of distributive justice is not ignored. For instance, in some parts of the world, changes to the climate mean that there is an over-abundance of water with flooding causing major concerns for countries like Britain and large swathes of South East Asia. Thus, social policy-makers need to be just as alert to possible injustices associated with, for example, flooding, as they are to those that come with drought.

One final point to make when considering the conceptual terrain concerns the relationship between the legal institution of property rights, on the one hand, and (i) systems of distribution and (ii) claims of distributive justice, on the other. Note that this is particularly important in the context of the central concern of this chapter. Any talk of distributive justice in water must acknowledge the role that property rights play in determining access to water. In general, to establish or allocate property rights – be they private or collective – over a set of resources has distributive implications. If resources are state-owned, then this will generate different distributive patterns and regimes than if the resources are privately owned. The privatization of water will have significant distributive effects because private property rights in any good typically include inter alia rights of use, control, transfer and exclusion (Honore, 1961, pp. 107–147). Those who hold private property in water will have a right to determine, within certain limits, how the water in question is used and who can and who cannot have access to it. Hence, granting private property is an *indirect* means of influencing the overall pattern of holdings across a society. To summarize:

(i)   Any system of property rights in water strongly determines a pattern of distribution of water across a society.
(ii) Establishing a system of private property rights in water will have significant effects upon the pattern of distributive access across a society.
(iii) Claims to private property rights in water are ultimately claims about what is 'distributively just'.

These three points are significant because many philosophical theories of justice, in their frameworks of evaluation, ignore private property rights and the impact those rights have on overall distributive patterns. However, as shall be seen in the next section, this is certainly not true of the work of Robert Nozick.

Nozickian historical claims

As was noted above, one strong undercurrent of contemporary thinking within water policy is the desire to transform water resources into private property and, through the establishment of water markets, to solve distributive problems.

This is part of a more general movement in favour of solving environmental problems through markets (Kolstad, 2011, pp. 67–82). Sometimes the establishment of such markets is justified via claims about the overall social utility of markets, whilst other justifications are historical in nature. Perhaps the most influential of these justifications is to be found in Robert Nozick's *entitlement theory* that he develops in *Anarchy, State and Utopia* (1974).

Nozick's theory must be understood in relation to, and in contrast with, the approaches of many modern theories of justice, at least up until the publication of Nozick's work. Most theories of distributive justice assume that there is a pool of social resources before us to which distributive principles need to be applied. Thus, the aim of the political philosopher is to formulate principles that determine how such goods should be allocated. So an egalitarian will argue (roughly) that social goods should be allocated equally, a meritocrat that they should be allocated to those who deserve them (on whatever criteria of desert are deemed appropriate) and a 'sufficientarian' that goods should be allocated in the first instance to ensure that all basic needs are met to an adequate level of sufficiency. Within what has been labelled the standard approach, the role of governments is thought to be to ensure that the favoured pattern of justice is realized. Such principles are typically ahistorical in the sense that the history of how the goods came into existence is irrelevant to the question of how they should be allocated.

It is the apparently *ahistorical nature* of such theories upon which Nozick concentrates much of his attention. Nozick is opposed to theories like Rawls's that begin with the question of how social resources should be distributed, as if they are part of a social pot waiting to be distributed. He claims that there is a fundamental flaw in accounts such as Rawls's and this is that they address the problem of distributive justice by asking, in the first instance, how the goods of a society should be justly distributed (Nozick, 1974, p. 198). But, in doing so, they assume that resources exist in a big social pot waiting to be allocated by some central authority. Nozick says it is *as if* goods 'fall like manna from heaven' and the job of the political philosopher is to work out how such manna should be distributed. According to Nozick, this is mistaken: goods do not fall like manna but instead are produced. Goods have a history. Indeed, Nozick argues that it is not possible to separate how goods are produced from how they should be distributed. According to Nozick, the production of goods and their proper distribution must not be treated as entirely separate matters.

His concern with history as a *category of analysis* is but one part of a larger taxonomy of theories of justice. Nozick classifies distributive theories using two distinctions between: (i) historical and end-state theories and (ii) patterned and unpatterned theories. An *historical theory*, in Nozick's picture, takes the historical origins of goods – that is, the tale of how they were produced and how they were transferred from producers to new holders – as central to understanding who should own and have productive control of any good. Nozick contrasts historical theories with end-state theories, like Rawls's. For Rawls, the choice of principles in the Original Position, behind a 'veil of ignorance' as it were, must be based on calculations about what people are likely to end up with under the various possible sets of principles – and assumes that there is no other way of choosing (Nozick, 1974, p. 202). Nozick suggests that if *any* historical entitlement theory is correct, Rawls's approach (and any other end-state theory) is wrong.

Nozick also distinguishes between patterned and unpatterned theories. According to a patterned theory there is a pattern of distribution that the social institutions such as the state should aim to realize in the future. For the egalitarian, the pattern is equality; for the utilitarian, the pattern is one of maximized utility and so on. Unpatterned theories, by way of contrast, are *indifferent* to the pattern of holdings across a society (Nozick, 1974, p. 157). Furthermore, Nozick argues that distributive theories should be *historical* and *unpatterned*: the theory he develops is one that assigns property rights to producers of goods and consequent rights of transfer to those who have property rights. Nozick's theory does not demand that the distribution resulting from just acquisitions, transfers and rectifications be patterned, i.e. correlated with anything else (such as moral merit, need, usefulness to society); people may be entitled to things obtained by chance or gift. Any distribution, irrespective of any pattern it may or may not have, is just, provided it has the appropriate history that it did, in fact, come about in accordance with the rules of acquisition, transfer and rectification.

According to Nozick, the reason why theories should be historical is that societies must respect the history of how goods were produced and subsequently transferred from legitimate owners through free exchange, and an end-result theory – such as Rawls's account – fails to respect that history. Distributive principles should be unpatterned because any attempt to impose a pattern impinges on our liberty. The very moment that people begin to make free use of resources and trade with others, any desired pattern will be disrupted. Nozick argues that the only way one can maintain a pattern is to limit the opportunities for people to produce and to exchange goods freely, but any such limitations involve a violation of fundamental rights. Accordingly, Nozick suggests that the proper principle of distributive justice should be '[t]o each as they choose, to each as they are chosen' (Nozick, 1974, p. 160). This principle is *unpatterned* and *historical*, in that there is no predetermined pattern of distribution that it holds to be correct (such as equality or maximizing utility) and that it respects the history of how the good came into being. Nozick also builds in a principle of rectification for instances where goods are acquired illegitimately, either through coercion or fraud (Nozick, 1974, pp. 151–152).

The difficult question for Nozick at this point is how people come to own un-owned things legitimately. How do un-owned things, such as for instance a stream, come to be owned? Transfer is, in one sense, easier to handle conceptually, for once ownership is established then how ideals of legitimate transfer might lead to patterns of exchange can be understood. This question of how ownership comes into existence is one that has puzzled political philosophers since at least the time of John Locke (1632–1704), who thought that by mixing one's labour with an un-owned object one came to own it. (This is sometimes referred to as the 'labour theory' of property.) Further, Locke argued that individuals have a right to commandeer private property by working on it, if and only if 'enough and as good [is] left in common for others' (Locke, 1967, p. 309). In his justification of the legitimacy of private property, Nozick makes use of both the Lockean idea of mixing one's labour in order to explain how un-owned natural resources might come to become private property and what he calls the Lockean proviso, which demands that enough and as good be left over (Nozick, 1974, p. 175).

There is – rather unsurprisingly – a remarkably extensive extant secondary literature that is critical of both the labour theory of property and the Lockean proviso (Varden, 2012, pp. 410–442). Indeed these two are perplexing ideas. For instance, some critics have asked if one mixes some fruit juice one has made one-self into the Pacific Ocean, does that mean one thereby owns the whole ocean? One might also argue, quite plausibly, that Nozick overstates the role of histori-cal considerations in questions of distributive justice. However, for the purposes of brevity, such critical questions about the system as a whole will be left to one side. Instead, the aim in the rest of this chapter is not so much to attack the theory itself (although there might well be many good reasons for being sceptical of it) but rather to explore its *applicability* to the issue of water distribution and the justification of private property rights in water. For argument's sake, it will be assumed that the Nozickian account of distributive justice is plausible and the focus will instead be on how well it legitimates private property rights in water. One general question to ask at this point about entitlement and property concerns the extent to which history only provides grounds for private property rights and not for other forms of property.

Indigenous land claims, cultural stewardship and Nozickian project

Nozick's entitlement theory was received with feverish enthusiasm by the free market right when it was published in 1974. They regarded *Anarchy, State and Utopia* as providing a sophisticated defence of free market capitalism as well as furnishing them with moral arguments that seemingly justified 'unpatterned' forms of inequality. Subsequent to the publication of Nozick's book, the history of a good's production and trade became the focus of much political philosophy. However, curiously and somewhat ironically, some leftists took the historical pro-ductivist arguments as inadvertently providing justifications of both the Marxist theory of exploitation and of Indigenous land claims. G. A. Cohen famously claimed in a number of works that Marx's theory of capitalist exploitation of the working class was in fact an historical entitlement theory (Cohen, 1995, p. 150), while David Lyons, in an article originally published in 1977 entitled 'The New Indian Claims and Original Rights to Land', extended Nozick's entitlement theory to contemporary Native American claims on land in the United States.

In this section, the focus will be upon the latter of these because there are a number of distinct politically significant historical claims associated with water that might be thought to be based on entitlement not need and yet which dif-fer markedly from those based on mixing one's labour. Here it is inter alia the claims of Indigenous groups to water resources and water systems they have traditionally lived on which is explored, as well as various riparian claims of other non-Indigenous people which are all clearly based on history. It will be argued that these historical claims differ markedly from those that are based on the idea of individual productive activity.

Lyons's article begins with an analysis of the claims made during the 1970s by Native American activists. After noting the injustices that have befallen Native Americans since the arrival of Europeans, Lyons suggests that the most natural arguments that might be advanced in favour of such claims centre on what might be called, following Robert Nozick, 'historical considerations affecting social justice'. He notes that the Nozickian language of 'justice in acquisition' and 'justice in rectification' fits neatly with the justifications for what were then known as the New Indian Land claims. Most of Lyons's article involves an exploration of the plausibility of using Nozick's framework to ground the Native American right to land originally owned by them.

Lyons investigates one obvious objection to such claims; it is an objection that concerns the historical rupture between the present time and the earlier period when Native Americans controlled the land. Lyons argues that this line of argumentation is underpinned by a conception of inheritance that fits our property regimes, but not those under which Native American groups operated during the time prior to the European invasion. Lyon suggests it involves an individualist conception of inheritance under which private property is transferred from one individual to another, and if there is a significant historical break then those rights of transfer disappear. But, as Lyons points out, the property regime prior to European colonization was based on the tribe and was communal rather than individual. Thus, the claim that inheritance rights are null and void because of the temporal distance between present day Native Americans and those who owned it originally cannot be sustained.

Ultimately, Lyons does concede that attempting to justify 'Indian Land Claims' on Nozickian grounds will inevitably fail because he believes that the Nozickian project – with its emphasis on socially unstable property rules – is flawed. According to Lyons, rights to property are always relative to social circumstances. Lyons argues that Native Americans have been systematically discriminated against in North American society and it is on this basis that rectification is justified. He argues that they have a valid claim to a fair share of society's resources as well as to social and economic opportunities. They also have a valid claim for compensation for unjust deprivation that the current generation have suffered.

Why then, given his ultimate rejection of Nozickian land rights claims, does Lyons bother making the initial extension? The answer is simple. He develops his arguments so as to make a more general critique of the Nozickian project itself. Nozick's focus on legal practices and rules in relation to resources fails to acknowledge the importance of those laws being properly understood as subservient to social circumstances. Now, whether or not one agrees that it is impossible to identify a right to property such as land or water that is not 'relative to circumstances', there are *independent* reasons to avoid basing Indigenous rights to resources like water on Nozick's mixing labour model. Doing so might well place the debate into an inappropriately 'productivist' model that undervalues the cultural connections with water and traditions of stewardship that are characteristic of many Indigenous groups.

The important point to take from this discussion – although Lyons does not make it directly – is that it is possible to distinguish between *distinct* species of historical claims. Indeed, talk of historical claims as a general class involves making sweeping generalizations across very different kinds of assertions. The historically based land claims of Indigenous peoples are based on cultural connections and traditions of stewardship, rather than on the notion of mixing ones's labour and therefore should not be understood in Nozickian terms. Moreover, the ownership rights in such cases, rather than individual and private, are typically collective or communal in nature. They represent a distinct category of historical claim (which might be labelled *cultural stewardist* as opposed to *productivist*) that deserve extensive analysis independently of the Nozickian approach. Accordingly, the critical arguments that are provided in the next section against the use of historical arguments to justify private property in water should in no way be thought to bear on discussions of Indigenous rights. These are entirely distinct matters. The arguments proffered here relate only to productivist claims like Nozick's.

Productivist historical claims and some politically salient features of water

In the previous section, a distinction was drawn between *productivist* and *stewardist* historical claims. In this section the focus will be upon the plausibility *in the case of water* of employing productivist historical claims to justify private property claims. In doing so it is essential that the analysis take into account salient features of water *as a distributive good*. All too often theorists of distributive justice apply allocative principles to different classes of goods in a uniform way without acknowledging morally salient specific features of the resource or good to which the principles are being applied; it is as if distributive theory is simply concerned with the abstract principles and the goods themselves are conceptually unproblematic.

Water – when considered as a distributive good – has a number of significant *idiosyncratic* features. Before reflecting more closely upon the productivist claims in greater detail, it is worth briefly noting some of these features. Firstly, as was noted earlier, water is both a benefit and a burden. Secondly, water has a multitude of uses and hence a diverse range of benefits associated with it. It is instrumentally useful in agriculture and industry and for recreation and domestic consumption as well as being a vital element of ecosystems considered as a whole. Thirdly, water has a multiplicity of meanings and values associated with it and, hence, any distributive decisions by policy-makers will involve not only access to a consumptive item – as would typically be the case if the allocation of a food staple like flour were the topic of discussion – but also will oftentimes have significant social and cultural effects (Linton, 2010; Strang, 2004).

However, perhaps the most significant feature of water is that much of it does in fact 'fall from heaven', as it were. Water is, in many cases, available for human consumption without human intervention. It is *typically* not produced in the Nozickian sense and, hence, the relevant history of how it came into existence does not include human labour. Nor does human intervention typically produce more of it. There are some notable exceptions to these claims of course. Desalination plants, for instance, produce potable water from seawater and in this way the history of the good and the labour that was mixed with the good becomes relevant to the determination of property rights. Equally, it is arguably true that underground bores make water available for human consumption as well as for industrial and agricultural uses that would not be available otherwise. The history of productive activity looks like it will be relevant here as well. But these cases are the exceptions. Most water over which disputes occur is not produced in a way that could justify private property rights in a Nozickian fashion. Hence, the 'social pot' approach to distributive justice (as Nozick describes views like Rawls's) is entirely appropriate for a large number of disputes over the allocation of water.

Further, not only is the mixing-one's-labour justification of private property less plausible when it comes to water, but in many instances privatizing water violates the second part of Nozick's justificatory framework, namely the Lockean 'enough-and-as-good' Proviso. It is simply not true that productive activity produces more and that privatization typically involves private enclosure of a scarce good rather than leaving 'enough-and-as-good'. Hence, productivist historical arguments for private property will fail in a great many of the contexts in which people wish to apply them in order to justify the privatization of water. To be sure, these criticisms do not bear on either the utilitarian arguments for private property regimes and markets, nor do they bear directly on stewardship arguments. The utilitarian arguments for privatization and commodification are based on claims about overall social benefit and hence the argument that water is not produced by human intervention is unproblematic for the utilitarian. They are simply concerned with realizing what they regard as the greatest overall social benefit from a resource and if that is achieved through private property regimes then private property in water resources are to be endorsed. For the defender of property rights based on historical stewardship, the preceding arguments about the nature of water are not relevant either since the historical claim is not based on any suggestion that the stewardship has transformed the resource, but simply that the group has a culturally significant historical relationship with the water resource and, further, has a history of stewardship that amongst other things involves non-instrumental reverent attention.

Concluding remarks on the role of historical claims in decisions about the allocation of water resources

This chapter began with a question concerning the role *historical considerations* should play in a theory of Just Water. Given the influence of Nozick's idea on public policy theory and, hence, on public policy-makers and his use of historical claims to justify trade (and ultimately marketization), questions regarding the role of historical considerations are important when exploring what is required for a theory of Just Water. However, closer examination of the theoretical terrain soon reveals that to speak of historical claims as if they were of a single or natural kind is mistaken. There are, in fact, a number of distinct kinds of historical claims, as the example of Indigenous communal water claims clearly demonstrates. More specifically, it is important – at least for the issues that animate this chapter – to distinguish between productivist and stewardist claims, with the focus here being on the former variety. Therefore, could the productivist claims be justified?

In response, it is argued that whatever the general intellectual merits of Nozick's approach to the problem of distributive justice, there are solid reasons for thinking that when it comes to water resources, *productivist* justifications cannot have the same kind of traction they might have elsewhere. With a few notable exceptions (such as desalinated water), productivism cannot provide justificatory grounds for private property rights in water since the history does not involve mixing labour, nor will 'enough-and-as-good' be left over after private appropriation. If policy-makers wish to justify the privatization (and the commodification) of water, then a utilitarian approach seems more plausible. The argument would be that property rights and markets in water are more efficient than other methods of allocation and it is the social benefits they bring that provide grounds for privatization. It is extremely unlikely that privatization of water can be justified on Nozickian grounds since, unfortunately for the productivist, water does regularly fall like 'manna from heaven'. If a more comprehensive account of the undesirability of privatization is to be developed, then a thorough going critique of the utilitarian approach is also required. However, for now it suffices to show that one of the most important justificatory routes to privatization does not work.

Finally, thinking imaginatively about the future of water policy requires, amongst other things, a clear-sightedness about the normative issues surrounding allocation. What social values should be promoted in public decisions that bear on water allocation? Water policy-makers need to be mindful of the centrality of questions of distributive justice if, in the longer term, they are to deal adequately with the challenges facing water policy across the globe.

References

Cohen, G. A. (1995). *Self-Ownership, Freedom and Equality*. Cambridge, England: Cambridge University Press.

Goudie, A. S. (2000). *The Human Impact on the Natural Environment: Past, Present and Future*. Oxford, England: John Wiley and Sons.

Honore, A. M. (1961). Ownership. In A. G. Guest (Ed.), *Oxford Essays in Jurisprudence*. Oxford, England: Oxford University Press, 107–147.

Kolstad, C. C. (2011). *Environmental Economics*. (2nd ed.). Oxford, England: Oxford University Press.

Linton, J. (2010). *What Is Water? The History of a Modern Abstraction*. Vancouver, Canada: University of British Columbia Press.

Locke, J. (1967). *Two Treatises of Civil Government*. P. Laslett (Ed.). Cambridge, England: Cambridge University Press.

Lyons, D. (1977). The New Indian Land Claims and Original Rights to Land. *Social Theory and Practice, 4*(3), 249–272.

Miller, D. (1976). *Social Justice*. Oxford, England: Oxford University Press.

Nozick, R. (1974). *Anarchy, State and Utopia*. New York, NY: Basic Books.

Rawls, J. (1972). *A Theory of Justice*. Oxford, England: Oxford University Press.

Strang, V. (2004). *The Meaning of Water*. Oxford, England: Berg.

Varden, H. (2012). The Lockean 'Enough-and-as-Good' Proviso: An Internal Critique. *The Journal of Moral Philosophy, 9*(3), 410–442.

Walsh, A. J. (1977). *A Neo-Aristotelian Theory of Social Justice*. Aldershot, England: Ashgate.

Walzer, M. (1983). *Spheres of Justice*. New York, NY: Basic Books.

4 Progressing from experience-based to evidence-based water resource management

Exploring the use of 'best available science' to integrate science and policy

Darren S. Ryder

Introduction

Human societies currently face major environmental crises and human-accelerated environmental change world-wide (Goudie, 2013). The past two decades have witnessed increasing global concern about the need for sustainable land and water management in the light of growing food and water insecurity. Human population increase, economic development, climate change and other drivers have altered the availability and use of water resources, resulting in an increased risk of extreme flow events (floods and droughts), threats to water quality and the demand for potable water outstripping renewable supply (Pahl-Wostl et al., 2013). Human water security is often achieved with little consideration of environmental consequences, and when water resource management seeks trade-offs between competing uses, the lack of scientific certainty around environmental water needs means the inclusion of science raises as many questions as it answers (Likens, 2010). In a recent study identifying the 40 top priorities for science to inform United States (US) conservation and management policy, the number one issue was water security and the need to document the quantity and quality of water required to sustain human populations (Fleishman et al., 2011). The debate that surrounds water resource management requires a substantial input from science because scientific evidence has played a major role in identifying the 'problem' as well as suggesting appropriate remedies. Arguably, there is a critical need for evidence-based information to contribute to water resource policy development, yet effective science-policy-management information exchange is still required for science to be best placed among the many socio-economic factors that inform water resources management and policy development.

Evidence-based practice requires policy development with the resultant management actions based on the latest objective information collected, using a hypothesis-driven scientific method to remove the subjectivity inherent in many experience-based approaches. An evidence-based policy and management process requires all disciplines involved seeking an improved understanding and integration of scientific knowledge. However, scientists have reported dissatisfaction with the lack of recognition of fundamental scientific principles by decision makers (see Richter, Warner, Meyer & Lutz, 2006; Lake, Bond & Reich, 2007; Likens, 2010; Nichols, Johnson, Williams & Boomer, 2015) and the deficiency of research-based evidence in policy development (Sutherland, Armstrong-Brown, Armsworth, Brereton & Brickland, 2006; Morgan, 2014). Policy makers in turn bemoan the lack of policy-relevant science, the time lag for its availability and the uncertainty and ambiguity that often accompanies scientific outcomes (Tomlinson & Davis, 2010). From the management perspective, poor and delayed communication of scientific outcomes has been identified as the major barrier to the effective uptake of science by on-ground managers of water resources (Webb, Miller, De Little & Stewardson, 2014). The management of aquatic systems requires an integration of knowledge from each of these perspectives, yet improving the scientific evidence base for policy and practice remains a challenge. However, the current state of scientific knowledge can also be an impediment, as existing data is often inadequate to answer essential policy questions (Lagacé, Holmes & McDonnell, 2008), or may not be directly applicable to specific management scenarios (Hillman, Crase, Furze, Ananda & Mayberry, 2005).

Decision-making processes for water resources management have increasingly sought the participation of affected communities (see Morton et al., 2009; Brandon, 2015), driven by recent examples of the need for community engagement to develop a social licence for water management linked to increased public and industry awareness of environmental issues (Dare, Schirmer & Vanclay, 2014). Management decisions made in the context of water scarcity and by necessity have been required to address the perceived dualistic needs for water supply by ecosystems and humans (Likens et al., 2009; Cosgrove & Rijsberman, 2014). Policy development is a rapid process that must embrace a diverse range of stakeholders (including science, agencies, industry and communities) in order to negotiate an acceptable compromise (Kramer & Pahl-Wostl, 2014). These constraints for management and policy decision-making and the increased role of industries and the public often lead to experience-based practice as the norm. These management actions and policy development decisions are based on familiarity rather than the latest objective scientific evidence, accepted practices of management known to succeed and institutional norms.

The Millennium Ecosystem Assessment (2005) suggested that such experience-based water resource management processes were a barrier to sustainable water use, and that water resources policy development should evolve towards a more holistic view of integrated water resource management to balance consumptive and environmental outcomes (Morgan, 2014). Thus, water resource policy in its broadest sense that follows an integrated path will involve a much wider range of stakeholders. For example, recent reforms in Australia have seen privatization, the introduction of water markets and an increased role for the Federal government (as opposed to State and local control), all contributing to the decentralization of decision-making and increasing the number of stakeholders (Koehn, 2015). As a result, the pathways for the contribution of science to the development and implementation of policy are typically neither static, direct nor transparent, blurring the boundaries of how personal and institutional experience and scientific evidence affect the decision-making process for water resource allocation.

Various institutions have sought to standardize the protection of natural resources by requiring that decisions be made based on the best available science (BAS) (Francis, Whittaker, Shandas, Mills & Graybill, 2005; Likens, 2010; Gerlach, Williams & Forcina, 2013; Green & Garmestani, 2012). One of the challenges facing freshwater scientists is to develop mechanisms for identifying and contributing BAS to the sound policy development and management of water resources. This chapter reviews the current understanding and use of BAS in water resource management from science, policy and management perspectives. A case study explores the use and misuse of BAS in the development and implementation of environmental flows, a highly contentious management and policy option that aims to improve the ecological health of regulated river systems, yet removes consumptive water that sustains rural populations. The challenges and opportunities to improve the use of scientific evidence in complex water resource management decisions are drawn from lessons learnt across different disciplines and countries each grappling with the need to integrate scientific evidence into collaborative policy and management development.

The mandate for 'best available science'

The compounding environmental and socio-economic loss associated with long-term water resource management has been paralleled by an acceleration in scientific data collection and knowledge generation, and the requirement to include scientific evidence in the development and implementation of policies to manage water resources. BAS is being increasingly used as a criterion to inform and prioritize the evidence used to shape well-informed resource management decisions. As a legal basis for natural resource policy development, BAS guides the implementation of the US Federal *Endangered Species Act 1973* (Doremus 2003; Lowell & Kelly, 2016; Murphy & Weiland, 2016), the US Federal *Magnuson-Stevens Fishery Conservation and Management Act 1976* (Adler, 2012; Wolters, Steel, Lack & Kloepfer, 2016) and the US Federal *Clean Water Act 1972* (Sullivan et al., 2006), and is a key element in the recent Paris Climate Consensus (United Nations, 2015) that sets long-term national climate action plans (Burleson, 2016).

BAS is an explicit requirement for water resource management in a number of countries and collectives such as the European Union. For example, the US Federal *Safe Drinking Water Act 1974* explicitly states that 'best available peer-reviewed science' is required to inform policy for protecting all waters actually or potentially designed for drinking use, whether from above ground or underground sources (Russell, 2016). The European Water Framework Directive (EWFD) in 2000 established a framework for the multi-jurisdictional protection of freshwater systems (Sommerwerk et al., 2010). It is characterized by a cyclical planning process, based on an iterative and adaptive approach for the inclusion of scientific evidence aimed at protecting and improving the ecological status of river basins and promoting sustainable use of water (Watts, Ryder, Allan & Commens, 2010; Green, Garmestani, van Rijswick & Keessen, 2013). Australia's National Water Initiative (Dovers, Grafton & Connell, 2005) advocates the use of 'best available scientific information' to underpin water resource management (Docker & Robinson, 2014). A guiding principle of the Australian national water reforms is the '[p]rovision of water for ecosystems should be on the basis of the best scientific information available on the water regimes necessary to sustain the ecological values of water dependent ecosystems' (Tomlinson & Davis 2010, p. 809).

Water management decisions will always involve trade-offs between consumptive, environmental and social objectives. Embedding scientific evidence in an adaptive management process is a dominant theme in water resources literature (Jacobs et al., 2016). The promise of adaptive management is that it facilitates the management of socio-ecological systems, despite their complexity, gaps in understanding and multiple, changing social goals (Allan & Stankey, 2009). However, the credibility of policy and on-ground decision-making relies on both the quantity and the quality of the evidence base, acknowledgement of the limitations of current scientific knowledge and the need to learn from decisions to fill knowledge gaps. Therefore, BAS provides an essential pathway to ensure that the ecological health and ecosystem services afforded by aquatic ecosystems are not compromised by the risks posed by water resource development.

What is best available science?

A limitation in the current legislative requirements is the absence of an explicit definition of BAS, and guidance on its practical application in the decision-making process (Ryder, Tomlinson, Gawne & Likens, 2010). Such a definition must include three elements: a clear definition of what constitutes 'science' as the basis for evidence, objective means to rate the quality of scientific information (i.e., to distinguish the 'best' science) and a determination of the availability of information, each of which is essential for successful integration of science and policy in the management of water resources. Similarly, Lowell and Kelly (2016) identify that the simple translation of the term 'best scientific and commercial data available' (US Environmental Protection Agency) and 'best available peer-reviewed science' (US *Clean Water Act 1972*), as well as 'best available scientific information' in the Australian *Water Act 2007 (Cth)* (Australian Commonwealth Government, 2007), to be collectively known as 'best available science' for implementation places restrictions on the 'science' included as evidence to inform policy development.

The many and varied types of scientific inquiry that can generate multiple forms of scientific outputs affect the evidence available for water resource management. Scientists can prioritize the questions they tackle, the approach taken and interpretations to test, often driven by the availability and longevity of funding. The scientific process depends on the elimination of false hypotheses, relying on the peer-review process and further testing to judge the veracity of their hypotheses. Science is therefore a self-correcting process, with new evidence emerging over time seeking to reduce the uncertainty inherent in the scientific process (see Gleick et al., 2010). Managers must transform scientific evidence from the realm of the general and abstract to information that can inform their decisions (Tress, Tress & Fry, 2006). This mismatch means that BAS poses the risk that it may include inaccuracies and uncertainty, which in turn may drive an unsustainable water resource policy agenda.

Identifying the 'best' science is also problematic given the diversity of approaches available to scientific enquiry. For example, Traditional Ecological Knowledge (TEK) (see Chapters 2 and 8, this volume) is not derived from hypothesis-based science, yet has the potential to provide a powerful evidence base for the conservation of bio-cultural diversity (Gratani, Bohensky, Butler, Sutton & Foale, 2014), the distribution of wildlife (Ziembicki, Woinarski & Mackey, 2013) and identify baseline conditions and an interpretation of long-term change in resource condition in highly variable environments (Wiseman & Bardsley, 2013). Gratani et al. (2014) observed the poor alignment between TEK and the western concept of knowledge that is based on short-term scientific evidence as a serious impediment to its integration into water resource policy and practice. Yet TEK is perceived as experience-based evidence, blurring the boundaries between evidence gathered by objective scientific enquiry and evidence gathered through natural resource management experience.

Scientific evidence is always open to interpretation, with value judgements regarding the validity and usefulness of science often required by end-users (Ryder et al., 2010). As such, a definition of the 'best' science will be specific to the management or policy issue at hand. The range of attributes that could define 'best' covers that which is academically credible, accurate or acceptable to support a user's position or agenda (Nichols et al., 2015). For example, a survey of scientists, managers and interest groups involved in US policy development and implementation for marine fisheries revealed that scientists valued the quality of methodology (following the scientific method), whereas managers rated experience and knowledge as the basis for defining the 'best' science for management decisions (Wolters et al., 2016).

The US Federal *Endangered Species Act 1973* has been a focal point in the published literature, defining 'best' as 'information that is collected by established objective protocols, properly analysed and peer-reviewed before released to the public' (Murphy & Weiland, 2016, p. 163). Peer-reviewed or published literature, expert advice and an acknowledgement that revision is necessary as uncertainties and/or limitations emerge are common attributes of 'best' scientific information in the literature. However, Gerlach et al. (2013) empirically tested the US Fish and Wildlife Service policy development, alarmingly finding that experience-based practice through normative isomorphism (seeking information from established organizational networks) and path dependency (information use based on historical preference) dominated the science selected for inclusion in management decisions. This process to satisfice the best available science based on familiarity and institutional norms is a reflection on both the quality of information, and the need for scientists to improve the availability of research and progress the currency of evidence-based science in management and policy decisions.

Following from the US Fish and Wildlife Service example, the concept of 'availability' is most relevant to end-users of scientific information and is a dynamic process through time. Interpretation and communication of science is key to its availability, yet is generally determined by the information providers rather than the needs of policy and decision makers as end-users. For example, the reward system for many scientists is based on peer-reviewed publications in high quality journals. Yet this is a protracted process that creates a temporal disjunct between the generation of evidence and its availability, as well as restricting its access by managers and policy makers who do not have access to journals (Murphy & Weiland, 2011). Alternatively, the rapid in-house production of a non-reviewed report that resides within agencies or institutions may be more available to decision makers but may compromise on quality of scientific evidence or relevance. Although there is a temporal continuum for availability, the accountability of both information providers and information users is determined by timely publication and dissemination of evidence (Haby et al., 2016). This highlights the challenge for an integrated process that incorporates measures of 'availability' and 'best' when reviewing the relevance of scientific evidence for use in water resource policy formulation or management actions.

Water management in a contentious landscape – The importance of defining and communicating BAS in environmental flow management

The multi-dimensional relationships required for sustainable water resource management are often based in legislation (Lowell & Kelly, 2016), yet human water security is frequently prioritized and achieved at the expense of the environment, with potentially harmful long-term implications for social-ecological systems reliant on water (Pahl-Wostl et al., 2013). Freshwater ecosystems can provide economic security, social security and ethical security. Thus, water allocated for the environment means indirectly supporting people through the maintenance of valuable ecosystem goods and services (Boulton & Ekebom, 2016). The Millennium Ecosystem Assessment (2005) identified that many ecosystems were being degraded or lost, with aquatic systems suffering disproportionately from the withdrawal of water for direct human needs for drinking, growing crops and supporting industry, with many impacts directly resulting from fragmentation by dams (Nilsson, Reidy, Dynesius & Revenga, 2005). The importance of a river's flow regime for sustaining biodiversity and ecological integrity is well established (Bunn & Arthington, 2002), and therefore the return of water to the environment is often a policy and management directive to sustainably manage human and aquatic ecosystem needs and values. Environmental flow is now a widely accepted term that covers the quantity, timing, duration, frequency and quality of water required to sustain freshwater, estuarine and near-shore ecosystems and the human livelihoods and well-being that depend on them (Acreman et al., 2014). For contentious water resource management issues such as the provision of environmental water to ecological assets, uncertainty around predicting scientific outcomes is very high, and therefore the risk of unacceptable management actions is also high. In such scenarios it is imperative to determine and successfully communicate BAS.

Successfully implementing the environmental flow concept requires dialogue among scientists, policy makers, water managers and users and local communities to balance the priorities among competing demands for water. However, scientific evidence around environmental flows is characterized by limited transferability among rivers, regions and countries due to the prevalence of site-scale studies and the inherent variability in natural systems (Pahl-Wostl et al., 2013). Environmental flow assessments have predominantly relied on scientific experts, because despite a large and rapidly increasing body of literature on the impacts of freshwater ecosystem degradation, the ability to generalize ecological effects of flow restoration is still poor (Stewardson & Webb, 2010). The literature is dominated by hypothesis-driven studies examining responses to individual environmental flow events and/or at a limited number of sites on individual rivers over short time frames, and often focused on responses of specific taxa such as fish or waterbirds. Such studies cannot be used to make defensible quantitative predictions of the likely ecological responses to flow restoration in new situations (Webb et al., 2015). Hence, while the scientific evidence base can provide the foundation for decisions, it transitions to more experience-based judgement the larger the scale of the environmental flow delivery. A key challenge in environmental flows is therefore to synthesize the objective hypothesis-driven science gained from individual case studies to support environmental flow decision-making at spatial and temporal scales relevant to policy and management decisions (Arthington, Bunn, Poff & Naiman, 2006).

The concept of environmental flows is embedded in the European Water Framework Directive (Acreman & Ferguson, 2010), South African (Tharme, 2003), United States (Poff & Zimmerman, 2010) and Australian water legislation, with each having a legislative requirement for the sustainable management of water resources based on best available science. Australia is a world leader in the implementation of large-scale environmental flow programs. Most notable among a number of initiatives, the Murray Darling Basin Authority's Basin Plan (Murray Darling Basin Authority, 2017) has passed into law the recovery of up to 2750 gigalitres (GL) per year to return as environmental water to the Murray and Darling River systems (Grafton & Connell, 2013). This contentious re-allocation of such a large volume of water from consumptive (i.e., agricultural) to environmental uses has caused great controversy among rural and regional communities and the industries that support them. Key to the reform is the legislative requirement to employ 'best available science' in policy development and management interventions (Tomlinson & Davis, 2010; *Water Act 2007 (Cth)* s21(4b)). One of the key points of contention in the development of the Basin Plan has been the credibility of the evidence base used to determine environmental flow allocations and particularly the use of expert opinion as the basis for best available science (Stewardson & Webb, 2010). Evidence-based practice relies on management actions and policy development derived from the implementation of the scientific process, and accepted by the scientific community through the peer-review process. In contrast, most current water resource management follows a murky path of expert opinion founded on a scientific evidence base, where decisions are made by managers and/or a small group of experts based on a synthesis of accumulated evidence and accepted practice that has been shown to work (Webb et al., 2014). However, it would be naïve to consider that a strong evidence base for predicting the ecological outcomes of environmental flows would resolve the controversy associated with large-scale water reforms such as the Murray Darling Basin Plan.

At the same time as the widespread call to increase the evidence base and improve the way it is used in water resource management, calls to increase the participation of stakeholders are becoming formalized in the decision-making processes used in water resource management (Felipe-Lucia et al., 2015). Strategies to incorporate a scientific evidence base into a participative decision-making process are reliant on the successful communication of research findings to various stakeholders, something that has proven difficult to date (Lacroix, Xui & Medgal, 2016). Moreover, if the scientific evidence base is used in a prescriptive, top-down fashion to guide the decision-making process, it is likely to impede participation and inclusion of other perspectives and can lose its legitimacy in the process (Rauschmayer, Berghöfer, Omann & Zikos, 2009; Haby et al., 2016). Thus, communication, stakeholder involvement and transparency are imperative to defining the attributes of BAS for use in policy development.

Clarity in the communication of BAS is essential where information may be subject to diverse and contested interpretations. This requirement includes careful definition of commonly used terms, such as those for fundamental concepts such as 'monitoring' and 'assessment', as well as acknowledgement of cultural differences in the scales at which science and resource management operate, and of the different interests, motivations and rewards of the various groups (Wolters et al., 2016). In particular, disagreement and uncertainty must be understood as fundamental to the scientific process, with dispute amongst scientists not meaning that the science is less credible or that the process is failing (Cullen, 1990). What it does require is the use of 'soft skills' to reach consensus among stakeholders (Allan & Stankey, 2009). These skills include organizing participatory planning sessions, facilitating inter-sectoral negotiation, coordinating communication and outreach programs or eliciting local knowledge, and require hands-on learning-by-doing approaches (Martin, Harrison-Atlas, Sutfin & Poff, 2014). This approach means that progress on the successful integrative management of water resources requires an a priori investment in the integration of communication and policy development activities from the outset of every applied science research program (Ryder et al., 2010). Investment in communication creates shared ownership and support for multiple program goals by all project participants (van Wyk et al., 2006), shared understanding of project-specific definitions of BAS and stakeholder levels of risk-aversion or risk-tolerance, and greatly enhances the delivery of research outputs that have practical application and of policies that reflect valid scientific inputs.

There are, however, concerns by scientists that a more participative approach will reduce the input of objective hypothesis-driven science into decision-making by subjecting policy outcomes to the whims of opinion (Bradshaw & Bekoff, 2001). However, the exposure of policy makers, stakeholders and the general public to scientific thought over long timeframes can influence decision-making via 'complex and iterative interactions that . . .require fundamental changes in belief systems, values and norms' (Lawton, 2007, p. 465). If it is used to expand the policy options available, and determine the likely outcome of those policy options to enable informed choice, it is conducive to a participatory decision-making approach (Pielke, 2007). If, however, in the first instance it is used to justify or prescribe a particular policy option without the consideration of alternative options, then other perspectives may be excluded from the participatory process. The latter strategy for adopting science in policy development covertly introduces evaluative judgments and subjective uncertainties, packaged as science, into policy development (Sarewitz, 2012), which reduces the credibility and legitimacy of science (Boully & Maywald, 2011).

At the heart of the environmental flows debate for successful science-policy-management integration is a legislative requirement that BAS informs policy and on-ground management. The ever-increasing challenge for these partnership-driven approaches is that policy and management often require information or pose questions for which science does not currently have unequivocal answers. Continued research to improve the mechanistic basis of ecological response models that provide evidence for links between environmental drivers and ecological responses that infers causality where experimental evidence is lacking was proposed by Webb et al. (2012) to manage the uncertainty surrounding the science underpinning environmental flows. The framework is termed 'Eco-Evidence' and outlines a transparent and repeatable method for assessing the strength of the available scientific evidence regarding particular management actions. However, if evidence is used to justify decisions rather than to provide options and likely outcomes of these options to stakeholders, its effectiveness can be undermined. This approach has been successfully used to assess the evidence concerning responses of vegetation, fish, macroinvertebrates and floodplain geomorphology to environmental flows in the Murray Darling Basin (Webb et al., 2011, 2012, 2013, 2014, 2015). Worldwide, there are many examples of similar scientific frameworks for environmental flow development and assessment that aim to synthesize the best available science (see Building Block method [King & Louw,1998]; physical habitat assessment or PHABSIM [Maddock, 1999]; ecological limits of hydrologic alteration or ELOHA (Poff et al., 2010); and Soil and Water Assessment Tool or SWAT [Zhang et al., 2012]). Adler (2012) went so far as to suggest that scientists need not propose or formulate changes in policy but simply provide good, reliable and candid information about the relative ecological benefits of environmental flows. Ultimately, the plethora of research articles from around the globe targeted at improving the integration of science, policy and management in environmental flow development and implementation, and the many examples of failed science-policy-management collaborations in water resource management, suggest little progress towards successfully achieving the mandate of incorporating BAS to achieve sustainable water use.

Challenges, opportunities and partnerships for effective use of BAS

As demand for water increases, it is becoming evident that there is insufficient water to meet all environmental, economic and social requirements, revealed in the concept of the 'planetary boundary' that identifies the tolerable levels of freshwater consumption for human and environmental survival (Gerten et al., 2013). By its very nature, water is a common property, yet management decisions must be made to allocate water among different users (see Chapters 3 and 5, this volume). The use of the BAS concept can promote truly integrative management that provides a conduit for greater awareness, communication, involvement, transparency and understanding among research, policy and management communities.

As policy develops and craves an objective evidence base for water resource management, managers worldwide are already grappling with the legislative requirement for using BAS in environmental decision-making. The US Federal *Endangered Species Act 1972* has been a focal point for researchers reviewing the challenges and successes of putting into practice the legislative requirements for BAS (Murphy & Weiland, 2011; Lowell & Kelly, 2016; Murphy & Weiland, 2016). Differential success in the use of hypothesis-driven science by agencies such as the US Fish and Wildlife Service (FWS) and the US National and Oceanic and Atmospheric Administration (NOAA) Marine Fisheries Service was attributed to three main issues. First, agencies were infrequently engaged in identifying and then appropriately employing objective scientific data specific to the issue at hand. Second, agencies used scientific uncertainty to validate the arbitrary, selective and uncritical use of scientific data. Finally, adaptive management was often used post hoc to address uncertainty in scientific evidence rather than as an a priori process to define the best science (Murphy & Weiland, 2011; Lowell & Kelly, 2016; Murphy & Weiland, 2016). The consistent message arising from these reviews is the need for structured trans-scientific processes to make the linkage between science inputs and policy outcomes that can take advantage of well-developed tools and employ these in modelling exercises that assess the ecological costs and benefits of the proposed action and alternative actions.

The successful joint delivery of an environmental flow to the Colorado River Delta negotiated by US and Mexican agencies highlights the potential for BAS embedded in a transparent and structured process to inform positive scientific, policy and management outcomes (Pitt & Kendy, 2017). The agreement specified a process, calling on a group of stakeholders, including federal, state and local water managers as well as non-governmental conservation organizations from both countries, to develop a flow delivery plan. The flow delivery plan was developed, approved and executed in an exceptionally short period of time, with limited scientific data, under numerous operational constraints. Success was driven by the close interaction between policy makers, water managers and scientists, driven by clear objectives for ecological outcomes, as well as overtly stating scientific uncertainties and the benefits of continued scientific studies to improve uncertainty (Pitt & Kendy, 2017).

The need for greater communication and collaboration among scientists, policy makers and managers to deliver timely scientific evidence remains one of the key challenges for the sustainability of water-dependent ecological and social communities. Lessons from the health sector that have developed rapid response processes to support evidence-informed decision-making in health policy and practice (Haby et al., 2016) serve to demonstrate that objective science can be used as the basis for BAS without the delays inherent in the peer-review publication process. A simple structured approach was used to identify the best methodological approaches for rapid critical reviews of the available research evidence, facilitate evidence-informed decision-making among stakeholders and identify how best to operationalize outcomes (Haby et al., 2016). Rapid evidence-based decisions framed around risk and uncertainty that are vital to the health care sector are just as relevant to water resource management where rapid responses to deliver environmental flows to support critically endangered communities must also be made. The multi-stakeholder basis for effective water management requires that these decisions are grounded in a governance framework and inclusive of interactions among affected communities, policy makers, relevant information providers (that includes scientists and non-scientists) and ultimately natural resource managers charged with delivering timely outcomes.

It is clear from these success stories that moving towards an evidence-based mode of management and policy development can be facilitated by clear a priori and collaborative governance processes. Neither simplistic approaches that ignore differences between disciplines, issues and stakeholders, nor approaches that assume each of these to be unique offer successful ways to proceed. Scientific evidence must be framed around guidelines that are sufficiently flexible to accommodate cultural differences among stakeholders, yet promote consistency in both the production and use of a diversity of scientific information with clear communication of the uncertainty in the scientific evidence base. Roux et al. (2006) stated that scientists cannot afford to remain detached experts who deliver knowledge to managers, but must assume the roles of collaborative learners and knowledge generators. Ultimately the successful inclusion of scientific evidence requires it is a key element in a collaborative decision making process, and acknowledgement by scientists that sustainable water resource management must be based on the best science available and not the best science possible.

References

Acreman, M. C., & Ferguson, A. J. D. (2010). Environmental flows and the European water framework directive. *Freshwater Biology, 55*(1), 32–48.

Acreman, M. C., Overton, I. C., King, J., Wood, P. J., Cowx, I. G., Dunbar, M. J., Kendy, E., & Young, W. J. (2014). The changing role of ecohydrological science in guiding environmental flows. *Hydrological Sciences Journal, 59*(3–4), 433–450.

Adler, R. W. (2012). *Restoring Colorado River ecosystems: A troubled sense of immensity.* Washington, USA: Island Press.

Allan, C., & Stankey, G. H. (2009). *Adaptive environmental management: A practitioner's guide.* Victoria, Australia: CSIRO Publishing.

Arthington, A. H., Bunn, S. E., Poff, N. L., & Naiman, R. J. (2006). The challenge of providing environmental flow rules to sustain river ecosystems. *Ecological Applications, 16*(4), 1311–1318.

Australian Commonwealth Government (2007). Australian Water Act 2007. Retrieved from https://www.legislation.gov.au/Series/C2007A00137.

Boully, L., & Maywald, K. (2011). Basin bookends, the community perspective. In D. Connell and R. Q. Grafton (Eds.) *Basin Futures: Water Reform in the Murray Darling Basin* (pp. 101–115). Canberra, Australia: ANU Press.

Boulton, A. J., & Ekebom, J. (2016). Integrating ecosystem services into conservation strategies for freshwater and marine habitats: A review. *Aquatic Conservation: Marine and Freshwater Ecosystems, 26*, 963–985.

Bradshaw, G. A., & Bekoff, M. (2001). Ecology and social responsibility: The re-embodiment of science. *Trends in Ecology and Evolution, 16*(8), 460–465.

Brandon, T. O. (2015). Fearful asymmetry: How the absence of public participation in Section 7 of the ESA can make the best available science unavailable for judicial review. *Harvard Environmental Law Review, 39*(2), 311–362.

Bunn, S. E., & Arthington, A. H. (2002). Basic principles and ecological consequences of altered flow regimes for aquatic biodiversity. *Environmental Management, 30*(4), 492–507.

Burleson, E. (2016). *Paris agreement and consensus to address climate challenge.* London: ASIL Insight.

Cosgrove, W. J., & Rijsberman, F. R. (2014). *World water vision: Making water everybody's business.* Washington, USA: Routledge.

Cullen, P. (1990). The turbulent boundary between water science and water management. *Freshwater Biology, 24*(1), 201–209.

Dare, M., Schirmer, J., & Vanclay, F. (2014). Community engagement and social licence to operate. *Impact Assessment and Project Appraisal, 32*(3), 188–197.

Docker, B., & Robinson, I. (2014). Environmental water management in Australia: Experience from the Murray-Darling Basin. *International Journal of Water Resources Development, 30*(1), 164–177.

Doremus, H. (2003). The purposes, effects, and future of the Endangered Species Act's best available science mandate. *Environmental Law, 34*(2), 397–450.

Dovers, S., Grafton, R. Q., & Connell, D. (2005). A critical analysis of the National Water Initiative. *Australasian Journal of Natural Resources Law and Policy, 10*(1), 81–99.

Felipe-Lucia, M. R., Martín-López, B., Lavorel, S., Berraquero-Díaz, L., Escalera-Reyes, J., & Comín, F. A. (2015). Ecosystem services flows: Why stakeholders' power relationships matter. *PloS one, 10*(7), e0132232.

Fleishman, E., Blockstein, D. E., Hall, J. A., Mascia, M. B., Rudd, M. A., Scott, J. M., & Vedder, A. (2011). Top 40 priorities for science to inform US conservation and management policy. *BioScience, 61*(4), 290–300.

Francis, T. B., Whittaker, K. A., Shandas, V., Mills, A. V., & Graybill, J. K. (2005). Incorporating science into the environmental policy process: A case study from Washington State. *Ecology and Society, 10*(1), 35.

Gerlach, J. D., Williams, L. K., & Forcina, C. E. (2013). Data selection for making biodiversity management decisions best available science and institutionalized agency norms. *Administration and Society, 45*(2), 213–241.

Gerten, D., Hoff, H., Rockström, J., Jägermeyr, J., Kummu, M., & Pastor, A. V. (2013). Towards a revised planetary boundary for consumptive freshwater use: Role of environmental flow requirements. *Current Opinion in Environmental Sustainability, 5*(6), 551–558.

Gleick, P. H., Adams, R. M., Amasino, R. M., Anders, E., Anderson, D. J., & Bax, A. (2010). Climate change and the integrity of science. *Science, 328*(5979), 689–690.

Goudie, A. S. (2013). *The human impact on the natural environment: Past, present and future.* Cambridge, Massachusetts: John Wiley & Sons.

Grafton, R. Q., & Connell, D. (2013). *Basin futures: Water reform in the Murray-Darling basin.* Canberra, Australia: ANU Press.

Gratani, M., Bohensky, E. L., Butler, J. R., Sutton, S. G., & Foale, S. (2014). Experts' perspectives on the integration of indigenous knowledge and science in Wet Tropics natural resource management. *Australian Geographer, 45*(2), 167–184.

Green, O. O., & Garmestani, A. S. (2012). Adaptive management to protect biodiversity: Best available science and the Endangered Species Act. *Diversity, 4*(2), 164–178.

Green, O. O., Garmestani, A. S., van Rijswick, H., & Keessen, A. (2013). EU water governance: Striking the right balance between regulatory flexibility and enforcement? *Ecology and Society, 18*(2), 24.

Haby, M. M., Chapman, E., Clark, R., Barreto, J., Reveiz, L., & Lavis, J. N. (2016). Designing a rapid response program to support evidence-informed decision-making in the Americas region: Using the best available evidence and case studies. *Implementation Science, 11*(1), 117–129.

Hillman, T., Crase, L., Furze, B., Ananda, J., & Maybery, D. (2005). Multidisciplinary approaches to natural resource management. *Hydrobiologia, 552*(1), 99–108.

Jacobs, K., Lebel, L., Buizer, J., Addams, L., Matson, P., McCullough, E., & Finan, T. (2016). Linking knowledge with action in the pursuit of sustainable water-resources management. *Proceedings of the National Academy of Sciences, 113*(17), 4591–4596.

King, J., & Louw, D. (1998). Instream flow assessments for regulated rivers in South Africa using the Building Block Methodology. *Aquatic Ecosystem Health and Management, 1*(2), 109–124.

Koehn, J. D. (2015). Managing people, water, food and fish in the Murray-Darling Basin, south-eastern Australia. *Fisheries Management and Ecology, 22*(1), 25–32.

Kramer, A., & Pahl-Wostl, C. (2014). The global policy network behind integrated water resources management: Is it an effective norm diffusor? *Ecology and Society, 19*(4), 11.

Lacroix, K. E. M., Xiu, B. C., & Megdal, S. B. (2016). Building common ground for environmental flows using traditional techniques and novel engagement approaches. *Environmental Management, 57*(4), 912–928.

Lagacé, E., Holmes, J., & McDonnell, R. (2008). Science-policy guidelines as a benchmark: Making the European Water Framework Directive. *Area, 40*(4), 421–434.

Lake, P. S., Bond, N., & Reich, P. (2007). Linking ecological theory with stream restoration. *Freshwater Biology*, 52(4), 597–615.

Lawton, J. H. (2007). Ecology, politics and policy. *Journal of Applied Ecology*, 44(3), 465–474.

Likens, G. E. (2010). The role of science in decision making: Does evidence-based science drive environmental policy? *Frontiers in Ecology and the Environment*, 8(6), 1–9.

Likens, G. E., Walker, K. F., Davies, P. E., Brookes, J., Olley, O., Young, W. J., & Arthington, A. (2009). Ecosystem science: Toward a new paradigm for managing Australia's inland aquatic ecosystems. *Marine and Freshwater Research*, 60(3), 271–279.

Lowell, N., & Kelly, R. P. (2016). Evaluating agency use of 'best available science' under the United States Endangered Species Act. *Biological Conservation*, 196, 53–59.

Maddock, I. (1999). The importance of physical habitat assessment for evaluating river health. *Freshwater Biology*, 41(2), 373–391.

Martin, D. M., Harrison-Atlas, D., Sutfin, N. A., & Poff, N. L. (2014). A social-ecological framework to integrate multiple objectives for environmental flows management. *Journal of Contemporary Water Research and Education*, 153(1), 49–58.

Millennium Ecosystem Assessment (2005). *Ecosystems and human well-being. Vol. 5.* Washington DC, USA: Island Press.

Morgan, E. (2014). Science in sustainability: A theoretical framework for understanding the science-policy interface in sustainable water resource management. *The International Journal of Sustainability Policy and Practice*, 9(2), 37–54.

Morton, S. R., Hoegh-Guldberg, O., Lindenmayer, D. B., Harriss Olson, M., Hughes, L., McCulloch, M. T., & Saunders, D. (2009). The big ecological questions inhibiting effective environmental management in Australia. *Austral Ecology*, 34(1), 1–9.

Murphy, D. D., & Weiland, P. S. (2011). The route to best science in implementation of the Endangered Species Act's consultation mandate: The benefits of structured effects analysis. *Environmental Management*, 47(2), 161–172.

Murphy, D. D., & Weiland, P. S. (2016). Guidance on the use of best available science under the US Endangered Species Act. *Environmental Management*, 58(1), 1–14.

Murray Darling Basin Authority (2017). *The Basin Plan*. Retrieved from https://www.mdba.gov.au/basin-plan.

Nichols, J. D., Johnson, F. A., Williams, B. K., & Boomer, G. S. (2015). On formally integrating science and policy: Walking the walk. *Journal of Applied Ecology*, 52(3), 539–543.

Nilsson, C., Reidy, C. A., Dynesius, M., & Revenga, C. (2005). Fragmentation and flow regulation of the world's large river systems. *Science*(5720), 308, 405–408.

Pahl-Wostl, C., Arthington, A., Bogardi, J., Bunn, S. E., Hoff, H., Lebel, L., & Tsegai, D. (2013). Environmental flows and water governance: Managing sustainable water uses. *Current Opinion in Environmental Sustainability*, 5(3–4), 341–351.

Pielke, R. A., Jr. (2007). *The honest broker: Making sense of science in policy and politics*. New York, USA: Cambridge University Press.

Pitt, J., & Kendy, E. (2017). *Shaping the 2014 Colorado River Delta pulse flow: Rapid environmental flow design for ecological outcomes and scientific learning*. Ecological Engineering.

Poff, N. L., Richter, B. D., Arthington, A. H., Bunn, S. E., Naiman, R. J., Kendy, E., & Warner, A. (2010). The ecological limits of hydrologic alteration (ELOHA): A new framework for developing regional environmental flow standards. *Freshwater Biology*, 55(1), 147–170.

Poff, N. L., & Zimmerman, J. K. (2010). Ecological responses to altered flow regimes: A literature review to inform the science and management of environmental flows. *Freshwater Biology*, 55(1), 194–205.

Rauschmayer, F., Berghöfer, A., Omann, I., & Zikos, D. (2009). Examining processes or/and outcomes? Evaluation concepts in European governance of natural resources. *Environmental Policy and Governance, 19*(3), 159–173.

Richter, B. D., Warner, A. T., Meyer, J. L., & Lutz, K. (2006). A collaborative and adaptive process for developing environmental flow recommendations. *River Research and Applications, 22*(3), 297–318.

Roux, D. J., Rogers, K. H., Biggs, H. C., Ashton, P. J., & Sergent, A. (2006). Bridging the science-management divide: Moving from unidirectional knowledge transfer to knowledge interfacing and sharing. *Ecology and Society, 11*(1), 3.

Russell, C. S. (Ed.) (2016). *Safe drinking water: Current and future problems*. Washington, USA: Routledge.

Ryder, D. S., Tomlinson, M., Gawne, B., & Likens, G. E. (2010). Defining and using 'best available science': A policy conundrum for the management of aquatic ecosystems. *Marine and Freshwater Research, 61*(7), 821–828.

Sarewitz, D. (2012). Beware the creeping cracks of bias. *Nature, 485*(7397), 149.

Sommerwerk, N., Bloesch, J., Paunović, M., Baumgartner, C., Venohr, M., Schneider-Jacoby, M., & Tockner, K. (2010). Managing the world's most international river: The Danube River Basin. *Marine and Freshwater Research, 61*(7), 736–748.

Stewardson, M. J., & Webb, J. A. (2010). Modelling ecological responses to flow alteration: Making the most of existing data and knowledge. In N. Saintilan and I. Overton (Eds.) *Ecosystem Response Modelling in the Murray-Darling Basin* (pp. 37–49). Canberra, Australia: CSIRO.

Sullivan, P. J., Acheson, J., Angermeier, P. L., Faast, T., Flemma, J., Jones, C. M., & Zanetell, B. A. (2006). Defining and implementing best available science for fisheries and environmental science, policy, and management. *Fisheries, 31*(9), 460.

Sutherland, W. J., Armstrong-Brown, S., Armsworth, P. R., Brereton, T., & Brickland, J. (2006). The identification of 100 ecological questions of high policy relevance in the UK. *Journal of Applied Ecology, 43*(4), 617–627.

Tharme, R. E. (2003). A global perspective on environmental flow assessment: Emerging trends in the development and application of environmental flow methodologies for rivers. *River Research and Applications, 19*(5–6), 397–441.

Tomlinson, M., & Davis, R. (2010). Integrating aquatic science and policy for improved water management in Australia. *Marine and Freshwater Research, 61*(7), 808–813.

Tress, B., Tress, G., & Fry, G. (2006). Researchers' experiences, positive, negative, in integrative landscape projects. *Environmental Management, 36*(6), 792–807.

United Nations (2015). Retrieved 22 April 2017 from https://treaties.un.org/Pages/ViewDetails.aspx?src=IND&mtdsg_no=XXVII-7-d&chapter=27&clang=_en.

van Wyk, E., Breen, C. M., Roux, D. J., Rogers, K. H., Sherwill, T., & van Wilgen, B. W. (2006). The ecological reserve: Towards a common understanding for river management in South Africa. *Water, 32*(3), 403–409.

Watts, R. J., Ryder, D. S., Allan, C., & Commens, S. (2010). Using river-scale experiments to inform variable releases from large dams: A case study of emergent adaptive management. *Marine and Freshwater Research, 61*(7), 786–797.

Webb, J. A., Little, S. C., Miller, K. A., Stewardson, M. J., Rutherfurd, I. D., Sharpe, A. K., & Poff, N. L. (2015). A general approach to predicting ecological responses to environmental flows: Making best use of the literature, expert knowledge, and monitoring data. *River Research and Applications, 31*(4), 505–514.

Webb, J. A., Miller, K. A., De Little, S. C., & Stewardson, M. J. (2014). Overcoming the challenges of monitoring and evaluating environmental flows through science-management partnerships. *International Journal of River Basin Management, 12*(2), 111–121.

Webb, J. A., Miller, K. A., King, E. L., Little, S. C., Stewardson, M. J., Zimmerman, J. K., & LeRoy Poff, N. (2013). Squeezing the most out of existing literature: A systematic re-analysis of published evidence on ecological responses to altered flows. *Freshwater Biology, 58*(12), 2439–2451.

Webb, J. A., Nichols, S. J., Norris, R. H., Stewardson, M. J., Wealands, S. R., & Lea, P. (2012). Ecological responses to flow alteration: Assessing causal relationships with Eco Evidence. *Wetlands, 32*(2), 203–213.

Webb, J. A., Wealands, S. R., Lea, P., Nichols, S. J., de Little, S. C., Stewardson, M. J., & Anderssen, R. S. (2011). Eco Evidence: Using the scientific literature to inform evidence-based decision making in environmental management. In Proceedings of MODSIM2011 International Congress on Modelling and Simulation (pp. 2472–2478). Perth, Australia.

Wiseman, N. D., & Bardsley, D. K. (2013). Climate change and indigenous natural resource management: A review of socio-ecological interactions in the Alinytjara Wilurara NRM region. *Local Environment, 18*(9), 1024–1045.

Wolters, E. A., Steel, B. S., Lach, D., & Kloepfer, D. (2016). What is the best available science? A comparison of marine scientists, managers, and interest groups in the United States. *Ocean and Coastal Management, 122*, 95–102.

Zhang, Y., Arthington, A. H., Bunn, S. E., Mackay, S., Xia, J., & Kennard, M. (2012). Classification of flow regimes for environmental flow assessment in regulated rivers: The Huai River Basin, China. *River Research and Applications, 28*(7), 989–1005.

Ziembicki, M. R., Woinarski, J. C. Z., & Mackey, B. (2013). Evaluating the status of species using Indigenous knowledge: Novel evidence for major native mammal declines in northern Australia. *Biological Conservation, 157*, 78–92.

5 Accounting for water

From past practices to future possibilities

Liz Charpleix

Introduction

Sustainable[1] management of water is an increasingly urgent and complex issue across the globe. The availability of water, in terms of quantity and quality, affects all living creatures and the natural environment. Accurate measurement of the quantity and quality of the freshwater that exists on Earth assists in its effective allocation and maintenance; to this end, humans have assumed responsibility for its management.

Mainstream approaches to water management calculate its value using the principles of financial accounting, which is a process devoted to the collection, categorization and analysis of information relating to resources, in order to describe the economic activity of an entity (Godfrey, Hodgson, Holmes, and Tarca, 2006, p. 94). Economics is 'the science of determining how available resources may be best used' (Agudelo, 2001, p. 3). As the economy 'discovers' water's usefulness, it figuratively transforms it into a resource, according to Evernden (1981, based on the analysis of Frankfort, Frankfort, Wilson, Jacobsen, and Irwin, 1946/2013). At this point, water changes from 'something possessed of innate value [a 'Thou'] to a neutral object [an 'It']' (Evernden, 1981, p. 148; Frankfort et al., 1946 (2013), lines 136–165). Resourcist paradigms package water as a commodity and humans and the environment as consumers.[2] As Evernden points out, 'By treating everything as homogeneous matter in search of a use, [resourcism] devalues all' (quoted in Berryman and Sauvé, 2016, p. 106). While, as Agudelo (2001) explains, 'Resources have economic value whenever users are willing to pay for them rather than do without' (p. 5), it is now true that in many places in the world, users have no choice about whether to pay for water. This is a result of what Snyder (2001) explains as the '"dilemma of common-pool resources" [which results in] the problem of overexploitation of "unowned" resources by individuals or corporations that are caught in the bind of "If I don't do it the other guy will"' (p. 478).

Consumers' desire, or need, to access resources that they perceive as useful to them – or in fact, vital, as in the case of water – has enabled the neoclassical economics paradigm to become dominant in Western economies (Agudelo, 2001; G8+5, 2007; UNEP, 2000, p. 105). This paradigm examines the behaviour of individuals who participate in the economy (Darity, 2008). Neoclassical economics is part of the neoliberal economy, in which the role of the State in society is secondary to that of the free market (Calhoun, 2002). The dominant rationalist, neoliberal economy, which excludes relational valuation systems through its demand for commensurability, is tenacious, according to Carruthers and Espeland (1991), due to its historical institutionalization during the rise of capitalism. Cahill and Paton (2011) agree, noting that 'the current structure of the economy is path dependent, institutionally bound, and locks in certain interest and behavioural norms' (p. 23).

These neoliberal economic practices, which incorporate exclusionary and privileging perspectives informed by Eurocentrism and anthropocentrism[3], accentuate rather than assist in solving existing water management problems. Due to the asymmetries inherent in these practices, they may obstruct the equitable allocation of water, mask the non-economic value of the environment, distort reciprocity between the human and non-human worlds and block non-resourcist knowledges of water.

In a resources-focused world, it has been considered appropriate to allocate water according to the demands of those who are able to express them, commonly referred to as 'users'. This term denotes a utilitarian approach that ignores intrinsic and relational qualities of water. Resulting inequities in allocation may be ameliorated by recognizing valid needs beyond the restrictions of utility, and by choosing non-resourcist language, for example 'recipient' rather than 'user'.

In seeking an alternative to this utilitarian view of water valuation, this chapter will explore how water is, has been and could be valued for its relational and existential qualities in addition to its value as a commodity (see also Chapter 6, this volume). Refinement of the valuation scheme, which forms the first step of the accounting process, would add useful complexity to the existing model. The next step is to account for its qualities and quantity in ways that are beneficial to both human and non-human recipients of water. A range of ontological conceptualizations of water will be examined. The examples included are selected from the wide range being implemented worldwide, with particular attention to Aotearoa/ New Zealand, where history has recently been made in respect to a river's legal status, and Australia, an example of a settler society with a long history of Indigenous culture based on a strong relationship with the environment. Possibilities for the creation of accounting systems that are able to report on a broad range of non-economic aspects of water will be evaluated. The aim is to enable improved accessibility, distribution and sustainability of water by a rebalancing of the water accounting paradigms that currently favour resourcist treatments of the substance. The conclusion is reached that the ontological approach offering the best likelihood of achieving these aims is a hybrid model of water accounting.

The word 'hybrid' is used in two ways in this chapter. One usage denotes a system synthesized from two or more origins. For example, water accounts compiled under United Nations guidelines use a 'combination of different types of units of measurement in the same accounts' (UNDESA, 2012, p. 65). In this case the combination is formed from commodified and relational waters, a distinction that essentially aligns with economic and non-economic approaches to valuation. The other usage of the word 'hybrid' is taken from Homi Bhabha's work on colonialism; through substituting commoditized water for the subjugated people discussed in Bhabha's theory of post-colonial[4] cultural politics. In this adaptation of Bhabha's theory, innovations arise in the spaces between commodified and relational waters, drawing on elements of both (see Charpleix, forthcoming). These usages are differentiated respectively as 'synthetic hybrid' and 'Bhabha-ian hybrid'.

Bhabha's exploration of post-colonial hybridity, whereby subjects are empowered by challenging the expectations and impositions of the colonizer, provides an analysis that can be applied to the study of water practices in settler societies. His work describes an ironic compromise he calls 'mimicry' which arises from 'the desire [by colonists] for a reformed, recognizable Other, *as a subject of a difference that is almost the same, but not quite*' (Bhabha, 1994, 2004, p. 122, emphasis in original). Nicholls (1997) explains that such mimicry 'opens up a gap between coloniser and colonised that disrupts the terms of this opposition, and as such becomes the site for power, agency and identity in the "post-colonial" world' (p. 5). New cultural meanings can best arise where space is made available for them:

> It is in the emergence of the interstices – the overlap and displacement of domains of difference – that the intersubjective and collective experiences of *nationness*, community interest, or cultural value are negotiated. (Bhabha, 1994, 2004, p. 2, emphasis in original)

Work in legal pluralism provides additional avenues for the application of this adaptation of Bhabha's theory of hybridity. For example, in her exploration of a legal pluralistic approach to Indigenous rights, Anker (2014) notes that 'the reification of [state] law is a denial of the human element, [and] of relation, interpretation, symbolism and embodied practice', whereas 'the minimum condition for the "recognition of difference" . . . [is] a plurality in the very nature of law' (p. 5). Plurality in all post-colonial institutions, not only legal systems, can be negotiated in the interstices referred to by Bhabha. The economy is one such institution.

In Aotearoa/New Zealand, a unique recognition of water's place inside and outside the economy occurred in 2012, when the legal representatives of the Whanganui River gained legal standing for the river as an entity in its own right (Office of Treaty Settlements, 2012), giving it 'the rights, powers, duties and liabilities of a legal person' (Office of Treaty Settlements, 2014). This was the result of a lengthy legal battle, culminating in a determination from the Waitangi Tribunal (Waitangi Tribunal, 1999), which placed non-economic value on the traditional connections between the river and its Māori peoples while also preserving the place the river occupies in the economy of the country. This legal recognition offers researchers into intrinsic value an example of what is possible when the limitations of Western economic and legal paradigms are challenged by alternative thinking, such as, in this instance, through Bhabha-ian hybrid negotiations.

What is water? An ontological analysis

Water is many things

A scientific approach describes water as a compound of hydrogen and oxygen. From a Western anthropocentric, neoliberal perspective, water is a resource with economic value. Water may be viewed as a flow and a stock, a good and a service, with intrinsic value (that is, according to Bonnett (2003), 'mak[ing] no reference to human satisfaction or respect in the experience . . . itself' (p. 630)) and with direct and indirect economic value. It is both a commodity, often under strategic control of human agencies, and a relationship, creating linkages between and within human and non-human communities. It has value when it is used and when it is not used. The 18th century economist Adam Smith cited water to illustrate the distinction between 'value in use' and 'value in exchange':

> Nothing is more useful than water; but it will scarce purchase anything; scarce anything may be had in exchange for it. A diamond, on the contrary, has scarce any value in use; but a very great quantity of other goods may frequently be had in exchange for it. (Smith, 1776, cited in White, 2002, p. 662)

Water as a commodity

Technological advancements and increasing human populations have revolutionized work, waste management practices and energy sources in the centuries since the Industrial Revolution. Simultaneously, nature has become separated from society, a development that Kaika (2005) calls '[m]odernity's Promethean project' to 'fight . . . for human emancipation through the domination of nature' (p. 12). In the process, water has become a modern 'container . . . of exchange value', a commodity (Kaika, 2005, p. 12) and a resource that needs to be managed; the result is described by Linton (2010) as 'modern water' (p. 30). The coining of the term 'hydrologic cycle' in 1931 coincided with the growth of state water agencies worldwide, 'rendering [modern water resources] "legible" for administrative purposes' (Linton and Budds, 2014, p. 171).

Modern water, on its journey through the pipes and processes of the supply chain that separate the source from the recipient, has become, in the general mindset, a placeless substance. This 'deterritorialization', meaning severance of the bond between specific groups of people and specific water bodies (Linton, 2010, p. 18), renders water commensurable, enabling its commodification (Linton, 2014, p. 113) and allowing its incorporation in the economy. Commodification also enables exclusion of non-economic recipients, thus limiting equitable and sustainable access to water.

Water as a relational agent

Linton (2010) also identifies water as 'a process of engagement . . . always taking form in relation to the entities with which it engages' (p. 18). Linton's discussion about the hydrosocial[5] cycle reverses the concept of the hydrologic cycle, now established in practice, which separates water from its social context. Instead, the hydrosocial cycle puts 'people and politics at the centre of all water issues' (Linton, 2014, p. 114), urging a strategy of resistance to the commodification of water. Proactive human input to this strategy is crucial to assist in freeing water from commodification.

Hydrosocial relationships have been explored by Tadaki and Sinner (2014), who subdivide relational ontologies into two categories. In the first, which is generally quantitative, water contains 'an essential 'value' (way and extent of mattering) universal to humans and largely independent of the context in which [it is] situated' (p. 143). In the second category, people imbue their environments with attributes unique to their lived experience. For example, '[p]eople in different social locations can read swimming "value" as composed of different attributes' (Tadaki and Sinner, 2014, p. 145).

Within a hydrosocial, or relational, world-view, the human and non-human co-constitute each other. Roe (2009) has described the assemblages produced by connections between humans and non-humans, including water, as an exercise of mutual configuration (p. 251). Throughout history, opinion has been divided about whether the sentience and agency required to create and maintain meaningful connections are absent from animals and the environment (Leopold, 1949; Shaw, 2013; van Bogaert, 2004). Agency in sociological terms is the capacity for autonomous social action (Calhoun, 2002), and sentience, according to Harris (1988), is 'the highly integrated combination of physiological processes in the organism [which registers and reacts to effects]' (p. 80).

There is support for the understanding that agency and sentience extends to non-humans. For example, actor-network theory recognizes agency through which both humans and non-humans can be instrumental in their connections: '*any thing* that does modify a state of affairs by making a difference is an actor' (Latour, 2005, p. 71). Strang (2014) quotes Tilley, after Appadurai: '[I]f things have their own biographies, it is but a short step to consider these things as having their own agency and actively having *effects* in relation to persons [and] other things/species' (p. 139, emphasis in original). For example, British archaeologist Matt Edgeworth notes that the agency of a river is derived from the solar and gravitational energy-driven 'flowing water [which] . . . actively and physically challenges human projects and intentions' (Strang, 2014, p. 159). Conversely, Strang's (2014) view on sentience is that water neither has it nor needs it to create networks across the human-nature divide, asserting that 'it is possible to excise intentionality from the equation [allowing] us to acknowledge the agentive capacities of things without proposing or implying a form of *faux* animism' (p. 165).

However, Indigenous world-views, such as those of the Australian Aboriginal peoples and New Zealand Māori, understand co-becoming (Bawaka et al., 2016, p. 2; see also Chapter 12, this volume), a concept which aligns with Bhabha-ian hybrid negotiations, rather than '*faux* animism', which indicates the dominance of anthropocentric hierarchies of culture. Bawaka et al. (2016) describe the concept of *gurrutu*, which, for the Yolŋu people of Australia, 'brings everything together within an infinite pattern of kinship, obligation and care. . . . *Gurrutu* reveals place/space as relational, as always emerging, as human and more-than-human, and as both bounded and constituted through flows and relationships' (p. 6). Cloke and Jones present this as a circulating temporality in which '[p]ast, present and future are continually reprocessed while the materiality of the landscape is worked by, and marks, this process' (Cloke and Jones, 2001, cited in Bawaka et al. 2016, p. 12), producing an emergent state of 'sentience' (Langton, 2002, cited in Bawaka et al., 2016, p. 12). The recognition of agency and sentience in Australian Indigenous perceptions of water, in contrast to the placelessness, anonymity and narrow utility of modern water, affirms its relationality. Similarly, for the Māori, '[t]he river, like the land, was transmitted from ancestors, from the original ancestress, Papatūānuku, the earth mother, through the first people to the current occupying tribes, and was bound to pass to the tribes' future generations' (Waitangi Tribunal, 1999, p. 48).

Strang (2014) confirms water's relational place in the world: '[W]ater's ubiquitous capacity to flow between [bodies and environments] articulates most clearly that . . . human-environmental relationships are composed of interactions between material and social processes' (p. 135). Such views on relationality complement, and refer back to, Aldo Leopold's 1949 call for human ethical behaviour to take account of the whole biotic community from a biocentric (placing equal value on all life, whether human or non-human) or ecocentric (replacing the primacy of humans with that of the earth, including all its living and non-living elements) rather than an anthropocentric standpoint (Arney, 2011).

An existential view of water

A further ontological conceptualization of water is one in which it is neither relational nor a commodity; it simply exists. Such existence is not meaningless, as Rolston (1988) notes: 'Nature is not inert and passive until acted upon resourcefully by life and mind' (p. 198). Since long before the evolution of humans as a species, non-anthropocentric water has followed a pattern of what he describes as 'negentropic constructiveness in dialogue with an entropic teardown . . . [T]his is nature's most striking feature' (Rolston, 1988, p. 199). In this ontology, humans are irrelevant to water, although there is no situation in which humans have no relationship with water: 'An ecology always lies in the background of culture' (Rolston, 1988, p. 3).

An existential ontology, by preceding (and perhaps, in the long term, succeeding) anthropocentric ontologies, also precedes Bhabha's theory of hybridity, which depends upon challenging entrenched human expectations. This gives the existential ontology both strength, through its pre-human primacy, and weakness, in the face of anthropocentric agency based upon intellectual capacity and egotistic hierarchies of expectations (see Chapter 6, this volume). Although, as Rolston (quoted by Oelschlaeger, 1991, p. 296) points out, 'nature itself – the evolutionary process – was enriched by species Homo sapiens', Passmore (quoted by Oelschlaeger, 1991, p. 297) laments that the development of human culture has 'paradoxically enabled behavior that impairs the integrity of nature'.

Water accounting

Systems of accounting for water can be categorized pursuant to these ontological approaches; innovation in the development of systems is required to meet needs identified by previously ignored ontologies. Before describing some of these systems, it is necessary to provide a brief background to the term 'water accounting'.

Why account for water?

Water accounting systems use quantitative and qualitative analysis as a basis for the preparation of reports to support water management decisions. Such analysis depends upon treating water as a commodity, in accord with the neoliberal economic model. Unregulated extraction in times of abundance prior to the development of formal water control institutions (for example, collecting water from free-flowing rivers in areas without municipal regulatory oversight) was a historically appropriate strategy, based on informal reporting mechanisms. Strategies which are implemented when regulatory systems spread into previously unregulated regions are more formal and scientifically informed. Qualitative reporting is based on non-financial measures, while quantitative analysis, upon which market-based economic systems depend, commonly relies on monetary values.

A limitation of the use of money as a medium of exchange is that it may not be able to appropriately acknowledge all the elements, such as cultural and environmental features, which give value to water. This is an insidious aspect of the dominant water accounting approaches, although some have attempted to find ways of overcoming the problem. Nevertheless, incommensurable approaches to valuing water, such as that outlined by Bhabha's theory of interstitial action, remain underexplored. The consequent imbalance of power exerted by the dominant economy tends to exclude some recipients from equitable access to water, both humans, of whom the United Nations currently records 783 million without access to clean water and 1.7 billion living in river basins where water use exceeds recharge (UN, 2016), and non-humans such as wetlands, flora and fauna.

What is water accounting?

Water accounting can be defined as 'combin[ing] the science of hydrology with financial accounting models to help ensure that adequate water measurement, monitoring and reporting systems are in place' (Australian Government Bureau of Meteorology [BOM], 2015d). Water accounting provides scope for information-gathering beyond traditional financial parameters, in order to 'improve understanding of how water resources are sourced, managed, shared and used[,] improve public and investor confidence in the amount of water traded, consumed, recovered and managed for environmental and other public benefit outcomes [and] inform users' decisions about the allocation of resources' (BOM, 2017). The models discussed below include the economic, which use financial measures (for example, to enable relevant costing to consumers) and the non-economic, which can be volumetric (accounting for quantities, for example, to establish resource stores to enable consumer allocations), quality-focused (accounting for condition, for example, to support implementation of conservation measures) or hybrid systems (synthetic and Bhabha-ian).

Figure 5.1 summarizes a range of current and potential water accounting models (see further Fanaian, 2013). Following this, descriptions of valuation methods and accounting systems using these valuation methods are provided. The dominant, commodifying systems are discussed first to describe existing practices, followed by a discussion of alternatives. Critical notes specific to the different models of valuation and accounting are included in this section, while a broader critique of the water accounting task is conducted in the following discussion section.

What is water valuation?

The first step in the water accounting process is the valuation of water. Witold Kula, quoted by Linton (2010), notes that the act of measurement (for valuation or other purposes) requires that 'we abstract from a great many qualitatively

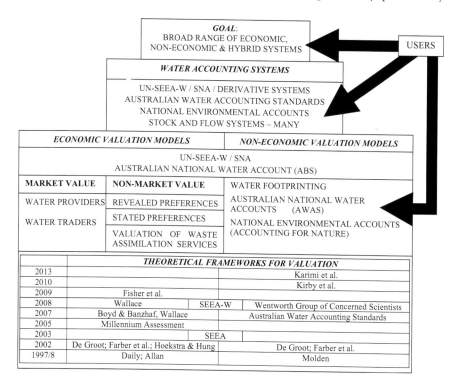

Figure 5.1 Water accounting pyramid: Summary of progression from research to valuation models to accounting systems.

Author's own.

different objects a single property common to them all, such as length or weight, and compare them with one another in that respect' (pp. 98–99). For accounting purposes, the goal is to equate the object being valued with an economic measure, enabling it to be traded in an economic market.

Since the 1960s, research into the value of water has explored ways of measuring incommensurable and relational aspects of water alongside the commodity values allocated by the neoclassical economic system. The understanding of water as a common good adds complexity to its valuation (see Chapter 6, this volume). Although Savenije (2001, p. 11) considers that its particular characteristics make it difficult to value on economic terms, water has found a place in the economy. Economic value is accepted in the framework of values found in Aotearoa/New Zealand's water governance system by Tadaki and Sinner (2014, p. 142). In their terms, this is 'value as magnitude of preference', which is a measurement of the quantity or intensity of preference for an outcome, or its exchange value; however, this may be expressed using non-financial quantifiers in some circumstances.

Economic quantifiers derive their assessments from the market as well as using non-market indicators to measure transactions based on their instrumental contribution to an economy. The United Nations' SEEA-W water accounting system includes 'shadow prices', which are estimates of value where the market price is absent or distorted (UNDESA, 2012, p. 121). However, this is not extended to identifying cultural or other incommensurable values. The quandary in accounting for water is in identifying quantifiers to describe such attributes. Relational assessments of intrinsic value, including models that account for water's contribution to human-identified goals, such as value as contribution to a goal, value as a system and values as 'ways that matter' (Tadaki and Sinner, 2014, p. 142), are generally excluded from economic water accounting models.

Valuation methods

Valuations of water (either its quantity or its quality) are made using economic methods, non-economic methods or a combination of the two. Commensurability is the goal of all valuation methods that incorporate economic quantifiers, including those that also attempt to recognize non-economic aspects of water. Anything can be traded in an economic market if a way can be found to equate it with a unit of currency.

Economic valuations

The method adopted for economic valuation is determined by the purposes for which it is being used, either to supply water in an economic market or to safeguard its existence as an ecosystem element (in which case its intrinsic value is subordinated to its perceived market value). Economic valuation methods include market and non-market valuations. Market valuations of water as a commodity price it according to the economic rules of supply and demand, with allowances being made for apportioning infrastructure construction, maintenance and replacement costs (Southern Water, n.d., p. 122). (See Table 5.1.) Non-market

Table 5.1 Market valuations

Market transfer facilitator	Type of valuation/ pricing	Purpose	Calculation method	Examples
Water providers	Fixed cost tariffs	Cost recovery for capital assets used in supply of water and sewerage services[i]	Calculated by providers based on their costs and forecasts (Southern Water, n.d., pp. 129, 136–139).	Municipal water provision; groundwater markets
	Usage tariffs	Recovery of cost of supplying water	Based on consumption	Municipal water provision; groundwater markets
	Other tariffs For other services provided	Other service charges e.g. sewerage or fire service connections (Southern Water, n.d., pp. 139–140)	Calculated by providers based on costs and forecasts	Sewerage, fire service
	Rising block tariffs	To encourage minimization of usage while meeting equity responsibilities	• First block of usage priced low to meet basic needs • Second block of usage at standard price for essentials • Third block at premium price for non-essentials (South West Water, n.d.)	Used in Israel (FAO, 2004); UK trials abandoned when no behavioural changes observed (Water Briefing, 2011)
Trading Brokers[ii]	Market value	Redistribution of water rights; Australian system includes: 1 Entitlements (perpetual or ongoing assets) 2 Allocations (volumes allocated in a given season according to entitlements held) (ACCC, 2012)	• Markets (separate for each state and territory) set prices based on supply and demand • Subject to oversight by government regulators	Water access right • The right to take water from a water source • Can include an obligation to contribute to fixed network losses (also known as conveyance losses) Water delivery right • A right to have water delivered by an operator • Irrigation right • A right to receive water that is not a water access right or a water delivery right

Source: Author's own

(i) Cost recovery is not usually directly aligned to per capita allocations of proportion of providers' costs; it may include adjustments for equity or historical reasons, for example converting between charging systems (Southern Water, n.d).

(ii) Water traders: operating in Australia, US, Chile, South Africa, Canada, China, India and other countries (Hadjigeorgalis, 2009); this discussion describes the Australian model (see Chapter 11, this volume).

Table 5.2 Imputed water valuation techniques

Valuation method		Explanation
Revealed Preference (based on observed market values)	Residual value	Marginal contribution of water to output, measured by subtracting all other costs from revenue
	Production function approach	Marginal contribution measured as the change in output from a unit increase in water input in a given sector
	Optimization models and programming	Marginal contribution measured as the change in sectoral output from reallocation of water across the entire economy
	Hedonic pricing	Service demand reflected in the prices people will pay for associated goods, e.g. houses closer to a natural amenity
	Opportunity Cost	Price differential compared to alternative
Stated Preference (based on Willingness to Pay [WTP][i] or Willingness to Accept [WTA][ii])	Contingent Valuation	Interview recipients; Service demand may be elicited by posing hypothetical scenarios that involve some valuation of alternatives, e.g. willingness to pay for increased fishing quotas
	Travel Cost	Interview recipients; Recreation areas attract distant visitors whose value placed on that area must be at least what they were willing to pay to travel to it
	Avoided Cost	Services allow society to avoid costs that would have been incurred in the absence of those services, e.g. flood control avoids property damages
	Replacement Cost	The cost of replacing ecosystem services with artificial systems, e.g. replacing wetlands with water treatment infrastructure
	Factor Income	Services that provide for the enhancement of incomes, e.g. increased water quality enhances commercial fisheries income
Valuation of water's waste assimilation services	Pollution damages avoided	Cost of harm that would occur if waste not assimilated
	Costs of preventing damage	Cost of measures taken to prevent pollution

Source: Farber et al., 2002, pp. 388–390; Lange, 2011

(i) Willingness to Pay is a measure of what a participant is willing to pay to acquire a good or service (De Groot, Wilson, and Boumans, 2002, p. 403).

(ii) Willingness to Accept is a measure of what a participant is willing to accept in compensation for forgoing access to a good or service (De Groot et al., 2002, p. 403).

valuations (including shadow pricing) require techniques that allocate economic value through observed or assessed inputs and outputs (see Table 5.2). These may be used in cost-benefit analysis of projects (Lange, 2011) and for assessment of social value not adequately captured by market valuation techniques, in other words, value conceived in terms of what a society would be willing and able to pay for a service or what it would be willing to accept in order to forgo that service (Farber, Costanza, and Wilson, 2002, p. 388–390). Non-market values include direct use values (the value of clean water to a recipient), indirect use values (the value of wetlands nutrient sequestration in reducing algal blooms), optional use values (the value associated with knowing that an asset is available for future use) and non-use values (passive, in the knowledge that an asset exists now, or bequest, in the knowledge that it will exist for future generations).

A category of economic valuations that combines both market and non-market systems is called ecosystem services. Research into ways of valuing the elements of the natural world from a neoclassical economic perspective has produced this term, which means 'contributions of the natural world which generate goods which people value' (Bateman et al., cited in Potschin and Haines-Young, 2011, p. 579). Valuable contributions include ostensibly non-use (public goods, see also discussion of non-resource instrumental values in Chapter 6, this volume) goods such as scenery, as well as those produced without human intervention or financial outlay. The Millennium Ecosystem Assessment Report (MA) in 2005 (cited in Fanaian, 2013) compiled a functional framework based on four categories:

- Provisioning Services (products obtained from ecosystems);
- Regulating Services (benefits obtained from regulation of ecosystem processes);
- Cultural Services (nonmaterial benefits obtained from ecosystems); and
- Supporting Services (services that protect other ecosystem services).

This has been finessed to avoid double-counting of services, to clarify definitions and characteristics of the services and to add a category of services provided to humans by the ecosystem (Boyd and Banzhaf, Wallace, and Fisher et al., all cited in Fanaian, 2013). More recently, a further category known as Blue Carbon has been defined, describing the carbon sequestered in the sediments of coastal wetlands (Boon and Prahalad, 2017).

Since the 1990s, ecosystem services applications called 'Payment for Ecosystems Services (PES)' and 'Markets for Ecosystem Services (MES)' have used the concept of market value to derive income from usage of ecosystems services (Goméz-Baggethun, de Groot, Lomas, and Montes, 2010, pp. 1214–1216). De Groot et al. (2002, p. 394) separated the values contained within ecosystems into three major types, adding ecological values (based on ecological sustainability) and socio-cultural values (based on equity and cultural perceptions) to the economic values outlined in Table 5.2. Ecosystem Services programs include water rights and carbon trading schemes (Coull and Valatin, 2008) and land stewardship programs (Tasmanian Land Conservancy, 2015).

The United Nations Statistical Commission has published a summary of current knowledge regarding accounting for ecosystems, 'SEEA Experimental Ecosystem Accounting' (UN, 2014). They note '[a]ccounting for ecosystems in physical (i.e. non-monetary) terms is a key feature' (p. 2) but there is no attempt to identify intrinsic environmental values. Within an Ecosystems Services context, assessing water for its value requires translation of the non-financial values into financial terms. Sullivan (2010) calls this 'the ideational transformation of "the environment" into new commodity fictions called "ecosystem services"' (p. 113), adding, '[w]hen nature health becomes converted into a dollar sign, it is the dollar not the nature that is valued' (Sullivan, 2010, p. 127).

Non-economic valuations

While modern economic usage concentrates solely on human benefits, a broader understanding of benefits on a relational basis, as described by Roe (2009), has existed since prehistoric times.[6] Contemporary researchers are pursuing relational outcomes through a range of non-economic valuation possibilities. Systems include water footprinting and national accounting systems, two Australian examples of which are described briefly below.

Water footprinting is an economic application developed by Hoekstra and Hung (2002, p. 15), built upon the volumetric concept identified by J. A. Allan in 1993 (Chapagain, 2006, p. 6). This model identifies virtual water for the pur-pose of understanding a country's dependence on internal or external sources of water. The underlying point of this theory is that it is possible for water-scarce countries to supplement their own water stocks by importing water within com-modities, rather than as bulk water. The approach has been criticized for having an over-simplified methodology that may limit its usefulness as an economic and policy tool (Frontier Economics, 2008).

A more complex water accounting model based on a financial accounting model (BOM, 2015a, p. BC10) is the Australian National Water Accounting system (Australian Water Accounting Standards – AWAS model). This has been produced as a tool to assist in decision-making about resource allocation (BOM, 2015a, p. 3). It uses volume as the most appropriate measure to pro-vide information to the public and to specialist users involved in the planning, monitoring, management and policy development for water as a resource (BOM, 2015a, p. BC8).

An alternative system, designed to assess the quality of water as well as that of other environmental elements, is the National Environmental Accounts (Australian model), designed by Wentworth Group of Concerned Scientists (2008, p. 6). This system measures the health of environmental indicators based on vigour (the level of productivity or 'pulse' of an ecosystem), organization (the structure or number of interactions within an ecosystem; healthy ecosys-tems have many interactions, whereas disturbed systems are highly simplified and have fewer interactions) and resilience (an ecosystem's ability to recover following disturbance). For water, the relevant factors under this system are its volume and its quality. The volume is evaluated based on a number of pre-existing frameworks, including the AWAS model of water accounts (Wentworth Group, 2008, p. 4), while the quality measurements for the trials have been pro-vided by Natural Resource Management institutions across Australia (Cosier, 2012). Based on the scientific practice of reference condition benchmarking, a non-monetary quantifier, the 'econd', was created to enable 'apples to be compared with oranges' (Cosier, 2012). Its purpose is to amalgamate qualitatively different indicators into a single quantifier for reporting purposes.

Synthetic hybrid valuations

Research into water valuation methods involving synthesis of economic and non-economic paradigms is ongoing. Two of these are briefly described below.

The National Water Account Australia, coordinated by the Australian Bureau of Statistics, is based on the United Nation's System of Environmental-Economic Accounts (SEEA), which 'bring into direct focus the relationship between the environment and well-being not revealed through traditional measures of economic activity, such as GDP and national income' (UNStats, n.d., p. 2). Supply and use data 'for physical and monetary volumes' (ABS, 2013) provide information about flows of fresh and desalinated water from the environment to the water supply industry and other economic activities, particularly agriculture, and flows of water from the water supply industry to households and businesses. Water use is linked to the economic data contained in the System of National Accounts, from which the headline indicator Gross Domestic Product is derived (BOM, 2015b).

Also related to SEEA, the United Nations' System of Environmental-Economic Accounting for Water [SEEA-W] (UNDESA, 2012) assesses a variety of aspects of water, including water quality, physical water supply and use, water emissions and assets. The physical data is provided to national statistical agencies by scientific agencies. The economic data is provided through national surveys by statistics bureaux. Valuation methods include versions of market and non-market valuation, with the addition of mathematical models, discussion of which is beyond the scope of this chapter. Temporal and spatial characteristics are also important.

Accounting systems

Water valuation techniques form a technical base for the water accounting pyramid. Above these sit water accounting systems (see Figure 5.1). Once the chosen aspects of water have been measured and valued, the resulting data is classified, analyzed and presented in a format aimed at meeting the needs of stakeholders (Godfrey et al., 2006, p. 17). The following examples are from the range of water accounting systems developed to date.

Since 2010, the volumetric (non-economic) National Water Accounts (Australian Water Accounting Standards – AWAS model) have been prepared annually, reporting on 'volumes of water traded, extracted and managed for economic, social, cultural and environmental purposes' (BOM, 2015c). Currently, there are ten significant Australian regions included in the National Water Accounts. More are to be added over time.

The National Environmental Accounts (Australian model) system is another non-economic method of accounting, but this is designed for integration into the dominant international economic system, through links with SEEA. The Wentworth Group (2014) stress the connection between the economy and the environment: '[W]e need to transform the economy so that a healthy environment becomes a partner to economic growth rather than a competitor' (p. 8). Trials have been completed successfully (Sbrocchi et al., 2015, p. 3). A state, territory and federal government joint project to create regional scale National Environmental Accounts is due to commence in 2017 (Wentworth Group, 2016).

Many countries have adopted the System of Environmental-Economic Accounting for Water (SEEA-W) model as a basis for preparation of water accounts, varied to suit national purposes. For example, the Southern Africa Development Community (SADC) Economic Accounting of Water Use Project is an accounting system based on SEEA-W, Australian water accounts (both ABS and AWAS systems), practical experience in SADC and pilot projects in southern Africa. Characteristics of this framework include livestock watering as an economic activity, valuation of the opportunity cost of time spent fetching water and the indirect and direct costs of water-borne diseases (Egis Bceom International, 2010).

Hydrological researchers are continuing to develop other volumetric water accounting systems that monitor and record flows and stocks through river basins. These have been used in a number of water catchments in south-east Asia, Nepal, Afghanistan, south-east and southern Africa (Water Accounting+, n.d.), eastern Africa, India, China, Iran and Sri Lanka (Karimi, Bastiaanssen, and Molden, 2013, p. 2461).

Kirby, Mainuddin, and Eastham (2010) describe a dynamic water accounting system based on Excel spreadsheets, which is a high-level, whole-of-basin level system, glossing over detail but providing rapid response analysis (pp. 6–7). Molden (1997) developed methods known as WA to account for water used in irrigation, from micro (recipient) to macro (basin) levels. Karimi et al. (2013) developed a system called WA+, building on and improving the WA system (pp. 2461–2463).

Water accounting systems are used by municipal water providers to monitor water stores to ensure maintenance of supply to domestic and industrial customers in their region. In Australia and Aotearoa/New Zealand, such providers include TasWater, the Tasmanian state water authority, and Horizons Regional Council, which supplies water to the Whanganui area. In accordance with their object of servicing customers from storages constructed, maintained and monitored in anticipation of customer demand, they account for water on an economic basis, setting tariffs priced on a cost-recovery basis. Wastewater processing is a related service (TasWater, 2014, p. 6). As these organizations are operated as commercial ventures, stakeholders include customers and others who have an interest in the business of water supply or asset provision. Responsible financial and operational management based on accurate accounting is therefore crucial.

Relational and existential models of valuation

Despite the various refinements and developments in water accounting practices (and water governance policy more generally), formal measurement of non-anthropocentric values are noticeably absent from the dominant accounting and valuation systems. Not only are they difficult for currently available accounting systems to recognize, accept and assess, but there is a lack of demand from ecocentric subjects (as we might perceive them) due to their literal voicelessness. Nature cannot express its concerns as specifically as humans do. Nevertheless, non-economic and hybrid accounting models, such as those negotiated in post-colonial domains where Bhabha-ian hybridity enables a fuller range of incommensurable values to be identified and incorporated, have the potential to illuminate values found in ecocentric water practices.

In Indigenous waters, value lies in such scales as the temporal rather than the economic. As Langton points out, 'The objectification and use of resources is part of Indigenous knowledge and philosophy, but this value of country does not dominate to the exclusion of all other values. Rather, resource-use decisions are made with reference to a range of other concerns, including dialogue with the ancestors' (Langton, quoted in Weir, 2009, p. 14). This is possible within the Indigenous view of time, which is described by Rose (2000) in the Australian context as 'a sound, a wave, a call that is pulled into specific places, social groups and ecosystems as it flows from the source [as] actions invigorate this out-pouring' (p. 295). By contrast, the Western view of time is linear, enabling schedules to be structured alongside cashflows. To paraphrase Baldwin's (2012) challenge to the West's response to climate change, water management 'warrants a different kind of analysis. Not one that seeks to expose how *past* colonial relations condition the present, but one that seeks to uncover how the *future-conditional* conditions the present' (p. 631).

Time scales provide a structure for the intrinsic qualities of nature, which, according to Rolston (1988), are 'pleasures good in themselves' (p. 110):

> From a short-range, subjective perspective we can say that the value of nature lies in its generation and support of human life and is therefore only instrumental. But from a longer-range, objective perspective systemic nature is valuable intrinsically as a projective system, with humans only one sort of its projects, though perhaps the highest. (Rolston, 1988, p. 198)

The observance of water's value in relation to the passing of time can assist in improving broader environmental sustainability. As Strang (2014) points out, this has been endangered by anthropogenically induced problems related to resourcism:

> [S]ustainability relies on the orderly movement of things, i.e. at a rate which permits the material renewal not just of resources themselves, but also of the other elements of the social and ecological systems upon which their production and use depend. ... This temporal stability is now being anthropogenically disrupted, with emergent effects on all scales. (p. 149)

Another angle from which possibilities for non-commodified water valuation arise is seen in Bonnett's (2003) differentiation between '*inherent intrinsic value*', where water holds value '*independently* of how it may serve or satisfy human beings', and '*derivative intrinsic value*', where 'its existence is dependent on satisfying a valuer' (p. 630, emphasis in original). Adhering to the view that water has agency and sentience, Rolston (1988) points out that '[t]he inventiveness of systematic nature is the root of all value' (p. 198), regardless of how it may or may not serve any external valuer. As Rolston (1988) explains, 'Against the standard view that all value requires a beholder[,] we now claim that some value requires only a holder, which can be an individual but can also be the [evolutionary] system that carries value to and through individuals' (pp. 222–223).

Discussion

Water management for stakeholders

The outcome of any water accounting task depends upon the integrity of valuation and accounting systems, their coherence with the needs of water recipients and innovations that meet evolving needs. The dominant, resourcist accounting approaches may be unable to ensure that all water management is equitable, effective and capable of meeting recipients' expectations. For example, TasWater produced a net profit of $33,154m (USD25,462) after tax in the 2014–2015 year, yet 26 (out of 76, compared to a target of 20) townships were on 'permanent boil water alerts' (Tasmanian Audit Office, 2015, p. 3; TasWater, n.d., p. 24). Only 46 per cent of its sewage treatment reached regulated compliance levels, for reasons determined to be beyond the authority's control (TasWater, n.d., pp. 25, 29).

TasWater's public reporting notes their environmental legislative compliance obligations, yet the environment is not specified as a stakeholder (TasWater, 2014). As long as the environment is excluded from the accounting process, it will be treated inequitably through inattention. By contrast, Horizons Regional Council (Aotearoa/New Zealand) displays an inclusive attitude to the environment: 'We must balance our desire to use water resources with a responsibility to ensure they can sustain our recreational use and their "slippery" inhabitants' (Horizons Regional Council, n.d.).

Water valuation for stakeholders

Both Tasmanian (Australia) and Whanganui (Aotearoa/New Zealand) water management authorities function on an economic basis in which water is the product, priced to recoup the costs of operations, supply and research. Unlike TasWater, however, Horizons sees beyond water as a resource with qualities and quantities to be maintained, to its store of relational value, such as mauri, 'the life-force and personality' (Waitangi Tribunal, 1999, p. 37) of the water body and bed. By communicating in the Māori language (as well as in English), the

Council makes efforts to communicate understanding of the importance of water management criteria to Māori communities (Horizons Regional Council, 2014). Horizons Regional Council (1997; 2003) considered but dismissed the possibility of valuing water economically:

> No attempt has been made to quantify the values of the river in dollar terms. This would be very difficult to do, and would entail the risk of argument focusing on the validity of methods of valuation rather than on the values the Council seeks to protect. The Council considers the cultural, intrinsic and amenity values outlined … are significant for the Region. (p. 5)

Nevertheless, recognition of the range of the river's values does give it some weight as a stakeholder in Horizons Council's operations.

Water accounting for stakeholders

There is utility for human stakeholders (such as farmers or corporations operated by humans) in the dominant water accounting systems, despite systemic problems at both macro and micro scales. Macro-level problems include choice of system and path dependency. Liebowitz and Margolis (1999, pp. 981, 984) point out that path dependency, which recognizes that 'where we go next depends not only on where we are now, but also on where we have been', means 'we lose the usual presumption that individual choices lead to an optimal outcome'. On a micro level, problems include reporting issues, the difficulty of obtaining reliable and relevant data and the exclusion of some recipients and processes because there is no uniformity about which elements of the hydrological cycle are included in the accounts (Karimi et al., 2013, p. 2469; Perry, 2012, pp. 217, 224, 229; see also Chapter 1, this volume).

The National Environmental Accounts aim to overcome some of these issues in their project 'to integrate the management of our environment into everyday economic decisions' (Sbrocchi et al., 2015, p. 4). In so doing, their purpose is to raise the status of the environment to approximate that of economic stakeholders in water management systems. While avoiding commodification, such aims reiterate the trope that the environment is important only insofar as it maintains economic relationships with the human-centred world.

Towards a recognition of the multidimensionality of water

The theories of researchers such as Bhabha (1994, 2004) and Linton (2014) provide guidance for recognition of the diversity of this planet's waters and their dynamic participation in hydrosocial processes. Inclusion of concern for welfare beyond the anthropocentric, subverting the common view of it as a solely human concept (see, for example, Daly, 2011), is a step towards overcoming the restrictions on water valuation of the neoclassical economic model. The words 'welfare'[7] and 'wellbeing'[8], which are usually understood to apply solely to humans, are derived from Middle English words that applied to human and non-human subjects alike. Before humans evolved, the environment managed its own welfare and wellbeing, through morphological processes such as the thermodynamic dialogue between negentropy and entropy, as described by Rolston (1988, p. 199).

Environmental welfare was a feature of the Whanganui River battle. After the centuries the claimants and their ancestors had spent living by a river that sustainably met their needs, they were concerned over its physical and spiritual degradation subsequent to colonization in 1840. They had become petitioners alongside any group wanting to have a say in the river's management, similarly subject to rules laid down by the colonial legal system. The [Aotearoa/ New Zealand] Resource Management Act 1991 is more inclusive than prior laws towards the sustainable management of the environment. Nevertheless, as the Tribunal judges noted, 'There is no process within the Act that does not leave the ultimate power and control in the hands of a regional or territorial authority' (Waitangi Tribunal, 1999, p. 343). For the sake of the wellbeing of the river as a taonga[9] to its Māori inhabitants, Waitangi Tribunal claim (Wai) 167 restored the responsibility for the river's welfare to its traditional owners.

Valuing the Whanganui River within the hegemony of neoliberal conventions, for example in the Ecosystems Services regime (Goméz-Baggethun et al., 2010, p. 1209), would dishonour the post-colonial progressiveness of Wai 167. However, accounting for 'condition' (Cosier and Sbrocchi, 2013, p. 2), or environmental wellbeing, using the Australian system of National Environmental Accounts, could be an appropriate method for valuation of the river's sustainability and water quality. Reconceptualizations of other qualities of the river, such as social, cultural, relational and political, still face the dilemma of how to persuade authorities to recognize and communicate incommensurable values.

Towards a relational system of water accounting

The water valuation and accounting systems described thus far might suit economically oriented applications but not the specific situation seen in the Whanganui River. In any settler society, there are both pre-colonial and colonial attitudes towards environmental features. Generally, the imported views have become predominant, but as the Whanganui River case shows, the pre-colonial views will not necessarily have been abandoned, and may have been retained and adapted over time. Thus the possibilities for future approaches to valuation of the Whanganui River (as a model for other water bodies) are several:

1   The colonizers' approach;
2   The Indigenous inhabitants' approach;
3   A parallel application of the two systems, each applying to different, predetermined circumstances;
4   A synthetic hybrid of the two systems;
5   A Bhabha-ian innovation, intervening between the two systems.

Option one – The colonizers' approach

The colonizers largely disregarded the views of the Māori who lived along the river and used the river without concern for its history or interdependencies. They adopted it as a resource for the supply of products and services for transport, infrastructure and power generation in their nation-building schemes. As a resource, the river has financial value in the national accounts. It can be economically valued for its inherent qualities, for example its aesthetics, through imputed value techniques.

Option two – The Indigenous approach

The Indigenous inhabitants lived with the river as an integral part of their community, alive and with unique needs, responsibilities and capabilities. They assimilated and responded to its changing presence according to their own cultural and legal understandings. Valuation using currently available indicators cannot encompass the intrinsic value of the river as a cultural and spiritual entity.

Option three – Parallel application of the colonizers' and the Indigenous approaches

The resolution of Wai 167 required the Waitangi Tribunal to respect the Indigenous approach to living on the river while simultaneously ensuring that accessibility by broader society to the river, including its economic and tourism benefits, was not lost. For example, an incorporated association's application for a Water Conservation Order was rejected on the grounds that its implementation would restrict the local council's ability to approve certain river uses (Waitangi Tribunal, 1999, p. 381) and also that it 'would be contrary to the claimants' claimed right of management' (Waitangi Tribunal, 1999, p. 382). Underlying this decision are the two perceptions of the river's value, which stem from it being *both* a resource (for the local council to administer) *and* an integral part of a network of geographically connected communities (to be valued as the communities determine). This action was taken before the role of guardian (Te Pou Tupua) was created to be the human face of the Whanganui River (Waitangi Tribunal, 1999, p. 382), a solution that is best described by Option Five.

Option four – Synthetic hybrid

SEEA-W is an example of a synthetic hybrid accounting system, creating economic reports from collections of quality and quantity data. Some views of this version of hybridity are skeptical of its form. Papastergiadis (2000) identifies 'the hybrid, which is born out of the transgression of [the] boundary [between 'us' and 'them'] . . . as a form of danger, loss and degeneration' (p. 174). Latour (2013) notes that the 'genetic metaphor' of the hybrid entity, 'an ambiguous, sterile mixture of two radically different and irreducible, pre-exi[s]ting entities', is not useful for removing the dualism from a '"hyphenated" view of Nature/Culture'

(p. 561). Referring to this genetic metaphor, Papastergiadis (2000) cautions that 'the hybrid may yield strength and vitality . . . [only if] . . . the boundary is marked positively – to solicit exchange and inclusion' (p. 174). SEEA-W benefits from positive boundaries defining its system components, but they are a restraint on non-economic applications.

Option five – Bhabha-ian innovation

In the conflict between the colonized Whanganui River Iwi[10] and the colonizing state, defined boundaries were an ongoing barrier to resolution of the conflict. Dissolving the boundaries allowed space for innovative thinking, from which emerged the river's guardian structure, Te Pou Tupua:

> Rather than speaking about the River as an inanimate commodity, the Crown, local authorities and others who interact with the River will be required to speak to the River and view it holistically as a living and indivisible whole. (Whanganui River Maori Trust Board, n.d., p. 24)

A river with its own voice is a situation that modern Western societies have not dealt with before. It offers intriguing opportunities to escape what Hill (2011) calls 'our enslavement to a manipulated, deceptively simple economic bottom line (when the absolute bottom line is bio-ecological)' (pp. 20–21). The river's incommensurable aspects, such as its cultural and ecological wellbeing, contribute to the bio-ecological bottom line, just as physical assets in a factory contribute to an economic bottom line. Given the absolute necessity of water to life, Hill (2011) points out that 'economic sustainability must serve . . . ecological sustainability . . . and not vice versa' (p. 22).

How is an interstitial valuation structure to be created? Bhabha (1994, 2004) urges a release of historical attachment in the creation of a new way forward based on his concept of 'mimicry':

> What is theoretically innovative, and politically crucial, is the need to think beyond narratives of originary and initial subjectivities and to focus on those moments or processes that are produced in the articulation of cultural differences. (p. 2)

Creating a new valuation structure outside economic markets while encompassing the inherent particularities and universalities of water requires lateral thinking. Crucially, in a post-colonial world, such a system cannot embody a wholly Indigenous nor a wholly Western paradigm, yet it must recognize both.

Te Pou Tupua's role is to speak from Bhabha's 'third space[, which] displaces the histories that constitute it, and sets up new structures of authority, new political initiatives' (Huddart, 2006, p. 126). An accounting system that recognizes and responds to the actions and needs of Te Pou Tupua should be in thrall neither to the commensuration of the Western economic system nor the incommensurability of the agential river. The Regulatory Impact Statement published in connection with the tabling of the Te Awa Tupua (Whanganui River Claims Settlement) Bill 2016 (Office of Treaty Settlements, 2016) places strong emphasis on the importance of the wellbeing and intrinsic values of the river. As yet Ngā Tāngata Tiaki o Whanganui, the Whanganui River Iwi's representative in Te Pou Tupua, have not published any mention of progress towards a new water valuation system (Ngā Tāngata Tiaki, 2016). However, the protracted legal settlement process could provide the opportunity to carve out an interstitial creative space, within which an innovative water valuation system can form.

Conclusion

Water accounting is a complex field of research and practice. The breadth of the range of water ontologies, from existential to relational to commodified approaches, guarantees that water accounting systems will never be limited to a single, standardized model. The incongruities between perceptions of water value may never be negotiated to the complete satisfaction of all parties; outcomes will vary according to differing needs.

Dominant water accounting systems are in most cases based on anthropocentric water management within resourcist markets and hold a strong position based on entrenched economic power. The path dependency inherent in this renders its neutralization by any alternative system difficult. However, water accounting research continues, with the goal of ensuring that the Earth is able to continue to support healthy life into the future.

With this aim in mind, the options reviewed here include economic, non-economic and hybrid accounting systems. Many economic systems are based on SEEA-W, which provide statistics on flow and stocks, including management of waste water, for inclusion in Systems of National Accounts and GDP calculations. Non-economic systems provide methods of calculating cost-benefit analyses, assessment of values not adequately captured by market valuations, quality assessments and flow and stock management. Inherent valuations can be attempted through non-economic valuations, but this results in commodification outcomes; otherwise inherent value cannot (yet) be assessed. Hybrid accounting systems provide potential either as combinations of elements of the other options, leading to primarily economic systems, or as experimental investigations into entirely new systems, for which Bhabha's post-colonial articulation provides an entry point.

The betweenness of Bhabha's interstitial space will allow water value assessment in a space unrestrained by the colonial structures that controlled the Whanganui River for so long. There is no precedent for an interstitial accounting system, but nor was there a precedent for the river's current legal status as an independent entity. The format will develop organically, just as the legal case developed.

If the inventiveness in the resulting accounting system matches the inventiveness of the outcome of Wai 167, it will be an exciting development in the field of water accounting. As Seed and Wright (2011) point out, '[Just as the fervour of displays of piety] to the reign of inquisitorial Christianity in the Middle Ages . . . looks bizarre to us today . . . [p]erhaps one day our own fealty to the hegemony of economics will look just as bizarre' (p. 247).

Notes

1 Sustainable development, according to the Brundtland Report definition, is being able to 'meet . . . the needs of the present without compromising the ability of future generations to meet their own needs' (WCED, 1987, p. 16).
2 See Oelschlaeger for a fuller discussion of the historical and religious roots of resourcism, which he describes as 'represent[ing] the transformation of modern people from Homo religiosus to Homo oeconomicus' (1991, p. 286).
3 Anthropocentrism (n): a value system that prioritizes humans over all other life forms (Anthropocentrism, 2016).
4 Post-colonial: from the work of Edward Said, whose 'practice of excavating the colonial past in the present' by investigating both the 'temporalities' of (post-)colonialisms (in the plural) and their 'spatial imaginaries' is distinctive due to 'the centrality he has given to the relationship between geography and empire' (Baldwin, 2012, p. 625).
5 Hydrosocial: 'a socio-natural process by which water and society make and remake each other over space and time' (Linton and Budds, 2014, p. 170).
6 Multilateral benefits, not necessarily related to cognitive capacities, become available through interaction between humans, plants and environmental features, for example through human alteration of watercourses to create swamps which supply food and shelter to humans, animals and vegetation (Gammage, 2011, p. 226).
7 'well', meaning 'in a good or satisfactory way', and 'fare', meaning 'perform in a specified way in a particular situation or over a particular period' (Welfare, 2016).
8 'well', as in note 7, and 'being', meaning 'existence, the fact of belonging to the universe of things material or immaterial' (Wellbeing, 2016).
9 'Taonga', meaning 'an ancestral treasure handed down' (Waitangi Tribunal, 1999, p. 46).
10 'Iwi', meaning 'people who share a common ancestor' (Waitangi Tribunal, 1999, p. 28).

References

ACCC. (2012). *A Guide to the Water Market Rules and Water Delivery Contracts*. Retrieved from http://www.accc.gov.au/system/files/A%20guide%20to%20the%20water%20 market%20rules%20and%20water%20delivery%20contracts_0.pdf

Agudelo, J. I. (2001). *The Economic Valuation of Water: Principles and Methods*. Delft, The Netherlands: IHE.

Allan, J. A. (1998). Virtual Water: A Strategic Resource. *Ground Water*, 36:4, pp. 545–546.

Anker, K. (2014). *Declarations of Interdependence: A Legal Pluralist Approach to Indigenous Rights*. England: Farnham.

Anthropocentrism. (2016). In *Oxford English Dictionary Online*. Retrieved from http:// www.oed.com.ezproxy.une.edu.au/view/Entry/336263?redirectedFrom=anthropocentr ism#eid

Arney, J. (2011). Land Ethic. In J. Newman (Ed.), *Green Ethics and Philosophy: An A-to-Z Guide*. Retrieved from http://knowledge.sagepub.com.ezproxy.une.edu.au/view/green ethics/n95.xml

Australian Bureau of Statistics (ABS). (2013). *Water Account, Australia, 2010–11: Explanatory Notes* (Cat. no.4610.0 0). Retrieved from http://www.abs.gov.au/AUS STATS/abs@.nsf/Lookup/4610.0Explanatory%20Notes12010-11?OpenDocument

Australian Government Bureau of Meteorology (BOM). (2015a). *Australian Water Accounting Standard 1: Preparation and Presentation of General Purpose Water Accounting Reports [AWAS1].* Retrieved from http://www.bom.gov.au/water/standards/documents/awas1_v1.0.pdf

Australian Government Bureau of Meteorology (BOM). (2015b). *Pilot National Water Account.* Retrieved from http://www.bom.gov.au/water/nwa/pilot_nwa.shtml

Australian Government Bureau of Meteorology (BOM). (2015c). *National Water Account.* Retrieved from http://www.bom.gov.au/water/nwa/index.shtml

Australian Government Bureau of Meteorology (BOM). (2015d). *National Water Account: Frequently Asked Questions.* Retrieved from http://www.bom.gov.au/water/nwa/faq.shtml

Australian Government Bureau of Meteorology (BOM). (2017). *Frequently Asked Questions.* Retrieved from http://www.bom.gov.au/water/standards/wasb/wasbFAQ.shtml

Baldwin, A. (2012). Orientalising environmental citizenship: Climate change, migration and the potentiality of race. *Citizenship Studies,* 16:5–6, pp. 625–640. doi: 10.1080/13621025.2012.698485

Bawaka Country, Wright, S., Suchet-Pearson, S., Lloyd, K., Burarrwanga, L., Ganambarr, R., Ganambarr-Stubbs, M., Ganambarr, B., Maymuru, D. and Sweeney, J. (2016). Co-becoming Bawaka: Towards a relational understanding of place/space. *Progress in Human Geography,* 40:4, pp. 455–475.

Berryman, T. and Sauvé, L. (2016). Ruling relationships in sustainable development and education for sustainable development. *The Journal of Environmental Education,* 47:2, pp. 104–117. doi: 10.1080/00958964.2015.1092934

Bhabha, H. K. (1994; 2004). *The Location of Culture.* New York: Routledge.

Bonnett, M. (2003). Chapter 6: Nature's intrinsic value. *Journal of Philosophy of Education,* 37:4, pp. 629–639.

Boon, P. I. and Prahalad, V. (2017). Ecologists, economics and politics: Problems and contradictions in applying neoliberal ideology to nature conservation in Australia. *Pacific Conservation Biology,* 23:2. doi: 10.1071/PC16035

Boyd, J. and Banzhaf, S. (2007). What are ecosystem services? The need for standardized environmental accounting units. *Ecological Economics,* 63:2–3, pp. 616–626.

Cahill, D. and Paton, J. (2011). 'Thinking socially' about markets. *Journal of Australian Political Economy,* 68:Summer 2011, pp. 8–26.

Calhoun, C. (2002). In *Dictionary of the Social Sciences.* Retrieved from http://www.oxford reference.com.ezproxy.une.edu.au/view/10.1093/acref/9780195123715.001.0001/acref-9780195123715-e-1781?rskey=KeByAE&result=1

Carruthers, B. G. and Espeland, W. N. (1991). Accounting for rationality: Double-entry bookkeeping and the rhetoric of economic rationality. *American Journal of Sociology,* 97:1, pp. 31–69.

Chapagain, A. K. (2006). *Globalisation of Water: Opportunities and Threats of Virtual Water Trade.* PhD Thesis, Delft, The Netherlands: UNESCO–IHE.

Charpleix, L. (Forthcoming). The Whanganui River as Te Awa Tupua: place-based law in a legally pluralistic society. *The Geographical Journal.*

Cosier, P. (2012). *Environmental Asset Condition Account Trials in Australia.* Paper presented at United Nations Statistics Division International Seminar *Towards Linking Ecosystems and Ecosystem Services to Economic and Human Activity.* Retrieved from http://www.wentworthgroup.org/uploads/UNCEEA%20NY%20Nov%202012.pdf

Cosier, P. and Sbrocchi, C. (2013). *Accounting for Nature: A Common Currency for Measuring the Condition of Our Environment.* Paper presented at *Our Place: State of the Environment 2013* conference, Auckland, New Zealand. [Paper provided by email from Peter Cosier].

Coull, J. and Valatin, G. (2008). *Payments for Ecosystems Services: Findings and Perceptions from the USA – Policy Summary.* Retrieved from http://www.forestry.gov.uk/pdf/PES_policy_summary_Jan08.pdf/$FILE/PES_policy_summary_Jan08.pdf

Crutzen, P. J. (2002). Geology of mankind. *Nature, 415:6867*, p. 23. doi: 10.1038/415023a

Daily, G. C. (1997). *Nature's Services: Societal Dependence on Natural Ecosystems.* Washington, D.C.: Island Press.

Daly, M. (2011). *Welfare.* Cambridge, UK: Polity Press.

Darity, W. A. Jr. (2008). In *International Encyclopedia of the Social Sciences 2nd ed., vol. 2.* Retrieved from http://go.galegroup.com.ezproxy.une.edu.au/ps/i.do?id=GALE%7CCCX3045300679&v=2.1&u=dixson&it=r&p=GVRL&sw=w&asid=da147bef9fa0d6b39c1071e35a6af961

De Groot, R. S., Wilson, M. A. and Boumans, R. M. J. (2002). A typology for the classification, description and valuation of ecosystem functions, goods and services. *Ecological Economics, 41:3*, pp. 393–408.

Egis Bceom International. (2010). *Economic Accounting of Water Use: Standardised Methodologies for Economic Accounting of Water Use in SADC.* Retrieved from http://www.sadcwateraccounting.org/_system/writable/DMSStorage/831en/SADC-EAW%20Methodologies.pdf

Evernden, N. (1981). The ambiguous landscape. *Geographical Review, 7:2*, pp. 147–157.

Fanaian, S. (2013). *Valuing Ecosystem Services Linked to River Flows in Lower Zambezi Basin, Mozambique.* MSc Thesis, Delft, The Netherlands: UNESCO–IHE.

FAO Natural Resources Management and Environment Department. (2004). *Pricing Methods.* Retrieved from http://www.fao.org/docrep/008/y5690e/y5690e06.htm

Farber, S. C., Costanza, R. and Wilson, M. A. (2002). Economic and ecological concepts for valuing ecosystem services. *Ecological Economics, 41:3*, pp. 375–392.

Fisher, B., Turner, R. K. and Morling, P. (2009). Defining and classifying ecosystem services for decision making. *Ecological Economics, 68:3*, pp. 643–653.

Frankfort, H., Frankfort, H. A., Wilson, J. A., Jacobsen, T. and Irwin, W. A. (1946; 2013). *The Intellectual Adventure of Ancient Man: An Essay on Speculative Thought in the Ancient Near East* (Kindle edition). Chicago: University of Chicago.

Frontier Economics Pty Ltd. (2008). *The Concept of 'Virtual Water' – A Critical Review.* Retrieved from http://www.dpi.vic.gov.au/__data/assets/pdf_file/0017/42650/Virtual-Water-The-Concept-of-Virtual-Water.pdf

G8+5. (2007). *Potsdam Initiative – Biological Diversity 2010.* Retrieved from www.bmu.de/N38948-1/

Gammage, B. (2011). *The Biggest Estate on Earth: How Aborigines Made Australia.* Sydney: Allen and Unwin.

Godfrey, J., Hodgson, A., Holmes, S. and Tarca, A. (2006). *Accounting Theory, 6th ed.* Milton, Qld: John Wiley and Sons Australia, Ltd.

Goméz-Baggethun, E., de Groot, R., Lomas, P. L. and Montes, C. (2010). The history of ecosystems services in economic theory and practice: From early notions to markets and payment schemes. *Ecological Economics, 69:6*, pp. 1209–1218.

Hadjigeorgalis, E. (2009). A place for water markets: Performance and challenges. *Review of Agricultural Economics, 31:1*, pp. 50–67.

Harris, E. E. (1988). *The Reality of Time*. Albany, USA: State University of New York Press.

Hill, S. B. (2011). Social ecology: An Australian perspective. In D. Wright, C. Camden-Pratt and S. Hill (Eds.), *Social Ecology: Applying Ecological Understanding to Our Lives and Our Planet*. Stroud, UK: Hawthorn Press.

Hoekstra, A. Y. and Hung, P. Q. (2002). Virtual water trade: A quantification of virtual water flows between nations in relation to international crop trade. *Value of Water Research Report Series No. 11*. Delft, The Netherlands: IHE Delft.

Horizons Regional Council. (n.d.) *Our Role and Goals*. Retrieved from http://old. horizons.govt.nz/managing-environment/resource-management/water/our-role-and-goals-water/

Horizons Regional Council. (1997; 2003). *Whanganui Catchment Strategy*. Retrieved from http://old.horizons.govt.nz/assets/publications/managing-our-environment/plans-and-strategies/Whanganui-catchment-strategy.pdf

Horizons Regional Council. (2014). *One Plan*. Retrieved from http://www.horizons.govt. nz/publications-feedback/one-plan

Huddart, D. (2006). *Homi K. Bhabha*, London: Routledge.

Kaika, M. (2005). *City of Flows: Modernity, Nature, and the City*. New York: Routledge.

Karimi, P., Bastiaanssen, W. G. M. and Molden, D. (2013). Water Accounting Plus (WA+) – A water accounting procedure for complex river basins based on satellite measurements. *Hydrology and Earth System Sciences, 17:7*, pp. 2459–2472. doi: 10.5194/hess-17-2459-2013

Kirby, M., Mainuddin, M. and Eastham, J. (2010). *Water-Use Accounts in CPWF Basins: Model Concepts and Description*. CPWF Working Paper: Basin Focal Project series, BFP01. Colombo, Sri Lanka: The CGIAR Challenge Program on Water and Food.

Lange, G-M. (2011). *Valuation of Water Resources*. Retrieved from https://unstats.un.org/unsd/envaccounting/workshops/seeawtraining/Valuation.ppt

Latour, B. (2005). *Reassembling the Social*. New York: Oxford University Press.

Latour, B. (2013). Is there an ANT at the beginning of ANThropology? A few responses to the subject matter of the collection. *Social Anthropology/Anthropologie Sociale, 21:4*, pp. 560–563. doi:10.1111/1469-8676.12053

Leopold, A. (1949). *A Sand Country Almanac and Sketches Here and There*. New York: Oxford University Press.

Liebowitz, S. J. and Margolis, S. E. (1999). Path dependence. In B. Bouckaert and G. De Geest (Eds.), *Encyclopedia of Law and Economics*. Ghent: Edward Elgar and the University of Ghent.

Linton, J. (2010). *What Is Water? The History of a Modern Abstraction*. Vancouver: UBC Press.

Linton, J. (2014). Modern water and its discontents: A history of hydrosocial renewal. *WIRES Water, 1:1*, pp. 111–120. doi: 10.1002/wat2.1009

Linton, J. and Budds, J. (2014). The hydrosocial cycle: Defining and mobilizing a relational-dialectical approach to water. *Geoforum, 57*, pp. 170–180.

Millennium Ecosystem Assessment. (2005). Retrieved from http://millenniumassessment. org/en/

Molden, D. (1997). *Accounting for Water Use and Productivity: SWIM Paper 1*. Colombo, Sri Lanka: International Irrigation Management Institute.

Ngā Tāngata Tiaki. (2016). *Ngā Tāngata Tiaki o Whanganui – Strategic Plan*. Retrieved from http://www.ngatangatatiaki.co.nz/?page_id=676

Nicholls, B. (1997). Disrupting time: Post colonial politics in Homi Bhabha's *The Location of Culture*. *Southern Review: Communication, Politics and Culture*, *30:1*, pp. 4–25.

Oelschlaeger, M. (1991). *The Idea of Wilderness: From Prehistory to the Age of Ecology*. New Haven: Yale University Press.

Office of Treaty Settlements. (2012). *Whanganui Iwi and the Crown: Tutohu Whakatupua*. Retrieved from https://www.govt.nz/treaty-settlement-documents/whanganui-iwi/

Office of Treaty Settlements. (2014). *Ruruku Whakatupua – Te Mana o Te Awa Tupua*. Retrieved from https://www.govt.nz/treaty-settlement-documents/whanganui-iwi/

Office of Treaty Settlements. (2016). *Regulatory Impact Statement: Te Awa Tupua (Whanganui River) framework*. Retrieved from http://www.treasury.govt.nz/publica tions/informationreleases/ris/pdfs/ris-justice-tatf-may16.pdf

Papastergiadis, N. (2000). *The Turbulence of Migration*. Cambridge, UK: Polity Press.

Perry, C. J. (2012). Accounting for water: Stocks, flows, and values. In *Inclusive Wealth Report 2012 – Measuring Progress Towards Sustainability*. Retrieved from http://www.unep.org/pdf/IWR_2012.pdf

Potschin, M. B. and Haines-Young, R. H. (2011). Ecosystem services: Exploring a geographical perspective. *Progress in Physical Geography*, *35:5*, pp. 575–594. doi: 10.1177/0309133311423172

Roe, E. J. (2009). Human-Nonhuman. In R. Kitchin and N. Thrift (Eds.), *International Encyclopedia of Human Geography*. doi: 10.1016/B978-008044910-4.00702-1

Rolston, H., III. (1988). *Environmental Ethics: Duties to and Values in the Natural World*. Philadelphia: Temple University Press.

Rose, D. B. (2000). To dance with time: A Victoria River Aboriginal study. *The Australian Journal of Anthropology*, *11:3*, pp. 287–295.

Savenije, H. H. G. (2001). *Why Water Is Not an Ordinary Economic Good*. Delft, The Netherlands: IHE Delft.

Sbrocchi, C., Davis, R., Grundy, M., Harding, R., Hillman, T., Mount, R., Possingham, H., Saunders, D., Smith, T., Thackway, R., Thom, B., and Cosier, P. (2015). *Evaluation of the Australian Regional Environmental Accounts Trial*. Sydney: Wentworth Group of Concerned Scientists.

Seed, J. and Wright, D. (2011). The religion of economics. In D. Wright, C. Camden-Pratt and S. Hill (Eds.), *Social Ecology: Applying Ecological Understanding to Our Lives and Our Planet*. Stroud, UK: Hawthorn Press.

Shaw, G. D. (2013). The torturer's horse: Animals and agency in history. *History and Theory, Theme Issue*, *52:4*, pp. 146–167.

Snyder, G. (2001). The place, the region, and the commons. In M. E. Zimmerman, J. B. Callicott, G. Sessions, K. J. Warren and J. Clark (Eds.), *Environmental Philosophy: From Animal Rights to Radical Ecology*. Upper Saddle River, New Jersey, USA: Prentice-Hall.

Southern Water. (n.d.). *Southern Water Price and Service Plan 2012–15*. Retrieved from http://www.energyregulator.tas.gov.au/domino/otter.nsf/LookupFiles/Southern_Water_Price_and_Service_Plan_2012-15_Approved_122439.pdf/$file/Southern_Water_Price_and_Service_Plan_2012-15_Approved_122439.pdf

South West Water. (n.d.). *Water Tariff Trial, Prices and Investment for 2009/10*. Retrieved from http://www.southwestwater.co.uk/index.cfm?articleid=6903&searchkey=water tariff trial

Strang, V. (2014). Fluid consistencies: Material relationality in human engagements with water. *Archaeological Dialogues*, *21:2*, pp. 133–150.

Sullivan, S. (2010). 'Ecosystem service commodities' a new imperial ecology? Implications for animist immanent ecologies, with Deleuze and Guattari. *New Formations*, 69, pp. 111–128.

Tadaki, M. and Sinner, J. (2014). Measure, model, optimise: Understanding reductionist concepts of value in freshwater governance. *Geoforum*, 51, pp. 140–151.

Tasmanian Audit Office. (2015). *Report of the Auditor-General No. 6 of 2015-16; Auditor-General's Report on the Financial Statements of State Entities, Volume 3: Local Government Authorities and Tasmanian Water and Sewerage Corporation Pty Ltd 2014-15*. Retrieved from http://www.audit.tas.gov.au/wp-content/uploads/Report-of-the-Auditor-General-No.-6-of-2015-16-Vol-3-2014-151.pdf

Tasmanian Land Conservancy. (2015). *Midlands Landscape Project*. Retrieved from http://tasland.org.au/programs/midlands-conservation-fund

TasWater. (n.d.). *Annual Report 2014–15*. Retrieved from http://www.taswater.com.au/About-Us/Publications

TasWater. (2014). *Corporate Plan: Financial Years 2015–2017*. Retrieved from http://www.taswater.com.au/About-Us/Publications

United Nations [UN]. (2014). *System of Environmental-Economic Accounting 2012 – Experimental Ecosystem Accounting*. Retrieved from http://unstats.un.org/unsd/envaccounting/seeaRev/eea_final_en.pdf

United Nations [UN]. (2016). *Global Issues: Water*. Retrieved from http://www.un.org/en/sections/issues-depth/water/index.html

United Nations Department of Economic and Social Affairs [UNDESA]. (2012). *SEEA-Water: System of Environmental-Economic Accounting for Water*. Retrieved from http://unstats.un.org/unsd/envaccounting/seeaw/seeawaterwebversion.pdf

United Nations Environment Program [UNEP]. (2000). *Decisions Adopted by the Conference of the Parties to the Convention on Biological Diversity at Its Fifth Meeting, Nairobi, 15–26 May, 2000*. Retrieved from http://www.cbd.int/doc/decisions/COP-05-dec-en.pdf

United Nations Statistics Division [UNStats]. (n.d.). *The System of Environmental-Economic Accounts (SEEA): Measurement Framework in Support of Sustainable Development and Green Economic Policy*. Retrieved from http://unstats.un.org/unsd/envaccounting/Brochure.pdf.

Van Bogaert, L-J. (2004). Sentience and Moral Standing. *South African Journal of Philosophy*, 23:3, pp. 292–301. doi: 10.4314/sajpem.v23i3.31399

Wallace, K. J. (2007). Classification of ecosystem services: Problems and solutions. *Biological* Conservation, 139:3–4, pp. 235–246.

Wallace, K. J. (2008). Letter to the editor: Ecosystem services: Multiple classifications or confusion? *Biological* Conservation, 141:2, pp. 353–354.

Waitangi Tribunal. (1999). *The Whanganui River Report: WAI 167*. Wellington: GP Publications.

Water Accounting+. (n.d.). *Projects*. Retrieved from http://www.wateraccounting.org/projects.html

Water Briefing. (2011). *South West Water Abandons Block Tariffs*. Retrieved from http://www.waterbriefing.org/index.php/home/company-news/item/3930-south-west-water-abandons-block-tariffs

Weir, J. K. (2009). *Murray River Country: An Ecological Dialogue with Traditional Owners*. Canberra: Aboriginal Studies Press.

Welfare. (2016). In *Oxford English Dictionary Online*. Retrieved from http://www.oed.com.ezproxy.une.edu.au/view/Entry/226968?rskey=JfnxKT&result=1&isAdvanced=false#eid

Wellbeing. (2016). In *Oxford English Dictionary Online*. Retrieved from http://www.oed.com.ezproxy.une.edu.au/view/Entry/227050?redirectedFrom=wellbeing#eid

Wentworth Group of Concerned Scientists. (2008). *Accounting for Nature: A Model for Building the National Environmental Accounts of Australia*. Retrieved from http://www.wentworthgroup.org/uploads/Accounting%20for%20Nature%202nd%20Ed.pdf

Wentworth Group of Concerned Scientists. (2014). *Blueprint for a Healthy Environment and a Productive Economy*. Retrieved from http://wentworthgroup.org/2014/11/blueprint-for-a-healthy-environment-and-a-productive-economy/2014/

Wentworth Group of Concerned Scientists. (2016). *Agreed Statement*. Retrieved from http://wentworthgroup.org/wp-content/uploads/2016/11/MEM-meeting5-statement.pdf

Whanganui River Maori Trust Board. (n.d.). *Whanganui River Settlement: Ratification Booklet for Whanganui Iwi*. Retrieved from http://www.wrmtb.co.nz/new_updates/Whanganui_River_Settlement_Ratification_2014_v2.pdf

White, M. V. (2002). Doctoring Adam Smith: The fable of the diamonds and water paradox. *History of Political Economy, 34:4*, pp. 659–683.

World Commission on Environment and Development [WCED]. (1987). *Our Common Future*. Oxford: Oxford University Press.

6 Rethinking the value of water
Stewardship, sustainability and a better future

Michael Allen Fox

This essay is in three Parts. Part 1 surveys some ideas about what water is – as a substance in its own right, and as an entity of major significance and symbolic importance. Part 2 explores basic considerations about the value of nature (the biosphere) and its components. Part 3 applies findings from Parts 1 and 2 to thinking about water, with reference to water management issues facing humanity today.

1. Appreciating water

The ancient Greek philosopher Thales of Miletus (c. 624–c. 546 BCE) attributed supreme importance to water in his cosmology. According to many sources, his first principle was that water is the origin of all things, that to which they will ultimately return, and perhaps even what they actually are made of. Some may consider this an extreme hypothesis; but whether it is or not, water has fascinated thinkers since the beginning of recorded thought, and we can easily see why. Nearly three-quarters of the planet Earth is covered by water. Water is not only a global climate regulator, as we all know from keeping track of the daily weather, but also, because of its 'transformation . . . between gas, liquid and solid phases[, water] is vital for the transfer of energy around the globe[,]. . . from the equatorial regions towards the poles' (Davie, 2008, p. 2). Water stores latent solar energy, and in addition, contains various forms of potential energy, such as that released when it turns to steam or by virtue of its expansive power when freezing. Water vapour in the atmosphere is critical to maintaining the warmth necessary for life on Earth to thrive. As a very efficient solvent, water is the key agent in transporting nutrients to growing things and in diluting and removing toxic substances from them. Because of its omnipresence and unique properties, water is essential to the bodily sustenance and self-regulation of human and many other organisms. From a structural standpoint, the human body is maximally adapted to absorbing water, 'thus ensuring proper hydration' (Applied-Water-Energy.com, 2011).

It is easy to see, then, that water is what fundamentally drives ecological and biospherical processes and is the vital factor in maintaining their momentum toward or away from a healthy, sustainable condition. (At least this is true of life 'as we know it' [Finney, 2015, p. 114].) Just as water penetrates and features predominantly in environmental change and stability, so too does it play a similar role within the bodies of organisms. From a hydrological standpoint (and from many others as well), living things are as much ecosystems as are those natural networks outside of them that usually go by this name. Or to put it better, organisms are ecosystems within larger ecosystems. Water is the dependency factor that unites these ecosystem levels. It governs the inner world of an organism's ecology as much as it does the outer ecologies of the biosphere as a whole. One source notes that:

> Virtually all chemical reactions in life processes take place in solution in water... Water is present both inside and outside cells. In the body of a mammal for example although it is about 70% water by weight, about 46% (approximately 2/3) is inside cells, and about 23% (approx. 1/3) is present outside cells in blood plasma and other fluids. (British Society for Cell Biology, n.d.)

And furthermore, micro-ecologies from the individual cell on up are nested within, and dependent upon, macro-ecologies with which they constantly interact.

Organisms know intuitively that they depend upon water to stay alive and to secure their wellbeing. Those who have lived through periods of water scarcity know this quite poignantly. Humans also know (or should realize) that they receive the highest level of benefits from inhabiting a watery planet, owing to the additional uses to which they have put this marvellous fluid – for washing, agriculture, industry, moving objects from place to place, and conducting trade, travelling, recreation, ceremonial rituals, and many other purposes. It is in no way surprising, therefore, that humans have attached spiritual significance to water and cultivated a reverential attitude towards it (Strang, 2015, pp. 41, 49). This larger sense of meaning lives on today and receives expression in various documents promulgated by international coalitions of Indigenous peoples. The following statement provides a good example:

> We, the Indigenous Peoples from all parts of the world assembled here, ... raise our voices in solidarity to speak for the protection of water... Our traditional knowledge, laws and ways of life teach us to be responsible in caring for this sacred gift that connects all life. (Indigenous Peoples Kyoto Water Declaration, 2003)

These broader cultural values of water often seem to be lost or buried in modern discussions of water policy (see Chapters 2 and 8, this volume).

In many traditions, water is considered essential to rituals of cleansing, purification, baptism, healing, and the veneration of water deities; it is also widely regarded as a source of life and bringer of death, and serves as a subject of profound contemplation and inspiration. The ancient philosopher Lao Tzu observed, in the *Tao Te Ching* (approximately 600 BCE), that water is the most beneficial force in nature, and a substance that exhibits opposing properties: quietness and compliance, on the one hand, and yet power and destructiveness, on the other. Being able to penetrate everywhere, water is like the energy and life-force of the universe itself (the Tao), and thus provides a model for whatever is good (Lao Tzu, 1995, verses 8, 78). In perceiving water's capacity to resist but also yield, Lao Tzu clearly grasps a fundamental property of water: its low viscosity. Davie (2008) explains that:

> Low viscosity comes from water molecules not being so tightly bound together that they cannot separate when a force is applied to them. This makes water an extremely efficient transport mechanism. When a ship applies force to the water molecules they move aside to let it pass! (pp. 3–4)

But even the most compliant substance can exert incredible forces of its own and – like water – erode, wash away, punish, and dissolve pretty much anything that stands in its way. It is well understood across the ages that water is both a benign, nurturing friend, but also under certain circumstances a ruthless adversary and leveler. What Lao Tzu adds to this knowledge is the perspective (found as well in Australian Aboriginal and other native cultures' Earth-defining narratives) that water symbolically expresses the meaning and conditions of life, and is not just the essential ingredient of it.

Low viscosity is just one of seventy-two features of water that set it apart. These are known as 'anomalies', or 'physical and chemical properties that are very different from [those of] other materials'. Another is that water molecules move faster under conditions of greater density or pressure. This has opened up a new discovery, made in 2001, that 'when water is confined within carbon nanotubes, the molecules form a single line in the centre. This allows them to flow a thousand times faster than expected' (Barbosa, 2015, p. 26). The importance of studying the foregoing property of water is that doing so may lead to better and lower-cost methods of desalination – an area of innovation much needed by the world's growing population, particularly those living in areas of relative water scarcity.

Improving the quantity and quality of water sources used for drinking and for irrigation is an essential condition of global political stability. But the problem of dryland salinity caused by generations of forest and other vegetation clearing and intensive agricultural practices (so-called 'secondary' or 'human-induced' salinity) cannot be properly addressed without the evolution of sustainable methods of food production (see Part 3). This is a serious issue in various parts of the world, including Australia, the US, Canada, Thailand, South Africa, Turkey, India, and Argentina (Pannell & Ewing, 2013).

All of this gives pause for reflection. Water is everywhere around us, in us, and intimately in touch with us, as we are with it. And yet, we still don't control it or necessarily even see it clearly. As Fishman (2011) observes:

Water is both mythic and real. It manages to be at once part of the mystery of life and part of the routine of life... Water has all kinds of associations and connections, implications and suggestiveness. It also has an indispensable practicality...

The good news is that most of what we know about water isn't really wrong, because we don't know that much. The bad news is that the invisibility of water in our lives isn't good for us, and it isn't good for water. You can't appreciate what you don't understand. You don't value and protect what you don't know is there. (pp. 2, 4)

Finney (2015) contends, on the contrary, that 'the structure, dynamics, and properties of water are in the main pretty well understood' (p. 115). But the rarified knowledge of physicists cannot be equated to the differently interactive understanding of laypersons, which depends on many strata of significance that have accumulated both experientially and historically. As an interdisciplinary approach to water reveals, there is much more to understand and appreciate than science alone can communicate (see also Chapter 5, this volume). Bearing in mind Fishman's caution above, therefore, let's step back for a moment and think about why and how we value things in the world around us. Then perhaps we can also render water more visible and discern its specific value more fully.

2. Valuing nature

Attributing value to nature (or to parts of it) has a very long and distinguished history. Many writers in ancient as well as more recent times have expressed identification with and love of the environment, an attitude of awe at natural processes, attachment to particular features of the natural world, the wish to modify nature for human benefit, the desire to personify natural forces, and so on. What is actually going on and what ideas are being posited when a person is engaged in valuing nature? Questions of this sort – which assume a second-order critical perspective on what normally goes unnoticed or is considered unproblematic – are frequently philosophical or conceptual questions. These call for some process of analysis before they can be answered. Several principal types of value have been identified in the literature of environmental ethics (or environmental philosophy). These have historically been in play, but it is only since roughly 1970 that scholars have attempted to clarify, categorize, and apply them in a more systematic fashion. In what follows, some basic value terms and their meanings that have been refined by this field of research will be brought into view.

Instrumental value

To speak of a human invention or artifact as being made for a certain purpose, which is its *raison d'être*, is to identify its instrumental value. A hammer is valued for its ability to accomplish the purpose of driving nails effectively, a car for being able to get one from A to B, and so forth. For obvious reasons, instrumental value is also referred to as use value, extrinsic value, or value attributable to that which serves as a means to an end. Things may possess instrumental value when employed to accomplish something, or (when in an idle state) as items recognized to be of potential use for achieving a certain result.

Humans have regarded nature as a whole in this way for millennia – as something placed at their command, to benefit them, to help realize their own interests, to do with as they wish, without accountability. The history of technology and of industrial capitalism has enshrined this ideology even further. But even earlier, long before the modern era, Biblical injunctions gave birth to what has come to be known as 'dominionism', the belief that humans are lords and masters over nature. Genesis 1:26 (King James version), for instance, asserts:

> And God said, Let us make man in our image, after our likeness: and let them have dominion over the fish of the sea, and over the fowl of the air, and over the cattle, and over all the earth, and over every creeping thing that creepeth upon the earth.

The same type of view was reinforced by the doctrine of 'the great chain of being', formulated by ancient Greek and Roman as well as medieval thinkers, according to which everything in creation has its place in a hierarchy extending from rocks to animate beings to spiritual beings, with God at the pinnacle of perfection. This systematic arrangement of existence was believed to justify, among other things, the exploitation of the natural world by humans. Another important milestone was the rise of empiricism and the scientific method in the seventeenth century, owing to the writings of Bacon (Bacon, 2000) and others, which emphasized gaining practical results from the disciplined and aggressive investigation of the natural world.

Non-resource instrumental value

Instrumental value is commonly cashed out (pun intended) in economic terms. 'What something is worth', within this framework, refers to its monetary value, its value in trade and the like. But what about values ascribed to the biosphere or some part or manifestation thereof that do not seem to fit this mould? Many people cherish what is or has come to be of historic, symbolic, scientific, aesthetic, psychological, experiential, anecdotal, cultural, or spiritual significance within the natural world – for example, places of spectacular beauty, formations that represent something (or someone) of importance and reverence to a particular group or culture, and environments that offer great recreational opportunities. One feels inclined to say that these aspects of nature should be viewed not just in economic terms but from within a wider frame of reference. For while it is true that economic value is, or may be, a measure of the relative worth something bears for individuals or for society at large, other value-measuring standards serve the same function. This is because what may be designated as non-resource instrumental value originates in them.

We have here a kind of hybrid appraisal indicator, which has the advantage of being both instrumental (because it singles out something that satisfies human desires and needs) and non-resource oriented (because it refers to what cannot be bought and sold). Unfortunately, the picture isn't as clear as it might seem or one might hope; for some argue that this sort of value too can and should be subsumed within a 'user pays' approach. Accordingly, the values mentioned earlier (historic, symbolic, and so on) will only attract consideration in social calculations about water use, land use, and other environmental components if people who wish to enjoy these values show a 'willingness to pay' a price that covers the costs of keeping them available in the form such users prefer and have become accustomed to. A leading environmental economist, for instance, contends that:

> Valuation is unavoidable; it is the essence of decision making. To pretend otherwise is irresponsible. It is better to be explicit about the valuations inherent in decision making and to seek to use the world's limited resources in the best way possible by recognising the inevitable trade-offs which all decisions imply. (Bateman, 2013)

It is difficult to fault the logic of this position in the abstract. But the implicit assumption at work here is that environmental use-decisions represent a zero-sum process, whereby gaining a benefit is a 'win' and forgoing another is a 'loss'. This reduces people's choices to their willingness to put a price on the desire to have or avoid certain consequences and their willingness to forgo (or to seek) certain benefits.

Some interesting questions now arise that we cannot resolve in this chapter, but which are relevant to the present inquiry. For instance, are there environmental goods to which people have a basic right and shouldn't have to buy in order to protect and enjoy (see Chapter 3, this volume)? There was a time when economists considered air and water to be free for the taking, and characterized the negative effects of industrial processes on them as collateral damage ('externalities') exempt from being included among the costs of manufacturing and marketing. This view fortunately does not dominate mainstream thinking in quite the way it used to, and the reason is that air and water are better understood as indispensable to the maintenance and minimum quality of life, for which people need some kind of guarantee, even if unfettered access for everyone is a thing of the distant past. So why should fundamental human rights not be extended to include possession of those things that are of non-resource instrumental value, and that also fulfill basic needs?

The monetary approach to value seems incompatible with economic thinking at an even deeper level. While it can be appreciated that at least part of the motivation for absorbing different kinds of value into an economic model is to enable factoring them into policy and decision-making, sometimes a line has to be drawn which says, 'These goods are not for sale or trade, because their goodness flows from relationships that have to be understood and appreciated in their own terms'. This insight is reinforced by a study of water use in Australia. Bark, MacDonald,

Connor, Crossman, and Jackson (2011) urge that '[a]s well as the direct economic uses, water-dependent ecosystems provide a myriad of ecosystem services of indirect economic value and they have intrinsic value beyond any economic considerations' (p. 21). Meanwhile, Prosser (2011) states more specifically that:

> Australians treasure their water for a range of economic, environmental, social, and cultural values… Australians also highly appreciate their rivers, lakes, estuaries, and wetlands because they are associated with a strong sense of place, and they have a desire to protect water resources and environments, both for future generations and for their intrinsic biological value. (p. 147)

As this team of authors points out, economics cannot capture all values of nature, and this ought to alter thinking patterns and open up a broader range of citizen participation in the policy arena.

The discussion thus far has assumed that attributing value to things requires a valuing (human) consciousness endowed with the capacity to judge and affirm what is of worth. But there are other ways of acknowledging value in nature that question (and also supplement) this approach. The following two concepts embody the idea that value may also exist independent of human consciousness, either in whole or in part (see also Chapter 5, this volume).

Intrinsic value

Deciphering the ontological status of value is a complex and challenging matter. Many environmental theorists and naturalists maintain that value in nature also exists independent of any valuing consciousness. According to different formulations of this point, nature (or parts or aspects thereof) possess value in and of themselves, in their own right, for their own sake, or inherently. Such value, in short, would continue as a feature of the world even if all valuing beings became extinct. Conceived in this manner, value is *something about the world*, not something that is only consciousness-dependent. It would follow that intrinsic value – to the extent that it exists – is found or discovered rather than invented, and is not just a feature of what is useful or experientially enriching to receptive minds. It is difficult to imagine a world without conscious beings in it, as this itself requires an act of consciousness. And it may be more difficult still to imagine that such a world contained intrinsic value. But these differences don't negate the concept of intrinsic value and its referential scope. Perhaps intrinsic value exists alongside instrumental value, as an essential supplement or complement, a point to which we will return presently.

Relational value

A different kind of challenge is posed by the view that value exists (or only exists) when a valuing subject encounters an object having a certain kind of potential to impinge upon it in certain ways. Relational value, thus understood, is more like an event than a property; it 'happens' or 'occurs' in the 'space' between consciousness and its object. The valuing event may be transitory and imperma- nent: for example, a musical composition or work of art or literature, according to the relational view, possesses aesthetic value only when an audience apprehends and interacts with it; its value exists in the experiential moment. The same could be said of a treasured encounter with some feature of the natural world. (So, for instance, 'wilderness' may be a state of experiential immersion within a pristine environment, rather than a place that no human has ever or has rarely set foot in.) But it would be a mistake to frame relational involvement with objects on an evaluative plane as merely the product of evanescent and/or fortuitous circum- stances. A more thoughtful, imaginative, ongoing engagement with an artwork or a phenomenon of nature may manifest a relationship of considerable duration or that depends on a relatively permanent structure being in place. We might also acknowledge that relationships sometimes exist even when we are unaware of them. Thus, for instance, we live our lives within an ecological context of co- dependency in relation to many things, and these are of value to us whether or not we know about the benefits we derive from them.

Valuable conclusions

Instrumental value and intrinsic value are usually defined in sharply contrasting terms; however, as hinted above, things can be said to have both kinds of value. Something can be of value in its own right or for its own sake (intrinsic value), yet also serve as a means to another thing's ends or betterment (instrumental value). Nature is full of such examples, as are the ways we talk about the world around us. This is a point to focus on for a moment. There is a tendency to suppose – especially within the Western mindset, supported by certain languages and religious/cultural outlooks – that things must be either this or that, but not both; or that ascribing some important properties to things automatically entails excluding others from being applicable. But on closer examination, it is evident that this is not necessarily so, as for example when one recognizes that utilitar- ian artefacts fashioned for everyday use by diverse human groups can also be objects of great beauty and expressive power. If this is so, then a shift in thinking is required that allows appraisal of the different orders of value people find in things as being compatible and complementary with one another and as layered or dimensional, rather than as oppositional qualities that negate or 'trump' one another when push comes to shove. If it is accepted that value judgments reflect different outlooks on and engagements with the world, then validating any of these to the disadvantage of the others is inappropriate. The task, on the con- trary, is to learn how to harmonize values that are revealed by rich and differently nuanced experiences.

All of the perspectives on value under review here raise, in one way or another, the time-honoured question whether value is subjective or objective, invented or discovered, 'in the mind of the beholder' or 'really out there'. It is not easy to set- tle this matter, or even to be sure that the question is clearly formulated. Nor is any proposed answer determined by which of the value types proposed earlier one primarily opts for. That being said, the purely subjective view is plainly incompatible with intrinsic value and relational value, both of which posit an objective basis of or contribution to value judgments. Relational value poses a problem for the subjective/objective dichotomy, in that it does not hinge on an exclusive either/or choice, moving instead toward a both/and viewpoint, in which what is paramount is the *activity* of valuing by beings immersed in the world.

Many issues remain, of course, such as: (a) whether we should regard values as *natural* properties (like something's being pleasurable or having a certain colour), *non-natural* properties (those that are neither identical to, nor analyzable in terms of, any natural property), *supervenient* properties (those that are derived from or caused by some natural property), merely *emotive expressions* and so on; and (b) whether there are degrees of value, merely different kinds, or both degrees and kinds. These issues also intersect with the above value type-distinctions in interesting and complex ways, which we cannot go into here.

One might object that it is hard enough to have to balance competing stakeholder demands when it comes to any given environmental issue, without having to reconcile a bunch of crazy value positions as well. But it could be replied that – crazy or not – these are the kinds of values that animate stakeholder positions in the first place, so that they're already in the mix rather than merely being introduced *ab extra* by the abstract theorizing of philosophers (O'Neill, Holland, & Light, 2008, chaps. 1, 8; see Chapters 11 and 12, this volume).

3. Thoughts on valuing water

Do these meditations on value have broader implications? Certainly, they necessitate transcending old habits of thought whereby humans stand apart from and superior to nature, as merely responsibility-free consumers of an inexhaustible storehouse of resources. We could locate historical approaches to the natural world by means of opposing evaluative terms, such as the following, and it might be an illuminating side-exercise to do so.

awe	unconcern
immersion	separation
identification	difference
relatedness	externality
kinship	foreignness

Most will be able to draw their own conclusions about theories and approaches they have encountered, in relation to this contrast of attitudes toward being in the world and being part of it. The majority of research projects in environmental philosophy and thoughtful nature writing of recent times clusters toward the left, or more inclusive, connective side of this scale. A lesson in humility may be unfolding in such works – a lesson that has been forced upon our species, and has to some extent been processed by those who are gifted in the art of seeing further and more clearly than their peers. In this spirit, a prominent poet, novelist, and creative nonfiction writer reflects (Kingsolver, 2010) that 'we are not important to water. It's the other way around. Our task [therefore] is to work out reasonable ways to survive inside its boundaries' (p. 49).

We are not important to water, yet water is monumentally important to us, whoever we are. But what kind(s) of value does water have? One could just say that the value of water is beyond any measure, that water is priceless. And there would be good reasons for taking up this stance. Would it be pragmatically helpful to do so? Probably not. Plainly, water has huge amounts of instrumental value. It also just as evidently has non-resource instrumental value, as is indicated by the Indigenous Peoples Kyoto Water Declaration quoted in Part 1, and by various other examples given in Part 2. Does it have intrinsic value? Here, even if controversial, one would have to answer in the affirmative as well. (Recent international examples of the 'rights of nature' becoming legally enshrined support this conclusion [Herold, 2017; Roy, 2017; Safi, 2017].) Let's see how all of these perspectives come together, thinking locally for a moment.

In the New England Tablelands region of New South Wales, Australia, there are numerous waterfalls strung out along a highway known as 'Waterfall Way' that runs from Coffs Harbour, on the east (Pacific) coast, inland to Armidale, on the Great Dividing Range. Many of the falls are enclosed within national parks. In this region too is the Oxley Wild Rivers National Park, the very name of which suggests that water has value worth preserving for its own sake (and the wilder, the better). Australia's national park system began with the establishment of Royal National Park near Sydney 'for public health and recreation, convenience, or enjoyment' (Government of New South Wales, 1861), and with an inherent guiding attitude of conservation (wise and cautious management of resources for present and future use). Since that time, in Australia (as elsewhere), the conception of national parks has broadened to include more preservationist or protectionist goals. For example, the current Parks Australia Sustainable Tourism mission statement endorses the principle that '[s]ustainable tourism guides the management of all resources in such a way that economic, social and aesthetic needs can be fulfilled while maintaining cultural integrity, essential ecological processes and biological diversity' (Director of National Parks [Australia], 2011). This combines the instrumental values of economic use with the non-resource instrumental values of 'social and aesthetic needs' and 'maintaining cultural integrity', and also with the intrinsic values of 'essential ecological processes and biological diversity'.

It is generally agreed among scientists that all the water that has ever been on Earth is still present on the planet in some form. Even though it is very unequally distributed and utilized, and precipitation waxes and wanes, it continues to cycle through nature as always. This is an interesting fact, but it obviously cannot be the source of either pleasure or displeasure in any practical sense – especially in areas periodically gripped by severe, record-breaking droughts. Each nation and region has to deal with its own lot, as it can never be expected or even rationally hoped for that someone else will come to the aid of those who are chronically water poor, in order to provide them with water security. For Australia, the largest and driest inhabited continent and island, the lesson of having to succeed on its own is self-evident. (This is not to say that the lesson is fully assimilated; for Australians are among the highest per capita water users in the world [Lehane, 2014].) And the moral of the tales about water reviewed earlier is that humans need to work with, not against water. People may realize this in their guts and in their historical records and cultural memories, but it is a much bigger project to determine where this idea should take them in today's complex and situation-specific sites, where conflicting interests are at play.

A good lesson for everyone is the advocacy of stewardship (caretaking or custodianship) that is woven into both mainstream religious systems and Indigenous peoples' worldviews. In Part 1, a consensus of indigenous outlooks was briefly noted. The presence of stewardly thinking in Western religious texts is also striking. Butkus (2002, p. 20), for instance, claims that there are no fewer than twenty-six references to stewardship in the Judeo-Christian Bible, and even the previously quoted dominionist account of God's assignment of humans' place in nature is sometimes interpreted as a prescription to tend the Earth in a measured and loving way. Islam paints a picture of creation in which the Earth is good and does not belong to humans; humans, on the contrary, are nature's caretakers, the vice-regents (or *khalifs*) of Allah (Qur'an 2:30), for whose glory all acts are performed. (The Qur'an also states: 'We made from water everything living' [21:30].) These are just two examples drawn from what is normally described as 'world religion'. Speaking more broadly, however, Pyari (2011) contends that '[e]nvironmental stewardship has been an integral part of nearly all religions of the world' (p. 26).

Stewardship rests fundamentally on the idea that the environment is an inheritance of a certain kind, a trust of sorts, which is being looked after by the present generation both for its own wise use and to ensure the welfare of generations to come. (But most discussions of responsibilities to future persons omit to mention that at any given instant, at least three generations already coexist. So, to cite a cliché, 'the future is now'.) Stewardship in the past was categorized as a weak form of anthropocentrism, and I once did so myself. This is not all bad; for it both succeeds in connecting stewardship primarily with such practices as responsible husbandry and prudent management and conservation of natural resources, and also allows room for the projects of promoting species diversity and pursuing sustainable development. However, the time has come to endorse an enhanced outlook that involves an act of imagination and calls for innovative responses. Here's how it might go: a new notion of stewardship would embrace the intrinsic value of the natural world and the two kinds of instrumental value ascribed to it in Part 2. It would also affirm the relational view of value presented there and celebrate its more personally meaningful manifestations that so many individuals experience. Nature – or more precisely speaking, the environment or biosphere – would be treated respectfully and protectively, with the preservation and flourishing of ecological systems being seen as a goal and an end-in-itself.

Balance and measure would be guiding ideals in relation to use and development, with restraint and responsibility checking overexploitation and wastefulness. The non-resource instrumental values that are so important to people would be placed at the forefront of deliberations and recognized as goods that cannot really be assigned a monetary value or be traded away. Stewardship will be understood to entail passing on to future persons a biosphere that is no less rich and healthy than when inherited from one's forebears (see also Chapters 5 and 7, this volume).

Needless to say, most societies today, individually and collectively, fall woefully short of this vision and the standards it embodies. As Faruqi (2015) observes, politicians 'have treated the environment as a dispensable commodity to be sacrificed, plundered or destroyed for short-term economic gain, or used as a bargaining chip' to solicit votes from recalcitrant parliamentarians, whose support is needed in order to retain control of government and to advance certain agendas (p. 18). Although this appraisal was made within a specific context, there is no denying that it has wide application – for example, in Australia, to the issues of coal seam gas exploration and coal mining in water catchments (such as Sydney's) that threaten priceless farmland and irreplaceable groundwater supplies.

Any major water supply source, in Australia or elsewhere, begins as a commons. That is, water presents in the first instance an invitation to free and uncontrolled use of itself, especially when and where it appears to be of unlimited abundance. But as those who have studied the behaviour of commons users have shown, even when an unhappy outcome can be foreseen, without regulatory mechanisms in place there is a tendency to overuse and deplete a resource, to everyone's detriment. Ecologist Hardin (2008) has cautioned that '[w]henever a distribution system malfunctions, we should be on the lookout for some sort of commons. . . Each nation still has the problem of allocating . . . rights among its own people on a noncommonized basis'. Although blessed with many natural wonders, including the Murray-Darling River Basin, 'the world's largest catchment system' (Feldman, 2012, pp. 75–76), Australia has nevertheless grappled with the problem of fair and practical water allocation for quite some time and still seeks solutions that will adequately respond to competing demands and take into account good land management practices, protection of wetlands, and other sustainability needs (Planning Institute Australia, 2015). According to the latest government figures, the agriculture industry is consuming sixty-two percent of the country's water supply (Australian Bureau of Statistics, 2015). 'Water use efficiency in agriculture is already relatively high in Australia', says a leading research group (Lehane, 2014). Even though this 'efficient' water-usage level seems quite high, it is still less than the overall figure of seventy percent 'global water withdrawals' for irrigation recorded by the United Nations (UN-Water, 2014). But without doubt, seriously improved efficiencies of water usage are going to be unavoidable in the future, given the effects of persistent drought and climate change. And with a dominantly urban population that is steadily increasing, it is difficult to imagine how Australia can avoid confronting some critical and painful water choices concerning utilization of the Murray-Darling and other waterways. These would include whether the livestock industry and traditional crops like cotton and rice are sustainable for a dry land over the long term. Some propose that the issue of exporting 'virtual water' to other countries (the water required to produce goods that can be regarded as 'embedded' in them) also needs to be factored into this debate (Feldman, 2012, p. 19; Rhemtulla, 2012; Barlow, 2013, p. 15). This is just a microcosm of the many water supply and use problems that confront a wide spectrum of nations on this fragile planet.

There is enough evidence now (and even more for those who can see that climate change is indeed occurring) for everyone to understand that water makes demands on us as much as we do on it. And as Kingsolver states (see first paragraph of Part 3 above), the ongoing challenge of water is to 'work out reasonable ways to survive inside its boundaries', where we may take 'reasonable' to include a commitment to fairness in our dealings with one another, respect for the environment, sustainability, efficiency, and due recognition of the many non-resource instrumental values of water.

References

Applied-Water-Energy.com (2011). Water cycle fundamentals. Retrieved from http://www.applied-water-energy.com/watercycle.html

Australian Bureau of Statistics (2015). 4610.0 – *Water account, Australia, 2013–14* (9th ed.). Main findings 2013–14. Released 26 November. Retrieved from http://www.abs.gov.au/AUSSTATS/abs@.nsf/ProductsbyCatalogue/49F854E3831E4294CA2580580015E2A6?OpenDocument

Bacon, F. (2000). *The New Organon.* (Orig. pub. 1620.) Trans. M. Silverthorne. Cambridge, UK: Cambridge University Press.

Barbosa, M. (2015). Tapping the incredible weirdness of water. *New Scientist, 226*(3015) (April), 26–7.

Bark, R., MacDonald, D. H., Connor, J., Crossman, N., & Jackson, S. (2011). Water values. In Prosser, I. (Ed.), *Water: Science and solutions for Australia* (pp. 17–27). Collingwood, VIC, Australia: CSIRO Publishing.

Barlow, M. (2013). *Blue future: Protecting water for people and the planet forever.* New York: New Press.

Bateman, I. (2013). Why should we value nature? Retrieved from http://www.valuing-nature.net

British Society for Cell Biology (n.d.). Water and cells. Retrieved from http://bscb.org/learningresources/softcell-e-learning/water-and-cells

Butkus R. A. (2002). *The stewardship of creation.* Waco, TX: Center for Christian Ethics at Baylor University. Retrieved from http://www.baylor.edu/ifl/christianreflection/CreationarticleButkus.pdf

Davie, T. (2008). *Fundamentals of hydrology* (2nd ed.). London: Routledge.

Director of National Parks (Australia) (2011). *Sustainable tourism overview 2011–2016.* Canberra: Commonwealth of Australia. Retrieved from http://www.environment.gov.au

Faruqi, M. (2015). Too much at stake to keep ignoring environment. *Sydney Morning Herald*, 6 May, 18.

Feldman, D. L. (2012). *Water.* Cambridge, UK: Polity Press.

Finney, J. (2015). *Water: A very short introduction.* Oxford: Oxford University Press.

Fishman, C. (2011). *The big thirst: The secret life and turbulent future of water.* New York: Free Press.

Government of New South Wales. Crown lands alienation act of 1861. Retrieved from http://www.austlii.edu.au/au/legis/nsw/num_act/claao1861n26270.pdf

Hardin, G. (2008). Tragedy of the commons. In Henderson, D. R. (Ed.), *Concise encyclopedia of economics* (2nd ed.). Library of Economics and Liberty. Indianapolis: Liberty Fund, Inc. Retrieved from http://www.econlib.org/library/Enc/TragedyoftheCommons.html

Herold, K. (2017). The rights of nature: Indigenous philosophies reframing law. *Intercontinental Cry*, 6 January. Retrieved from https://intercontinentalcry.org/rights-nature-indigenous-philosophies-reframing-law

Indigenous Peoples Kyoto Water Declaration (2003). Third world water forum, Kyoto, Japan (March). Retrieved from http://www.waterculture.org/uploads/IPKyotoWaterDeclarationFINAL.pdf

Kingsolver, B. (2010). Water is life. *National Geographic* (special issue on water), *217*(4) (April), 36–49.

Lao Tzu (1995). *The tao te ching.* (Written approx. 600 BCE.) Trans. B. B. Walker. New York: St. Martin's Griffin.

Lehane, S. (2014). Australia's water security part 2: Water use. *Future Directions International*, 13 November. Retrieved from http://www.futuredirections.org.au

O'Neill, J., Holland, A., & Light, A. (2008). *Environmental values.* Abingdon, Oxon, UK: Routledge.

Pannell, D. J. & Ewing, M. A. (2013). Managing secondary dryland salinity: Options and challenges. Rev. 16 June. Retrieved from http://dpannell.fnas.uwa.edu.au/dp0403.htm

Planning Institute Australia (2015). Policy: Water and planning. Retrieved from http://www.planning.org.au/policy/water-and-planning

Prosser, I. (2011). Conclusions. In Prosser, I. (Ed.), *Water: Science and solutions for Australia* (pp. 147–52). Collingwood, VIC, Australia: CSIRO Publishing.

Pyari, D. (2011). Environmental stewardship and religion. *International Journal of Educational Research and Technology*, 2(1), 26–35. Retrieved from http://soeagra.com/ijert/vol3/5ijert.pdf

Rhemtulla, S. (2012). Cotton and virtual water in the Murray-Darling basin. A report for Friends of the Earth, Australia. (September.) Retrieved from http://www.foe.org.au

Roy, E. A. (2017). New Zealand river granted same legal rights as human beings. *The Guardian* (Australia edition), 16 March. Retrieved from https://www.theguardian.com/world/2017/mar/16/new-zealand-river-granted-same-legal-rights-as-human-being

Safi, M. & agencies (2017). Ganges and Yamuna rivers granted same legal rights as human beings. *The Guardian* (Australia edition), 21 March. Retrieved from https://www.theguardian.com/world/2017/mar/21/ganges-and-yamuna-rivers-granted-same-legal-rights-as-human-beings

Strang, V. (2015). *Water: Nature and culture.* London: Reaktion Books.

UN-Water (The United Nations Inter-Agency Mechanism on All Freshwater Related Issues, Including Sanitation) (2014). Statistics. Retrieved from http://www.unwater.org/statistics/en

7 Stewardship arrangements for water

An evaluation of reasonable use in sustainable catchment or watershed management systems

Mark Shepheard

Introduction

Integrated governance arrangements to implement sustainability attempt to redefine the nature of human-environment interaction and foster more environmentally and socially beneficial outcomes from the use of resources. In this chapter, stewardship is examined as a foundation for resource management arrangements that connect natural resource use with landscape ecological limits and the social wellbeing of communities. The objective of the chapter is to evaluate catchment or watershed management approaches in four case studies that define stewardship expectations regarding resource user rights and interests. The evaluation adopts a model of stewardship to interrogate and extend resource user accountability for sustainability in a defined catchment or watershed. Reasonable use is a key concept in this analysis as it represents a standard of performance or appropriate behaviour for the circumstances of a sustainable catchment or watershed. The case studies demonstrate how existing arrangements for sustainable catchment or watershed management anticipate a reimagined standard of reasonable resource access and use for more sustainable catchment or watershed systems.

The analysis reflects concerns about the extent to which collaborative catchment wide approaches to natural resource management can guide resource use practice to protect the environment from the impacts of agriculture (Comptroller & Auditor General, 2010). It extends the analysis of this issue by defining underlying notions about rights to resource use that limit the effectiveness of efforts to implement outcomes anticipated in catchment or water shed plans. The discussion is an important one given background evidence from Britain that farmers question the contribution of their resource use practices to diffusing source pollution of waters, despite the significant threat this poses to compliance with environmental accountabilities (Environment Agency, 2011). In Switzerland, it is also recognized that protecting waters from the impacts of agricultural production is a challenge as resource users attempt to meet societal expectations, including a high sensitivity to ecological issues and a strong desire for clean and unspoiled water bodies (Prasuhn, 2011; Stadelmann, 2009; Wagner et al., 2002).

Stewardship in the context of this chapter is considered as an organising concept (Lange & Shepheard, 2014), with the capacity to motivate behavioural change in natural resource use practice. A steward holds a position of responsibility for guardianship of a place (Pearsall & Trumble, 2001). Stewardship anticipates prudence to avoid environmental harm (Lee, 2005) and appreciates decision-making as a mixture of foresight, morals, and self-understanding (Barnes, 2009; Jacob, 1995). Stewardship combines with praxis to represent a level of skilled conduct by resource users. Praxis brings insight and in-depth acquaintance with practical considerations to the notion of stewardship in a way that enables timely, creative, and skilful intervention (Birden, 2009). This involves wise judgement, practical wisdom, prudence, and deliberation (Schwandt, 2010). In particular, praxis concerns the competence, sense, and sensitivity in knowing what is right and good to do given the circumstances and demands of a particular situation, i.e. transforming meaning into an immediate social context (Brown, 2007). Stewardship and praxis combine in the notion of stewardship accountability.

The four dimensions of stewardship accountability are derived from the author's research on liability of natural resource users for environmental harm. The dimensions are: norms of practice, limits on exploitative freedoms, legitimacy, and trust. Collectively these provide a basis from which to redefine underlying notions of what is reasonable natural resource use in the context of expectations defined in catchment or watershed plans. The four dimensions combine to establish a link between strategic level statements about resource use, formal and informal resource use rules, landscape systems, and the social systems in which natural resource users are accountable. Thus, the interpretation of responsibility delivered by stewardship accountability is more likely to meet the integrated and holistic local outcomes sought by collaborative resource management arrangements, making performance expectations more reasonable for sustainable resource use in a catchment or watershed.

Collaborative catchment or watershed management is being used to redefine expectations and adjust resource access and use arrangements for more sustainable systems. Participatory approaches to performance accountability are also being used as an attempt to improve integration across sectors and reduce the risk of conflict between natural resource users (Rey & Müller, 2007). An inherent question for natural resource users is how to implement the policy and regulatory expectations of performance developed by natural resource managers and specified in catchment plans.

The application of stewardship accountability is reviewed here using the four case studies to evaluate the effectiveness of existing catchment or watershed management planning and performance outcomes. The case studies are: the Canterbury Region of New Zealand, the Swiss Midland Lakes Region of Lucerne, the Anglia Region of England, and the 'du Nord' watershed in the Canadian Provence of Québec. Analysis of governance arrangements for sustainable catchment or watershed management in these case studies demonstrates how strategic pursuit of sustainable catchment or watershed systems seeks to adjust the exercise of rights and interests of natural resource users. The cases also illustrate some limitations in the implementation of accountability for sustainable resource use, and demonstrate how stewardship could provide a suitable basis for reimagining reasonable resource access and use behaviour.

The chapter is structured as follows:

1   The discussion commences by identifying the case studies, their characteristics, and how they are used in this chapter;
2   Sustainable catchment or watershed management is then described as a strategic process applied to achieve change in resource access and use practice;
3   The discussion then turns to describing limitations to the extent that resource users can reasonably be held to account for their actions. Identifying these limitations reinforces the need to reimagine what a reasonable standard of behaviour might be in a sustainable catchment or watershed system;
4   The model of stewardship accountability is then described as a foundation on which strategic water governance arrangements at a catchment or watershed scale can more effectively encourage change in resource use practice; and,
5   The chapter concludes by describing how the model supports the redefinition of reasonable resource use behaviour in sustainable catchment or watershed systems.

Case studies in catchment and watershed management

The concept of stewardship accountability has been developed in earlier research on the liability of natural resource users for environmental harm; particularly in relation to farming as a sector responsible for natural resource use decision-making across substantial areas of catchments or watersheds (Lange & Shepheard, 2014; Shepheard, 2011; Shepheard & Lange, 2013; Shepheard & Norer, 2013; Shepheard, 2010). The further development of stewardship accountability as a model to enact sustainability through catchment or watershed plans has been performed through case studies conducted between 2010 and 2013 and reported in this chapter. The case studies were selected to test the implementation of resource user accountability for performance outcomes developed through catchment or watershed planning. This involved interviewing a variety of stakeholders involved with catchment or watershed planning, and analysing the relevant policy and legal mechanisms adopted to implement resource user responsibility for environmental harm in each case study jurisdiction.

Four case studies are used here to provide an empirical perspective about the accountability of resource users (see Figure 7.1) (Blatter & Haverland, 2012). This relies on expectations that are documented in a collaborative resource management strategy associated with each case study, that then needs to be translated into on-ground accountability for resource use. The four case study regions offer a

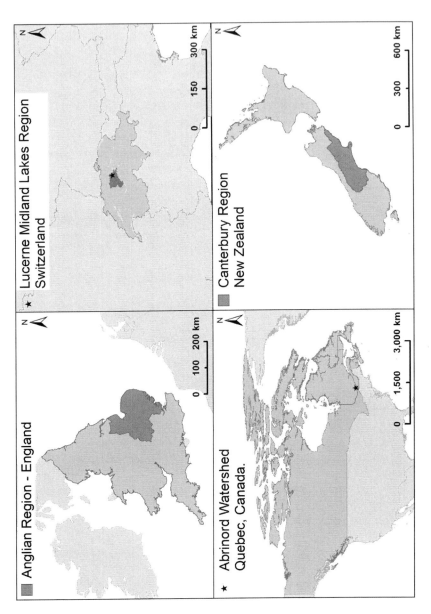

Figure 7.1 Locations of the case study catchments.

comparison between how strategic expectations about sustainable use of natural resources are specified and implemented in different legal regimes (New Zealand and England from a common law regime, compared with Switzerland and Québec from a civil law regime). Each case study region contains agricultural use of natural resources that presents practical challenges for stewardship of water in particular. The case studies correspond with regions where the author has carried out postdoctoral research projects. The choice of multiple case study sites widens the analysis by increasing the potential to identify variation in views about how access to and use of resources is affected by common good interests within a broader catchment and society (Yin, 2009). The practical effect of catchment or watershed plans is to broaden the scope of accountability of resource users to include expectations associated with ecological performance and social impacts. Achieving this requires adjustment of resource use rights and interests in pursuit of sustainable catchment or watershed systems. A brief introduction to each case study is provided below.

Canterbury Region, New Zealand

Resource use conflict in Canterbury reflects competition between uses of water and unresolved debates about water allocation (Shepheard, 2011). Irrigated agriculture accounts for at least 83 percent of water used in the region (Morgan et al., 2002). At the same time, an increasing interest in renewable energy sources is compounding pressure to make greater use of freshwater in hydropower development. Conservation and recreation interests highlight the failure of government to protect water resources from extraction and development (New Zealand Conservation Authority, 2011). There are claims that water resource management is not providing an adequate safeguard for the life-supporting capacity of water, its potential to meet future needs, or provide remedy for adverse environmental impacts (Sheppard, 2010). The uncertainty associated with entrenched conflict has provided impetus to find a better way forward (Canterbury Mayoral Forum, 2009; Land and Water Forum, 2010).

Lucerne Midland Lakes Region, Switzerland

Resource use issues for water in the Midland lakes Region of Lucerne are predominantly related to water quality (Shepheard & Norer, 2013). From the 1950s to 1980s nutrient levels continued to rise in the Swiss Midland Lakes (Lake Baldegg, Lake Hallwil and Lake Sempach), causing algal blooms (including toxic cyanobacteria), decline of underwater biota, and the de-oxygenation of water and lake sediments (Stadelmann, 2009). Discharge of fertiliser and waste affected drainage water from agricultural land are a prominent diffuse source of pollutants, including: rainfall runoff from the soil surface, leaching of fertiliser into subsurface drainage, soil erosion, over fertilised soils, high stocking rates, excessive dung production, and small effluent storage capacity (Stadelmann, 2009).

Due to the intensity of livestock grazing, the potential for nutrient contributions from farmland in the Midland Lake catchments is great. The region is one of the highest carriers of livestock per hectare of agricultural land, having greater than 1.79 units per hectare (Swiss average is 1.2) (Stadelmann, 2009). It supports almost 60 percent of the nation's livestock, led by pigs, followed in decreasing order by poultry, dairy cows, and other ruminants (Stadelmann, 2009). At least 22 percent of soil erosion from farming in Midland Lake Catchments ends up in the water (Prasuhn, 2011). As a result, economic and ecological costs are borne by downstream communities due to non-point source pollution, eutrophication, sedimentation, and more frequent high water events (Weisshaidinger & Leser, 2007).

Anglian Region, England

The Anglian region is characterised by water scarcity, stiff competition for agricultural abstraction of water, and the need to protect ecologically significant sites (Lange & Shepheard, 2014). The basin contains one of the world's most productive agricultural areas (Environment Agency, 2009a). Over half of the area (more than 700,000 ha) is used for agriculture – producing over a quarter of England's wheat and barley, half the nation's sugar beet, numerous vegetable, salad, and fruit crops, and one third of the nation's potatoes (Environment Agency, 2009b). The basin also contains substantial wetland habitats and the largest nationally protected wetland in Britain (Environment Agency, 2009b). There are licences to abstract 160,681 mega-litres (ML) of water per year for agriculture in the Anglian Basin (Environment Agency, 2009a). These abstractions represent, on average, 5 percent of the available water in the basin from both surface and subsurface sources although this can rise to 60 percent during the hottest and driest periods of summer (East of England Rural Forum, 2007).

Abstraction is a significant water management issue for the basin (Environment Agency, 2009b). Fifty-nine percent of catchments are over-abstracted or over-licensed during low flow (summer) periods, when demand for irrigation is at its highest (Environment Agency, 2009c). Pressure upon waters is only predicted to increase with climate changes (East of England Rural Forum, 2007). For example, by 2050 the available water resource could be reduced by up to 15 percent in an average year with summer river flows down by between 50 and as much as 80 percent (Environment Agency, 2008).

The Abrinord Watershed, Québec, Canada

Water resource conflict in Québec typically concerns the impacts of farming on water quality. Diffuse pollution of waters from agriculture is a challenge that brings water use and management into the community spotlight (Bureau d'audiences publiques sur l'environnement Québec (BAPE), 2000). Farming in Québec has become more specialised and intensive since the 1950s, with a greater use of monocultures and heavy chemical use corresponding to pressure on the environment from water pollution, soil degradation, biodiversity loss, and degradation of waterways (Bureau d'audiences publiques sur l'environnement Québec (BAPE), 2000, p. 3 Vol. 2).

The Abrinord (the *Agence de bassin versant de la rivière du Nord*) watershed covers the 'du Nord' river basin in Québec's Laurentian Region. Its first watershed plan (plan directeur de l'eau' or PDE) was approved in 2007 (Abrinord, 2007). The 2007 plan identified the high risk of nuisance associated with agricultural activity (Abrinord, 2007, p. 211), even though agriculture accounts for only 13 percent of land use in the basin (Abrinord, 2007, p. 6). Overall, the diagnosis was of a watershed suffering from poor farming practices, including bare ground on floodplains, absence or lack of sufficient riparian buffer strips, livestock access to waterways, increasing land drainage, and land leveling (Abrinord, 2007, p. 211).

Sustainable catchment or watershed management as strategic level guidance

Sustainable catchment or watershed management is a strategic process applied to help achieve behavioural change in the exercise of access and use rights over natural resources. However, an underlying notion of reasonable resource use associated with private use of natural resources limits how effectively watershed or catchment plans implement sustainability strategy into practice. This sets the context for why reasonable resource use needs to be reimagined.

Management arrangements for sustainable catchment or watershed systems tend to favour collaboration and attempt to share decision-making authority between community members and formal institutions. Processes to achieve this need to (Nowlan & Bakker, 2010):

- Enable local communities to articulate their interests;
- Allow for input from various interests;
- Enable management decisions to be made and acted on; and
- Provide a means to hold decision makers to account for their management choices.

An important aim promoted by these arrangements is to increase responsibility to meet the long-term goals within a community of interest in a way that is integrated across resource uses. Developing and supporting the most desirable behaviours to achieve this must address the challenge of optimising economic production within the constraints of environmental limits, with fair allocation of costs and benefits, and the provision of public goods from landscapes (Martin, Williams, & Kennedy, 2012). Meeting this challenge requires governance arrangements that provide clear signals about the norms of sustainable behaviour and how performance will be managed to secure their attainment (Martin & Gunningham, 2011).

Thus conceived, sustainability is a social project where by governance arrangements interact with individuals and communities to challenge and transform behaviour (Choquette, 2008). Social systems are the site for interpretation and implementation of behavioural change that (Ellis, 2008):

1   Is equitable within generations (now) and between generations (the future);
2   Recognizes the economy as geared to the needs and objectives of society and respect for environmental limits; and
3   Provides constraints on human activity to ensure the continued wellbeing of ecosystems into the future.

Catchment or watershed management is being implemented to help meet sustainability commitments in several countries. The catchment or watershed provides a territory in which a collaborative planning process is used to define behavioural outcomes to meet these needs. The process anticipates individual contributions to the ecological, community, and social wellbeing of the entire catchment or watershed (Howarth, 2009; Québec Ministry for the Environment and Water, 2002; Water Agenda 21, 2011). The resulting plan, in effect, defines a general interest in preserving resources for posterity and protecting resources from harm (Barnes, 2009).

The catchment or watershed is the core concept at the heart of these arrangements where expectations are developed by people representing various interests about the use and management of resources within the catchment area (Lucy & Mitchell, 1996; Raff, 2005; Rodgers, 2009; Stallworthy, 2002). The process has two principal social aims: First, it seeks reduced conflict and greater integration of interests (Rey & Müller, 2007), through deliberation and negotiation of the plan (Mollinga, Meinzen-Dick, & Merrey, 2007; Rangeon, 2008). Second, it defines a relationship between resource users and a community through sustainable practice outcomes to reduce the risk of harming the general interest from private use of resources (Barton, 2010; Choquette, 2008). Overall, the process attempts to define what sustainability requires of resource managers to meet the general interest within a particular area.

In this way the plan is a type of social contract, identifying expectations to change behaviour with the objective of developing a sustainable catchment system (Gunningham, Kagan, & Thornton, 2002; Lynch-Wood & Williamson, 2007; Turnbull Group, 2009). There are multiple meanings of anticipated behaviour that need to be appreciated to craft the standard of behaviour expected. The process generally points to maintaining a standard of reasonable resource management practice, as defined in a catchment or watershed plan. In theory, a natural resource manager in the catchment (such as a farmer) ought to be confident of not breaching their responsibility by following the expectations of the relevant watershed plan. Confidence in the outcome stems from trust in both the planning process, and between interests to act as agreed (Bowmer, 2002). This is likely to be reinforced when the process has built strong collaboration among stakeholders, providing a forum for reconciling differences, the development of local solutions based on particular circumstances, and shared responsibility for integrated governance and resource protection (Gangbazo, 2004).

Sustainable watershed or catchment governance processes have the potential to establish a standard of reasonable natural resource behaviour as part of a collaborative governance vision. This is part of an increasing push for governance arrangements made in civil society to achieve a more integrated approach to regulating environmental harm (Woods & Macrory, 2003). Strategy and rules are essential parts of the governance vision (Fisher, 2009) because they attempt to link private interests in the access and use of resources to agreed community expectations for a particular catchment area (Dimple, 2011; Geneva, 2008). What is sought is good practice relative to the general interest in achieving a sustainable watershed system. This is borne out in practice as governance arrangements for sustainable catchment or watershed management that have typically developed as strategic planning arrangements in the case studies.

Limitations to achieving behavioural change through sustainable watershed or catchment planning

Limitations to achieving the strategic intent of catchment or watershed management plans are identified from the case studies. Limitations arise because strategic level processes lack an appreciative system to foster learning and behaviour change. The link that transforms strategy to practice is indispensible if sustainable catchment management is to effectively guide collective action (Woodhill & Röling, 1998).

There are a number of ways that strategic catchment or watershed management plans can be limited in their capacity to generate behaviour change. Seven limitations have been identified from the case studies and are discussed in this section. These are:

1   Varying standards;
2   Lack of legal responsibility for socially constructed expectations;
3   Demonstrating connectivity between action and impacts;
4   Impractical expectations;
5   Financial incentives that foster a minimum standard of environmental performance;
6   Status of watershed plans as binding instruments for compliance; and
7   Legitimate commercial expectations.

These seven limitations illustrate how common practice is favoured as a default standard of reasonable resource use behaviour in the four case studies reviewed by this research. The risk associated with these limitations is that they limit the extent that catchment or watershed management plans are likely to achieve change to a more sustainable standard of resource use practice. Addressing the risk requires a clear link to be made between strategic level expectations in a plan and behavioural change at a practical resource user level.

Varying standards

Resource use and management by the farm sector is increasingly subject to public concern about sustainable use, and the effect of intensive production processes on the environment (Legg & Diakosavvas, 2010). Such concerns may be adopted and applied in regulation as obligations to protect (or at least not to cause harm) and to manage resources in a particular way (Fisher, 2009). The result can be a range of standards that can be applied to a particular catchment or watershed area. For example, in the Swiss Canton of Lucerne, there are three tiers of standard that may be relevant to water protection by farmers.

First, there is a statutory duty to avoid harmful effects on waters under Article 3 of the Swiss Water Protection Law 1991 (SR 814.20 *Water Protection Law 1991* (CH), 1991). Within this there is ample scope for farming practices to generate harm through diffuse source pollution (Shepheard & Norer, 2013). Although not a positive obligation to actively prevent the threat of a possible impact, the duty does provide a normative benchmark against which the discharge of specific water protection obligations can be assessed (Wagner Pfeifer, 2006). Second, there are general ecological performance requirements for farming. These are the ÖLN standards (proof of ecological performance) that are mandatory under Article 70 of the Swiss Agriculture Act for farmers to receive direct payments (SR 910.1 *Agriculture Act 1998* (CH), 1998). Ecological performance includes: animal friendly husbandry; balanced fertiliser use; use of ecological compensation areas (set aside); suitable crop rotation; soil protection measures; and targeted application of pesticides.

Third, there are specific standards of performance on farms within protected lake catchments, specified in a lake contract (Seeverträge). The contract is non-negotiable and contains obligations linked to soil nutrient and surface conditions that farmers are required to maintain in order to receive additional funding (see Shepheard & Norer, 2013, p. 131 for further details).

Sustainable watershed governance arrangement seeks to add socially constructed stakeholder expectations as a further dimension to existing performance standards. These are now discussed in the following section.

Lack of legal responsibility for socially constructed stakeholder expectations

The absence of formal legal responsibility for farm sector compliance with environmental regulation in some jurisdictions, and the lack of enforcement mechanisms for watershed governance in particular, has fostered a minimum standards approach to resource use accountability. For example, statutory ambitions for sustainable water governance in Québec are less likely to be realised when contemporary government decision-making has demonstrated reluctance to put requirements in place for farmers to change their practices (*Letter from Watershed Organisation 2 to Author (December 18)*, 2012). A 1997 regulation that included measures to reduce diffuse source nutrient pollution of waters from farms (Government of Québec, 1997) was amended in 1998 (Government of Québec, 1998), and again in 1999 (Government of Québec, 1999). These amendments introduced exemptions from earlier requirements to prepare agro-environmental plans and to cease or minimize the spreading of livestock manure and farm compost in protected areas. The amendments also extended the compliance timeframes for meeting more stringent nutrient discharge standards. Such regressions suggest ongoing challenges for the maintenance of statutory aspirations for sustainable water governance (Cantin Cumyn, 2010).

Another challenge exists where watershed planning organisations have no regulatory power to enforce a plan. In Québec, once a plan is produced, it is circulated to all relevant government departments, metropolitan regions, and municipal councils. However these bodies only have to consider the plan when exercising their powers and duties (*Collective Nature of Water Resources and Increased Water Resource Protection Act*, R.S.Q., 2009). This is a source of tension for watershed organisations and stakeholders, because their efforts are not binding upon other administrative bodies (Abrinord, 2012). Lack of enforcement capacity is a significant obstacle in converting collaborative effort on integrated sustainable watershed governance planning to action (Baril, Maranda, & Baudrand, 2006). This means that sustainable watershed planning processes may generate little more than aspirations for resource use accountability.

Identifying connectivity between action and impacts in a sustainable catchment or watershed

Sustainable catchment management expectations are generally based on notions of systems connectivity in minimising harm from actions that may generate impacts elsewhere in the system (Prasuhn, 2011; Stadelmann, 2009). Diffuse pollution of water from agriculture is an illustration of the challenge connectivity brings for the achievement of sustainable catchment management (Comptroller and Auditor General, 2010; Wagner et al., 2002). This is a concern in both England and Switzerland.

The problem of connectivity may also be viewed in terms of who bears the cost of service provision. For example, whether downstream recipients or upstream providers should bear the cost when making use of upstream rural land in England to provide periodic floodwater storage and reduce the risk of flooding downstream (Shepheard & Lange, 2013). Cost bearing for this sort of problem has been addressed in Québec, where, in 2003, the Superior Court refused to endorse a municipal council levy imposed on all landholders in a watershed to pay for downstream de-silting work in a river when there was no evidence of a connection between upstream practices and the need for the downstream work

(*Municipalité régionale de comté les jardins-de-napierville c. Municipalité régionale de comté le haut saint-laurent* [2003] CanLII 15536 [Superior Court of Québec, QC CS]). In this case, sustainable watershed governance was used as a means to try and argue for the imposition of a levy. Rejection of this by the court highlights that a general notion about good practice in the implementation of sustainable management cannot be substituted for evidence when seeking to apportion liability to remedy a downstream harm.

Expectations that lack practical relevance to the farm sector

Watershed obligations may hold little sway for farmers affected by other pressures. For example, in England, green production standards implemented through industry supply chains may play a more important role in defining on-farm resource use behaviour. These production standards arise from various sources, including: the farming industry, independent certification bodies, such as the UK Soil Association, as well as supermarkets and food processors to whom farmers sell their produce. Such standards can link farmer conduct to a perceived societal concern for sustainable catchment management through product market chains, to hold farmers to account for their practices.

Production standards may promote expectations for sustainable catchment management to varying extents. For example, the Red Tractor standard for potato production, in contrast to the LEAF Marque standard, prioritises quality attributes and market access over broader benefits derived from water conservation (Burrell, 2011). Alternately, LEAF Marque standards are acknowledged for fostering greater awareness of the environmental impacts of farm water use (Lange & Shepheard, 2014).

Supermarket standards appear to be more influential for farm resource management decision-making. These tend to reinforce interest in maintaining access to markets (Herzfeld & Jongeneel, 2012) and provide a product certification label as a marketing edge for the product (Martin & Verbeek, 2006). These standards also seek to assure buyers that farmers have met quality and environmental performance requirements (Burrell, 2011; Herzfeld & Jongeneel, 2012). In this way, the standards can (theoretically) steer farmers towards resource access and use practice that prompts consideration of catchment impacts. Examples include the reference to 'water as an integral part of any farming system that fosters a healthy and diverse ecosystem and enhances farm production' (Soil Association, 2012, p. 85); 'maintaining access to water while having a positive impact in the communities and watersheds we source from' (Pepsico UK & Ireland, 2012); the identification of the 'responsible use of water' as a component of performance accountability (John Lewis & Waitrose, 2012, p. 69); and the charity Linking Farming and the Environment (LEAF), which addresses farmers, consumers, and food businesses and refers to farmers as 'stewards, producing nutritious food in an environmentally and socially responsible way' (Linking Environment and Farming, 2012). In practice, however, food processor and supermarket quality standards can reinforce resource use based on product appearance regardless of broader environmental impacts.

Production standards, in this sense, reflect a problematic mix of economic and ethical motivations for water resources management, as well as a contradiction between reducing and expanding water use on vegetable farms. A crisps manufacturer may aim to reduce the water impact of crops it sources from water stressed areas by 50 percent over five years from 2012 (Pepsico UK & Ireland, 2012), however, in order to sell their potatoes, farmers do not just have to demonstrate reduced water use, they also have to produce good skin finish, which is achieved by using more water than is needed in order to minimize blemishes (Lange & Shepheard, 2014). Thus, efficient use of water can be limited by the business reality of needing to maintain access to markets for high-value crops. These standards can prove more significant than catchment plans and abstraction licence requirements in shaping how resource access and use rights are understood and qualified in the general interest (Lange & Shepheard, 2014).

Financial incentives that foster compliance with a minimal reasonable standard

Funding is an important incentive for resource managers to meet expectations in a catchment or watershed plan and assume responsibility for broader impacts. In Québec the 'Prime-Vert' programme is run by the Ministry of Agriculture to promote good agricultural practices in support of production that respects the environment, and meets broader expectations from consumers and society (Government of Québec, 2013). Funds are available to farmers who agree to follow an approved agri-environment plan. The scheme is independent of sustainable watershed planning processes.

Prime-Vert funds are available to support watershed management actions, through competitive bids. However the funding is limited and competition is fierce. There are five watershed organisations within the Montérégie region of southern Québec that compete with each other, and eligible farmer and industry based groups for the limited amount of money available in a single funding call per year (*Letter from Regulator to Author (22 October)*, 2014). This supports findings from Switzerland that in a harsh fiscal climate where public budgets are under pressure, initiatives like this amount only to the allocation of scarce public funds to a limited number of applicants who emerge as successful in a competition (Shepheard & Norer, 2013). This is hardly a robust financial platform on which to base sustainable catchment or watershed management action.

Status of sustainable watershed plans as instruments for compliance

The process of sustainable watershed governance in Québec provides a watershed organisation (the administrative body) with limited responsibility to plan outcomes for sustainable resource management in its territory. Stakeholders are then asked to voluntarily enact the agreed plan in the general interest (Baril et al., 2006; Cohen & Davidson, 2011). Watershed contracts or agreements are used by organisations to specify arrangements and responsibilities for local projects within a catchment (Choquette, 2008). Their purpose is to document particular management activities based on the agreed 'needs' of the watershed in a plan. In practice, a watershed contract has more moral than legal value: It is 'a moral agreement to engage with the plan and take action with no legal responsibility to act, and no consequences for not following through on what was agreed' (*Interview with Regulator* [Longueil, Québec, 28 February], 2013).

An effective watershed contract requires specific actions if it is to be used to apply and enforce particular management standards (Choquette, 2008). A contract that only provides non-specific statements of principle is little more than acknowledgement of the need to protect the watershed (Bouchard & Clavet, 2009). Such agreements fail to recognize a commercial reality for farmers: that they are affected and respond more to the needs of the economic commodity chain within which they are embedded rather than general expectations about improved management (Cohen & Davidson, 2011; Shepheard & Lange, 2013).

Legitimate commercial expectations

Water access and use can be a critical factor for primary production (Allan, 2011; Alliance for Water Stewardship, 2010), where access and use rights allow abstraction of water to be used as a private good with legitimate commercial expectations (Chartres & Varma, 2011). The private goods concept reflects the status and value of water through its connection with land value and economic production that bring expectations of certainty of supply and protection from expropriation (Rochford, 2011). In New Zealand, the High Court has identified that an access entitlement to water is akin to a right that brings legitimate expectations for exclusive use (*Aoraki Water Trust v Meridian Energy Limited* [2005] 2 NZLR 268, 2005) but also that a narrow preoccupation with property rights is out of keeping with popular and holistic notions of sustainability (Justice Barker in *Falkner v Gisborne District Council* [1995] 3 NZLR 622, 1995). This contrast emphasizes the contested nature of sustainable resource management in an institutional framework that favours both collaborative governance to define expectations and upholds freedom of private property rights (Skelton & Memon, 2002).

Sustainability expectations and transforming reasonable practice through stewardship accountability

Stewardship accountability as a conceptual model provides a basis for transforming strategic planning arrangements into effective change in resource use. It achieves this by defining the critical components of stewardship accountability underpinning resource use decision-making. This is instrumental to achieving reasonable resource user behaviour that is more closely aligned with sustainable catchment or watershed planning expectations.

Stewardship accountability does this by providing a model for resource user appreciation of strategic catchment or watershed arrangements and placing them into a practical resource use context (Lange & Shepheard, 2014). As an organising concept, stewardship accountability helps foster change in resource management decision-making in the interests of sustainability (Rodgers, 2003), by helping redefine the notions of performance that underpin sustainable use of natural resources (Rodgers, 2009). Stewardship accountability does this by fostering notions about reasonable resource use: i.e. limiting the exploitive freedom implicit in property rights, helping to form norms of conservation practice, and protecting legitimacy and social trust in return for sustainable practices (Shepheard & Martin, 2009). These are the four dimensions of stewardship accountability that enable a resource user understanding of reasonable use to be reset.

Stewardship as an appreciative process promotes collaborative social approaches to critical reflection about current practice and, where needed, to challenge assumptions and practice relative to real threats (Morris, Marzano, Dandy, & O'Brien, 2012). This is more likely to promote sustainable resource use practice as a habit of action (Walsh & Shepheard, 2011), more effectively connecting resource governance arrangements with ecological limits and the social wellbeing of a community. Emphasising individual behaviour as part of a broader eco-socio-legal context allows for appreciating the role of private decision-making to help deliver a range of outcomes for a more sustainable system (Lange & Shepheard, 2014).

Stewardship accountability is focused on creating a new reality for resource access and use, driven by commitment to wellbeing and respect for others through defining the necessity for change, what the options are, and making that part of lived experience (Brown, 2007). This connects individuals with their broader catchment or watershed community by empowering people to decide on changes that are best suited to their context within a catchment, and then identify and/or develop the tools needed to effect those changes. The dimensions of stewardship accountability are identified and discussed in the following section as a system to underpin reasonable resource use decision-making that is more likely to meet sustainable catchment or watershed level expectations of performance.

Four dimensions of stewardship as an appreciative system for reconceptualising resource user rights and interests

Stewardship accountability provides a means for adjusting resource use rights and interests by reconceptualising strategic expectations about practice in catchment or watershed plans. Stewardship accountability enables expectations of sustainable management to be defined and understood relative to particular resource use circumstances. This is important if strategic expectations expressed in sustainable catchment or watershed plans are to be effectively expressed as resource use accountabilities (Iza & Stein, 2009). The four dimensions of stewardship accountability are (Shepheard & Martin, 2008):

1   Limits to the exploitative freedom of private rights;
2   Norms of practice;
3   Legitimacy; and,
4   Trust.

These four dimensions are used to redefine resource user behaviour in a particular context. Contextualising strategic expectation is more likely to result in collaboration between interests to define workable resource use sustainability for a particular catchment or watershed. The result is greater clarity about responsibilities, and clarification of the extent that formal and informal resource use expectations form specific requirements for resource users to meet sustainability obligations (Lee, 2009).

The four dimensions of stewardship accountability are identified below in Table 7.1. The table presents each dimension as a range between private (self) interest on the left, and general interest on the right. The four dimensions are shown in the centre column of the table. The range from left to right also shows

Table 7.1 Stewardship: A continuum between private and general interests as reasonable practice in natural resource use

Reasonable Practice as a Private Interest (less sustainable)	*Dimensions of Stewardship Accountability*	*Reasonable Practice as the General Interest (more sustainable)*
Private interest reinforced. Justice as private freedom. Exclusive physical possession.	**Rights of access and use**	Private interest inclusive of public interest concerns. Justice as other regarding action in context of the general interest. Flexible tenure options.
Exploitation Resources as private capital. Efficient use as reasonable. Efficient use the only goal.	**Norms of resource use**	**Conservation** Resources as shared part of natural capital. Equitable use as reasonable. Water shared within society to meet production, social, and environmental needs.
Stakeholder expectation unbounded or poorly defined in practical terms.	**Legitimacy**	Strong connection with collaborative planning processes and responsibility clearly bounded. Collaborative governance networks. All legitimate uses acknowledged.
Low degree of trust Shield from regulation (maximize self-regulation) not inclusive of broader expectations, e.g. right to farm.	**Trust**	High degree of trust. Operations inclusive of formal and informal expectations for water use, clearly bounded, e.g. social licence.

(Source: After Shepheard, 2010)

a transition of reasonable resource use practice from a less to more sustainable standard. Breaking stewardship accountability into these four dimensions enables a reimagined version of reasonable use to be visualised and understood.

Conclusions

This chapter set out to discuss how stewardship accountability provides a basis for learning about sustainable management of natural resource systems and redefining reasonable resource access and use behaviour. It has identified that sustainable catchment or watershed management arrangements in four case studies exist primarily as strategic processes specifying desired resources use outcomes. Converting those processes into effective resource use decision-making is limited by the following issues:

1   Varying standards;
2   Lack of accountability for socially constructed expectations of performance;
3   Difficulty of establishing connectivity between practice and impact in a catchment;
4   Lack of relevance to farm resource use practice;
5   Conflicting interests and perverse financial incentives that undermine or counter the measures and objectives in the catchment of watershed plan;
6   The lack of legal status and enforceability of a catchment or watershed plan; and,
7   Legitimate commercial expectations that impact on interpretation of a reasonable exercise of resource access and use rights.

Stewardship accountability highlights the importance of having a catchment or watershed planning process to connect stakeholder collaboration to strategic expectations, and effectively incorporate those into resource use decision-making. Many strategic level planning processes lack this link to ensure that expectations are understood and can be applied by resource users.

Stewardship accountability provides the conceptual workings that are critical to converting strategic intent into practical resource use decisions and performance outcomes. To achieve strategic collaboration and planning, processes need to identify clearly what they are trying to achieve, and how objectives should be pursued in order to more effectively achieve behavioural change so that natural resource users positively contribute to development of a sustainable working system. This involves private access and use rights being subject to adjustment in the interests of sustainability: limiting exploitative freedoms (implicit in resource property and access rights), helping to form norms of sustainable resource use practice, and protecting legitimacy and social trust in return for practices that do not degrade the environmental, social, or economic wellbeing of a place. These dimensions ensure that strategic governance processes for sustainable catchment or watershed management enable development of expectations that are defined, appreciated, and understood relative to particular circumstances.

Acknowledgements

This research was made possible by: the Social Sciences and Humanities Research Council of Canada; the British Academy; Centre for Socio-Legal Studies at Oxford University; the Foundation for Law and Justice at Wolfson College, Oxford; the Swiss National Science Foundation; and the Australian Endeavour Postdoctoral Awards. I am honoured to have received this support and been privileged to work with inspiring colleagues, including: Professor Jaye Ellis at McGill University; Dr Bettina Lange at Oxford University; Professor Roland Norer at the University of Lucerne; and Dr Anne Brower at Lincoln University (NZ). Particular thanks to Catherine MacGregor at UNE for cartographic assistance in this chapter.

References

Agence de bassin versant de la rivière du Nord. (2007). *Portrait et Diagnostic du bassin versant de la rivière du Nord*. Saint-Jérôme, Québec: Abrinord.

Agence de bassin versant de la rivière du Nord. (2012). *Diagnostic de la zone de gestion intégrée de l'eau d'Abrinord, version préliminaire*. Saint-Jérôme, Québec: Abrinord.

Agriculture Act 1998 (Switzerland) SR 910.1.

Allan, T. (2011). *Virtual water: Tackling the threat to our planet's most precious resource*. London: I.B. Tauris & Co.

Alliance for Water Stewardship. (2010). What is water stewardship? Retrieved from: http://www.allianceforwaterstewardship.org/about-aws.html-what-is-water-stewardship

Baril, P., Maranda, Y., & Baudrand, J. (2006). Integrated watershed management in Québec: A participatory approach centred on local solidarity. *Water Science & Technology*, 53(10), 301.

Barnes, R. (2009). *Property rights and natural resources*. Oxford: HART Publishing.

Barton, B. (2010). Property rights created under statute in common law legal systems. In A. McHarg (Ed.), *Property and the Law in Energy and Natural Resources*. Oxford: Oxford University Press.

Birch, J. (2007). *Water resources position paper*. East of England Rural Forum. Retrieved from: www.eerf.org.uk

Blatter, J., & Haverland, M. (2012). *Designing case studies*. Basingstoke, UK: Palsgrave MacMillan.

Borden, S. (2009). Praxis, in *Encyclopedia of the Social and Cultural Foundations of Education*. Thousand Oaks, CA: SAGE Publications.

Bouchard, D., & Clavet, M.-E. (2009). Gestion de l'eau: une Politique nationale de l'eau à mettre en œuvre ou à revoir? (Water management: A national water policy to implement or review?). *Développements Récents en Droit de l'environnement*, 300, 117.

Bowmer, K. (2002). *Learning from existing practice. Reflections on developing a water sharing plan*. Paper presented at the Fenner Conference, Agriculture for the Australian Environment, Johnstone Centre, Charles Sturt University, Albury.

Brown, K. M. (2007). Praxis, in *Encyclopedia of Activism and Social Justice*. Thousand Oaks, CA: SAGE Publications.

Bureau d'audiences publiques sur l'environnement Québec (BAPE). (2000). *L'eau, ressource à protéger, à partager et à mettre en valeur (Water, a resource to protect, share and develop)*. *Rapport de la Commission sur la gestion de l'eau*. Retrieved from Québec City: http://www.bape.gouv.qc.ca/sections/archives/eau/

Burrell, A. (2011). 'Good agricultural practices' in the agri-food supply chain. *Environmental Law Review*, *13*(4), 251–270.

Canterbury Mayoral Forum. (2009). *Canterbury water management strategy strategic framework*. Christchurch, NZ: Canterbury Mayoral Forum.

Cantin Cumyn, M. (2010). Recent developments to the law applicable to water in Québec. *Vermont Law Review*, *34*(4), 859.

Chartres, C., & Varma, S. (2011). *Out of water: From abundance to scarcity and how to solve the world's water problems*. Upper Saddle River, New Jersey: FT Press.

Choquette, C. (2008). Le contrat de bassin: un outil de gestion à géométrie variable (The basin contract: A management tool with variable geometry). In C. Choquette & A. Létourneau (Eds.), *Vers une gouvernance de l'eau au Québec (Towards Water Governance in Québec)*. Quebec, Canada: Éditions Multimondes.

Cohen, A., & Davidson, S. (2011). An examination of the watershed approach: Challenges, antecedents, and the transition from technical tool to governence unit. *Water Alternatives*, *4*(1), 1–14.

Collective Nature of Water Resources and Increased Water Resource Protection Act, 2009 (Quebec) R.S.Q. c C-6.2.

Comptroller and Auditor General. (2010). *Environment agency: Tackling diffuse water pollution in England*. London, UK: National Audit Office.

Dimple, R. (2011). *Ecological goods and services: A review of best practices in policy and programing*. Retrieved from Winnipeg, Mannitoba: http://site.ebrary.com/lib/mcgill/ Doc?id=10497735&ppg=17

Ellis, J. (2008). Sustainable development as a legal principle: A rhetorical analysis. In H. Ruiz Fabri, R. Wolfrum, & J. Gogolin (Eds.), *Select Proceedings of the European Society of International Law*. Oxford: HART.

Environment Agency. (2008). *Potential impacts of climate change on river water quality*. Science Report SC070043/SR1. Bristol, UK: Environment Agency.

Environment Agency. (2009a). *River basin management plan Anglian River Basin District. Annex G: Pressures and risks*. Peterborough, UK: Environment Agency.

Environment Agency. (2009b). *River basin management plan Anglian River Basin District*. Peterborough, UK: Environment Agency.

Environment Agency. (2009c). *Water resources strategy: Regional action plan for Anglian Region*. Peterborough, UK: Environment Agency.

Environment Agency. (2011). *Catchment sensitive farming: ECSFDI phase 1 & 2 full evaluation report*. Bristol, UK: Environment Agency.*Falkner v Gisborne District Council* (1995). *New Zealand Law Reports*, Vol 3, 622.

Fisher, D. E. (2009). *The law and governance of water resources. The challenge of sustainability*. Cheltenham: Edward Elgar.

Gangbazo, G. (2004). *Gestion intégrée de l'eau par bassin versant: concepts et application. Direction des politiques de l'eau (Integrated watershed management concepts and application: The direction of water policy)*. Retrieved from: http://www.mddep.gouv.qc.ca/eau/ bassinversant/concepts.pdf

Geneva, R. (2008). *Thinking outside the fence: International land stewardship policy options for the Canadian agricultural sector.* Retrieved from Calgary, Alberta: http://site.ebrary.com/lib/mcgill/Doc?id=10237935&ppg=11

Government of Québec (1997). *Regulation respecting the reduction of pollution from agricultural sources (made under the Environmental Quality Act RSQ c. Q-2).* Order in Council 742-97, 4 June 1997.

Government of Québec (1998). *Regulation to amend the regulation respecting the reduction of pollution from agricultural sources (made under the Environmental Quality Act RSQ c. Q-2).* Order in Council 737-98, 3 June 1998.

Government of Québec (1999). *Regulation to amend the regulation respecting the reduction of pollution from agricultural sources (made under the Environmental Quality Act RSQ c. Q-2).* Order in Council 247-99, 24 March 1999.

Gunningham, N., Kagan, R. A., & Thornton, D. (2002). *Social licence and environmental protection.* London, UK: Centre for Analysis of Risk and Regulation, London School of Economics.

Herzfeld, T., & Jongeneel, R. (2012). Why do farmers behave as they do? Understanding compliance with rural, agricultural, and food attribute standards. *Land Use Policy, 29*(1), 250–260.

Howarth, W. (2009). Aspirations and realities under the water framework directive: Proceduralisation, participation and practicalities. *Journal of Environmental Law, 21*(3), 391–417. *Interview with Regulator (Longueil, Québec, 28 February).* (2013).

Iza, A., & Stein, R. (2009). *Rule: Reforming water governance.* Gland, Switzerland: IUCN.

Jacob, B. E. (1995). Ancient rhetoric, modern legal thought, and politics: A review essay on the translation of Viehweg's 'Topics and Law'. *Northwestern University Law Review, 89*, 1622–1677.

John Lewis & Waitrose. (2012). *Sustainability report.* Retrieved from: http://www.johnlewispartnership.co.uk/csr/our-latest-report.html

Land and Water Forum. (2010). *Report of the land and water forum: A fresh start for fresh water.* Land and Water Forum, Wellington, NZ. Retrieved from: www.landandwaterforum.org.nz

Lange, B., & Shepheard, M. (2014). Changing conceptions of rights to water? An eco-socio-legal perspective. *Journal of Environmental Law, 26*(2), 215–242. doi:10.1093/jel/equ013.

Lee, M. (2005). *EU environmental law: Challenges, change and decision-making* (Vol. 6). Oxford: Hart Publishing.

Lee, M. (2009). Law and governance of water protection policy. In J. Scott (Ed.), *Environmental Protection: European Law and Governance* (Vol. XVIII/3). Oxford: Oxford University Press.

Legg, W., & Diakosavvas, D. (2010). *Environmental cross compiance in agriculture.* Retrieved from: http://www.oecd.org/tad/sustainable-agriculture/44737935.pdf *Letter from Regulator to Author (22 October).* (2014). *Letter from Watershed Organisation 2 to Author (December 18).* (2012). On file with author.

Linking Environment and Farming. (2012). LEAF – For consumers. Retrieved from: http://www.leafuk.org/leaf/consumers.eb

Lucy, W. N. R., & Mitchell, C. (1996). Replacing private property: The case for stewardship. *The Cambridge Law Journal, 55*(3), 566.

Lynch-Wood, G., & Williamson, D. (2007). The social licence as a form of regulation for small and medium enterprises. *Journal of Law and Society, 34*(3), 321–341.

Martin, P., & Gunningham, N. (2011). Leading reform of natural resource management law: Core principles. *Environment and Planning Law Journal, 28*, 137.

Martin, P., & Verbeek, M. (2006). *Sustainability strategy*. Sydney: The Federation Press.

Martin, P., Williams, J., & Kennedy, A. (2012). Creating next generation governance: The challenge for environmental law scholarship. In P. Martin, Z. Li, T. Qin, & A. Du Pessis (Eds.), *Environmental Governance and Sustainability*. Cheltenham: Edward Elgar.

Mollinga, P. P., Meinzen-Dick, R. S., & Merrey, D. J. (2007). Politics, plurality and problemsheds: A strategic approach for reform of agricultural water resources management. *Development Policy Review, 25*(6), 699.

Morgan, M., Bidwell, V., Bright, J., McIndoe, I., & Robb, C. (2002). Canterbury strategic water study, report 4557/1. Lincoln, NZ: Lincoln Environmental.

Morris, J., Marzano, M., Dandy, N., & O'Brien, L. (2012). *Theories and models of behaviour change*. UK: Forestry, Sustainable Behaviours and Behaviour Change Review, Forest Research.

New Zealand Conservation Authority. (2011). *Protecting New Zealand's rivers*. Retrieved from Wellington, NZ: http://www.doc.govt.nz/publications/getting-involved/nz-conservation-authority-and-boards/nz-conservation-authority/protecting-new-zealands-rivers/

Nowlan, L., & Bakker, K. (2010). *Practicing shared water governance*. UBC Program on Water Governance. Vancouver, Canada: University of British Columbia.

NZ Environment Court (2005). Aoraki Water Trust v Meridian Energy Limited. *New Zealand Law Reports*, Vol. 2, 268.

Pearsall, J., & Trumble, B. (Eds.). (2001). *Oxford English Reference Dictionary* (2nd revised ed.). Oxford: Oxford University Press.

Pepsico UK & Ireland. (2012). *Environmental sustainability – Water*. Retrieved from: http://www.pepsico.co.uk/purpose/environmental-sustainability/water.

Prasuhn, V. (2011). Soil erosion in the Swiss midlands: Results of a 10-year field survey. *Geomorphology, 126*, 32–41.

Québec, Department of Agriculture, Fisheries and Food. (2013). *Prime-Vert: Programme d'appui en agroenvironnement 2013–2018 (Prime Vert: Agri-environment support programme 2013–2018)*. Québec: Government of Québec.

Québec Ministry for the Environment and Water. (2002). *Water our life our future: Québec water policy*. Que. Can.: Government of Québec.

Raff, M. (2005). Toward an ecologically sustainable property concept. In E. Cooke (Ed.), *Modern Studies in Property Law*. Oxford: HART Publishing.

Rangeon, F. (2008). L'intérêt général et les notions voisines (The general interest and related concepts). *La sante et le bien commun, 19*.

Rey, P., & Müller, E. (2007). *EC water framework directive and Swiss water pollution control legislation. A comparison*. Retrieved from Bern: http://www.bafu.admin.ch/publikationen/index.html?lang=de

Rochford, F. (2011). Limits on the human right to water – The politics of social displacement. *Australian Journal of Human Rights, 16*(2), 109.

Rodgers, C. P. (2003). Agenda 2000, land use, and the environment: Towards a theory of 'environmental' property rights? In J. Holder & C. Harrison (Eds.), *Law and Geography: Current Legal Issues 2002*. Oxford: Oxford University Press.

Rodgers, C. P. (2009). Nature's place? Property rights, property rules and environmental stewardship. *Cambridge Law Journal, 68*(3), 550.

Schwandt, T. A. (2010). Praxis. In S. Mathison (Ed.), *Encyclopedia of Evaluation*. Thousand Oaks, California: SAGE Publications.

Shepheard, M. (2011). The potential for improved water mangement using a legal social contract. *Water Law, 22*(2/3), 95.

Shepheard, M., & Lange, B. (2013). *Is there still an economic right to water? An analysis of the intersection of rights and regulatory regimes.* Retrieved from Oxford: http://www.fljs. org/content/is-there-still-an-economic-right-to-water

Shepheard, M., & Norer, R. (2013). Increasing water stewardship responsibility: Water protection obligations and the watershed management policy affecting farmers in Lucerne, Switzerland. *Environmental Law Review, 15*(2), 121–138. doi:10.1350/ enlr.2013.15.2.181

Shepheard, M. L. (2010). *Some legal and social expectations for a farmer's duty of care.* Irrigation Matters Report (02/10). Richmond, Australia: Cooperative Research Centre for Irrigation Futures.

Shepheard, M. L. (2011). *The potential for improved water management using a legal social contract* (26). Retrieved from Christchurch, NZ: http://hdl.handle.net/10182/3819

Shepheard, M. L., & Martin, P. (2008). Social licence to irrigate: The boundary problem. *Social Alternatives, 27*(3), 32–39.

Shepheard, M. L., & Martin, P. (2009). Multiple meanings and practical problems: The duty of care and stewardship in agriculture. *Macquarie Journal of International and Comparative Environmental Law, 6*, 191.

Sheppard, D. F. (2010). *Reaching sustainable management of fresh water.* Paper presented at the Resource Management Law Association Conference 'Sustainable Freshwater Management: Are we there yet?' Resource Management Law Association, NZ. Retrieved from: http://www.rmla.org.nz/wp-content/uploads/2016/09/address_session_3_v6_2.pdf

Skelton, P., & Memon, A. (2002). Adopting sustainability as an overarching environmental policy: A review of section 5 of the RMA. *Resource Management Journal, X*(1), 1–10.

Soil Association. (2012). *Soil Association organic standards: Farming and growing.* Retrieved from: http://www.soilassociation.org/LinkClick.aspx?fileticket=l-LqUg6iIlo%3d& tabid=353

Stadelmann, F. (2009). *Reduction of phosphorus inputs.* Paper presented at the COST Action 869, Working Group 4, Evaluation of projects in example areas. The Swiss Midland Lakes, Nottwil, Switzerland.

Stallworthy, M. (2002). *Sustainability, land use and environment. A legal analysis.* London: Cavendish Publishing Limited.

Superior Court of Québec. (2003) *Municipalité régionale de comté les jardins-de-napierville c. Municipalité régionale de comté le haut saint-laurent.* Québec, Canada: Superior Court of Québec.

Turnbull Group. (2009). *Governance of water.* Retrieved from Wellington: http://www. waternz.org.nz/documents/comment_and_submissions/090730_governance_of_ water.pdf

Wagner Pfeifer, B. (2006). *Environmental Law II* (2nd, added and expanded ed.). Basel: Sculthess.

Wagner, W., Gawal, J., Furumai, H., Pereira De Souza, M., Teixeira, D., Rios, L., & Hemond, H. F. (2002). Sustainable watershed management: An international multi-watershed case study. *Ambio, 31*(1), 2–13.

Walsh, A., & Shepheard, M. (2011). The role of virtue in natural resource management. In J. Williams & P. Martin (Eds.), *Defending the Social Licence of Farming. Issues, Challenges and New Directions for Agriculture.* Melbourne: CSIRO Publishing.

Water Agenda 21. (2011). *Watershed management. Guiding principles for integrated management of water in Switzerland*. Bern, Switzerland: Government of the Swiss Confederation.

Water Protection Law 1991 (Switzerland) SR 814.20.

Weisshaidinger, R., & Leser, H. (2007). Switzerland. In J. Boardman & J. Poesen (Eds.), *Soil Erosion in Europe*. Hoboken: John Wiley & Sons.

Woodhill, J., & Röling, N. G. (1998). The second wing of the eagle: The human dimension in learning our way to more sustainable futures. In N. G. Roling & M. A. E. Wagemakers (Eds.), *Facilitating Sustainable Agriculture*. Cambridge: Cambridge University Press.

Woods, M., & Macrory, R. (2003). *Environmental civil penalties. A more proportionate response to regulatory breach*. Retrieved from London: http://www.ucl.ac.uk/laws/environment

Yin, R. K. (2009). *Case study research design and methods* (4th ed.). Thousand Oaks: SAGE.

8 Water knowledge systems

Jacqueline Williams, Patricia Please and Lorina L. Barker

Introduction

Our research focuses on how to empower and recognize traditional water knowledge systems. With this in mind, we questioned how best we could share our collective conversation about the relationship with water from three different perspectives: an Indigenous perspective, a depth-psychology perspective and a stewardship perspective. As we explored our relationship and synergies in water knowledge systems, we realized that adopting an Indigenous research framework utilizing the conversational method was the best way to share our collective story. We each bring particular cultural and disciplinary approaches to water knowledge system issues. Dr Lorina Barker is a Wangkumara and Muruwari woman from northwest NSW and emphasizes the many examples where Aboriginal Elders are conveying traditional Indigenous water knowledge to government agencies and the wider public, to ensure the cultural, spiritual, social and emotional well-being of people, place and the environment. There are many complex reasons why Elders are sharing these water stories, and why they have chosen multimedia as the vehicle for that transmission. The inclusion of traditional water knowledge in water governance recognizes the importance of Indigenous water knowledges. Dr Patricia Please considers the questions at the centre of this project from the perspective of public engagement with, and participation in, natural resource management, using an integrated holistic approach that accommodates the importance of empathy, affect-emotion and eco-psychology. Dr Jacqueline Williams explores environmental stewardship from an enviro-social perspective, as a white Australian rural landholder and as an environmental scientist. She identified globalization as one of the main barriers to the recognition of traditional water knowledge systems, suggesting it is best understood as another wave of colonization. Our chapter is presented in the form of a narrative where each author presents their perspectives through an on-going dialogue from different cultural and disciplinary backgrounds.

Lorina   My name is Lorina Barker, and I am a Wangkumara and *Barker (LB):* Muruwari woman from northwest NSW and a descendant of the Adnyamathanha (Flinders Rangers SA), the Kooma and Kunja (southwest QLD) and the Kurnu-Baakandji (northwest NSW). I have always had a deep connection to land and water; they are my life, and my identity. My language, culture, kinship ties and knowledge of and respect for the natural and spirit worlds are intertwined and interconnected. We are the earth and the water; we are one and the same. My formative years were spent growing up on the *Ngarntu* (Culgoa River) at a place called Weilmoringle in northwest NSW. Weilmoringle forms part of Muruwari Country, my father's Country. It was here, during the Dreaming (well before I was born) that the Muruwari ancestors, the great creator beings like *Bida-Ngulu* the sky spirit, *Giyan* the moon and *Mundaguddah* the rainbow serpent traversed the earth, and with their creative powers the natural and spirit worlds came to be, and are forever interconnected and intertwined with the Muruwari people. *Bida-Ngulu* breathed life into people and where he ventured landforms, like the sand hills, clay pans and great plains, were created and sprung with life. After *Giyan*'s mischievous travels, he dove into the water and when he emerged he rose into the sky and became the moon. The *Mundaguddah* tracks became the rivers, waterholes, billabongs and creeks (Barker, 2014; Matthews, 1977). Embedded in these Dreaming stories is the lore that governs the way people engage with water and their intimate relationship with Country (Barker, 2014; Matthews, 1977). This long history and deep knowledge was captured in my PhD research (2014), and underpins and sustains connection to Country and all living things in both the natural and spirit worlds. Water storylines shape the interrelationship between people, place, water and other living spirit beings that protect the water places. This is what Deborah Bird Rose calls 'a non-human-being-centred view of the world' (Rose, 1992; Rose, 1996, p. 3).

Patricia   I am a migrant to Australia of twenty-nine years. I came from *Please (PP):* the United Kingdom and spent my early years in the United

States of America. My connection to land and water issues comes from having worked professionally on a range of land and water problems from different perspectives as part of my life experience in Australia. I arrived in 1988 and initially worked as a scientist/hydrogeologist on salinity and water matters. While that was enjoyable and challenging in many ways, my occupation did not satisfy certain deeper concerns I held in relation to human psychological dimensions of life, and I further trained and worked as a psychotherapist. This provided me with the conceptual understandings, words and skills required to engage empathically and to listen deeply about human issues and to explore personal narratives and their relationship and emotional content. In this text, I shall use the term 'depth psychology' to characterize this approach to understanding human experience. At a later date I merged my two discipline backgrounds to become a psychological/social scientist working on a range of land and water issues, including pest management, horticulture, salinity, marine reserves and commercial fishing.

Jacqueline Williams (JW):   My journey from birth as a white Australian woman (European ancestry) to the present has been underpinned by a family foundation that instilled a strong connection with and respect towards nature. My career pathway encompasses nursing and social welfare prior to environmental science studies that were embedded in action-orientated community work, including forest activism, Landcare and regional natural resource management (NRM) groups prior to entering academia where my research now focuses on natural resource governance. My adult life has involved extensive experience as a rural landholder including landscape rehabilitation and organic food production; and so I bring to this dialogue experience in the practical application of environmental management and restoration from my life journey with a particular interest in environmental stewardship.

LB: Part of my PhD research was about the *Ngarntu* (Culgoa River) in northwest NSW, about the knowledge systems and the way in which the river changed with colonization. That is, the name change of the river and the flow of the river, once the weirs, bore drains and ground tanks[1] went in, technologies that created 'new water' my community has adapted to, including the creation of water storylines for these man-made structures. This was an important way of maintaining cultural beliefs and practices in an ever-changing environment. My community adapted to the technologies of 'new water', and when descendants of the Kooma, Yawalaraay and Gamilaraay moved to Weilmoringle they took on the Muruwari songlines about the river and the waterways and, together, they created these things called water storylines, where events happen. It is a way of holding on to your traditions by creating these contemporary stories of the river and other water places. Stories are told to children so you know where to swim, where you can't swim; it's the way in which you relate to the river system traditionally and then it is adapted and changed in a contemporary sense. Thus, when I am talking about a contemporary sense I am talking about places where you couldn't swim in a particular part of the river because most likely there were undercurrents that were too strong; these places also form part of the songlines of Country (water and land). There are particular areas where *Giyan* went into the water and rose as the Moon and they are significant places that only certain people can swim or fish. I don't exactly know where *Giyan* went in and come up as the Moon, because this story wasn't explained in detail to us. Instead, we were told you can't swim around that bend or in that section of the river and so we avoid these places. There are other places where accidents occurred or an event took place you weren't allowed to swim there for a long period of time and when you swam there, you did so with caution and that story was always told to you. For instance, where someone has drowned or somebody got sick if they went into the water, they had a swim and they came out and got sick, or somebody hit a rock that was submerged and became a paraplegic—all these places form the contemporary water storylines—these places are known to everyone. Where someone had camped or where someone's traditional hut like, Granny Byno's place, we knew that it was her place because that's where Granny Byno lived by the river and she was one of the matriarchs of the community. It's the continuity of traditional mourning customs of avoiding places where people lived or died so as not to disturb the spirit of the place and it is a way of always remembering the individual. There are also places at the river where the *Guwinji Jugi* the water dog, who acts as a sentinel for different sections of the river. They have been sighted at the Bridge, at *Wamparinj*, 'the big hole and sometimes standing on the riverbank' (Barker, 2014, p. 90). This is what David Anderson (2000) calls a 'sentient ecology' where people, place and animals are interconnected in a symbiotic relationship (p. 116). In this case, it is the relationship between people, places and the animal spirit *Guwinji Jugi* the spirit dog. Also, when you have been relocated from elsewhere, you take your traditional beliefs and practices and adapt them to your new place and environment. It's the way in which Aboriginal People have maintained our cultural connections to the water even though it may not be our traditional body of water. This gives you a sense of what I have been doing in regards to water, so mapping the storylines, the songlines of the river and peoples use of technologies of 'new water'—like weirs, bore drains and ground tanks. I collected oral histories of Muruwari, and descendants of the Kooma, Yawalaraay and Gailmaraay peoples who shared with me stories about important places on the river: their special swimming and fishing places where they take their children and grandchildren and introduce them to place and re-tell the story about that particular section of the river. The river has changed, over generations that same body of water, when I was a kid we swum in a place we called 'the big hole' *Wamparinj*, a sacred resting place of the *Mundaguddah*, the rainbow serpent and his travelling route. The *Mundaguddah* travelled under the river through a system of tunnels that connected all the waterholes in the *Ngarntu* (Culgoa River). These waterholes never run dry; they are permanent, even when there is a drought and the bed of the river is dry. There are about a hundred waterholes in the *Ngarntu* (Culgoa) and each have their special names and they are all the paths that the *Mundaguddah* the rainbow serpent travels. There are certain family members, certain Traditional Owners, who couldn't swim in those waterholes, but because I live and walk in two camps, I am part of the Traditional Owners and am also a descendant of the People who have a long historical connection to Weilmoringle. Therefore, I and my extended family swam at 'the big hole' all the time and we knew the stories and we knew when we couldn't swim there and the rules of how we had to behave in the water. When I went back as part of my PhD collecting these stories, I was told that nobody swims there now, the reasons they told me they don't swim at 'the big hole' are because the scrub has grown up and the banks are too steep now from erosion so its not as accessible as it once was. Today, the kids, young people and family members have got their own special places that they go, they swim down at the sandy bar or sandy bank that's nice and accessible. I bring an Aboriginal perspective, a relationship with and connection to water, the environment and all living things in both the natural and spirit worlds.

PP: In conversation with Lorina and Jacqueline, I came to realize that I have an absence of an embodied and lived experience of the Australian landscape. However, I do have a series of lengthy scars on my body resulting from a serious motor vehicle accident that occurred on the bend of a dirt road near Charleville, Queensland whilst undertaking hydrogeological field work with two colleagues (sadly one of my colleagues died in this accident)—so I do have an intense embodied experience of the Australian landscape of a different kind. My different experience prompted me to reflect more deeply on what I could bring to this conversation on water management. And in turn that reflection pointed to my depth psychology training and work experience that shifted me from taking the more conventional physical, empirical, scientific approach to water management to taking an approach that values personal stories—emotion, relationship and the use of empathy and focussed listening to more deeply understand the range of perspectives that can be taken on any one issue. So I would like to share with you a story of the ideas and influences that I have come across in my travels that have helped me to understand and engage in other peoples' experiences of land and water issues in this country.

The practices and principles from depth psychology, which are slightly off centre from what we learn about in mainstream education, provided me with a different way to explore different knowledge systems that are often placed at the margins of professional and public discourse at best—and more often left out of the discourse completely. I have been struck by the writings of Syme and Hatfield-Dodds (2007, p. 12) in which they emphasize the need for those working on water issues to take on board a more nuanced approach that recognizes the 'multiple currencies of value' and 'expand the negotiation space and craft win-win solutions, rather than framing the entire process in terms of trade-offs between opposing values'. Whilst their selection of words in that quote is not necessarily how I would phrase the issue, they did point out, in language that makes more sense to me, that government water agencies have an underdeveloped and limited capacity to integrate expert and local knowledge with social science expertise. As well, they have observed that professional managers and policy advisors dealing with water management and reform 'often resist genuine community engagement and attention to social values and equity issues' (Syme & Hatfield-Dodds, 2007).

From where I stand, it is this 'limited capacity' and 'lack of genuine engagement' that results in a lack of recognition of Traditional Ecological Knowledge (TEK) systems, and thus a lack of relatedness resulting in increased disconnection, which can then lead to frustration and anger, and thus escalate to extended conflict. A good example of such conflicts is illustrated in the Liverpool Plains in New South Wales where concerns were raised by the Gomeroi Traditional Custodians in 2016 that Aboriginal heritage and cultural sites were overlooked by the Australian Government in the conditional approval granted for a coal mine (Australian Broadcasting Corporation, 2016). Similarly, in the Galilee Basin in central Queensland, Wangan and Jagalingou Traditional Owners raise concerns that a massive government-supported coal mine would destroy their ancestral homelands and waters, cultural landscape and heritage (Robertson, 2016). There are a range of approaches that can be taken to increase the skills and capabilities of key people working at the interface of government—non-government organizations (NGOs), industry groups and other local communities. These approaches can be defined in many ways, but at the heart lies the need for a more integrated holistic approach, and an emphasis on public participation and community engagement. These approaches have been identified as two of the five common principles emerging in water management in western countries (Hussey & Dovers, 2007, p. 152); however, their 'emergence' as a new social and professional norm in Australia appears to be limited and patchy.

JW: Spending time with First Nations Peoples, and hearing about and seeing the ongoing impact of the first wave of colonization by Europeans in Australia, only to be reinforced and entrenched today by neoliberalism (often considered as another wave of colonization [Choudry, 2003; Venn, 2009; Wikaire and Newman, 2014]), I was introduced to decolonizing theories and practice (Smith, 2012), which led me to other disciplines and cultural views of the state of the planet and humanity. Hugh Mackay (2017) recently highlighted the problems for Australia in the 2017 Gandhi Oration, entitled *The State of the Nation Starts in Your Street*, in which he shows that, despite our high standard of living, wealth and opportunity, we are less open and tolerant, highlighting the current lack of trust in all of our human institutions and the preoccupation of the 'all about me' culture. Current water governance systems place humans at the centre of nature, with water largely considered a commodity for human exploitation and use, a common assumption reinforced more recently by neoliberal ideology. This disconnect from our true place as being dependent on nature and stewards with ethical obligations (Fischer et al., 2015) is reinforced through globalization, which perpetuates unsustainable consumerism as the norm, relying on short term technological fixes that perpetuate the entrenched dominant colonial paradigms since the Doctrine of Discovery in the 15th century. Gómez-Baggethun, Corbera and Reyes-García (2013) define Traditional Ecological Knowledge (TEK) as the body of knowledge, beliefs, traditions, practices, institutions and worldviews developed and sustained by Indigenous, agrarian and local communities in interaction with their biophysical environment. These systems, underpinned by cultural knowledge, managed water sustainably over many thousands of years (as illustrated in Chapter 5 in this volume), yet current water governance fails to recognize these important knowledge systems and practices. Water governance systems in the neoliberal world of globalization have rendered water as a commodity to be traded and consumed, despite the rhetoric of environmental laws, which have largely resulted in the empowerment of developers rather than the stewards of the environment (Boer, 2000; Preston, 2011; Verschuuren, 2012). Western governance systems are broadly defined as the exercise of power, decision-making and implementation of decisions constituted by laws, rules, institutions and processes (Moore et al., 2011), and are considered to rely on mono-cultural knowledge systems (in particular western science) within top-down centralized power structures. In comparison, Indigenous and community-based governance systems are holistic, relying on diverse knowledge systems that are communal in nature (Aseron et al., 2015). Western governance systems devalue the cultural practices of the non-dominant cultures and marginalized groups, which we refer to as TEKS (Tang & Gavin, 2016). How has western governance continued to dominate globally and marginalize TEK? Given the range of challenges as present, I, like so many others, have been confronted with various questions such as: how can traditional water keepers self-determine within colonized systems of water governance?

PP: One does not need an expensive research project to show that western education and governance systems do not give explicit recognition to the felt experience of people and their relationship to the natural environment. Spend some time in a government job that involves land and water issues and you will quickly realize that many managers will make great efforts not to engage on this level, let alone sanction any projects that would facilitate studying it in any psychological depth. I have worked in and alongside several government agencies and know this to be the case. Whilst most work with the best of intentions, I would suggest that many people working in these systems may not even know how to start going about doing this in their professional roles. A key skill required is 'empathy'.

'Empathy' is not just a 'feel good' term to be used in private life. Heinz Kohut, an eminent psychoanalyst best known for his development of self psychology, provides a useful definition of empathy: '[I]t is the capacity to think and feel oneself into the inner life of another person' (Kohut, 1984, p. 82). Empathy has many functions, such as enhancing the feeling of connectedness to others and enabling the processing of emotional experience (Mollon, 2001). It is also a key tool to acquiring data on peoples' subjective experience. Most people have the capacity for empathy but it is an approach and a skill that is not usually given prominence in the professional world of water management. Kohut made the point that training and learning can make a difference in widening the range and scope of one's basic capacity for empathy. He used empathy in his treatment of people with narcissistic personality disorders.

My depth psychology training points to the fact that if we want to integrate more diverse voices and perspectives into water management, including people associated with TEKS, we need to learn to think and feel our way into an Indigenous way of thinking and feeling—particularly how they think and feel about their land and water resources. And for that we need to adopt an explicitly empathic approach—bring it to the foreground, use it effectively and bring what we learn from it into our discussions, planning and policies.

In the mid-2000s I was fortunate enough to be funded for a social psychological PhD project that provided me with the opportunity to apply my depth psychology training on the issue of dryland salinity. A key part of my research involved applying my empathic listening skills to explore the affective, feeling and emotional experience of working on dryland salinity by a range of stakeholders—scientists, landholders, extension and policy officers and an Indigenous NRM officer (Please, 2011). My main motivation for undertaking this research was the realization that there is a lack of explicit recognition of people's feelings in decision-making linked to government policy, funding and management. It is known that conscious recognition of this aspect of human experience can lead to improved engagement, increased motivation and creativity. Moreover, taking this innovative approach and acquiring knowledge of people's felt experience has the potential to be very beneficial in dealing with multiple value systems and perspectives associated with water and land management issues. My PhD research into salinity, and the associated affective and emotional experiences of salinity management stakeholders, found a home as a chapter in a book titled *Placing Psyche* (San Roque et al., 2011). This book took the Jungian notion of a 'cultural complex' (Singer & Kimbles, 2004) and applied it to Australian cultural themes. The concept of 'cultural complex' provides a framework for the analysis of affects, feeling and emotions, and is defined as follows: 'Intense collective emotion is the hallmark of an activated cultural complex at the core of which is an archetypal pattern' (Singer & Kimbles, 2004, p. 6). At the heart of what the range of stakeholders interviewed were talking about was a shared, deep caring and love for the land (including water). Phrases like 'I am doing something to help the landscape', 'I feel content being in my country and working in my country for its benefits', 'I guess it has always motivated me. That sort of spiritual side of it. Looking after the land . . . I love it. It is part of me.'— these were some of the deeply felt expressions from those I interviewed. These expressions were very much linked to the growing collective awareness in the Australian psyche of the uniqueness and fragility of the physical environment in this country and how we need to recognize the uniqueness of the land and water as well as individuals who understand it well.

I have a couple of reflections on this research that I think are of relevance to our story: my sense was that it was other professionals involved in depth psychology relating to Australian environmental issues who had an interest in this research. The general NRM-land and water research and government community did not appear to be very interested because, as I saw it, it did not seem to fit in with the dominant discourse on land and water issues at the time. To be exploring the world of affect and emotions in relation to a land management issue didn't easily fit in with the dominant positivist, reductionist approaches of the time. Mind you, dryland salinity dropped off the radar as a significant Australian environmental issue at about the same time—primarily due to a long-term drought that resulted in declining groundwater levels that led to a reduction in the overall salinity threat. But water issues didn't drop off the radar. My perception was that I was using methods, concepts and literature that were not 'mainstream' enough for others to feel comfortably interested with. Most of it came from the world of psychotherapy and mental health, which does not sit alongside land and water issues with ease. Discipline perspectives, such as 'social science' and 'behavioural science', tend to be acceptable. But 'depth psychology', 'analytic psychology', 'psychodynamic', 'psychotherapy' or 'relational' perspectives remain located at the fringe. And yet these approaches aim to explore the heart and soul of the human psychological experience.

David Tacey (2014) has written that Australian society has a pervasive fear of the unconscious—'We dare not set foot in the interior world, lest we go mad, lose our bearings, or risk being disintegrated by unruly forces' (p. 283). He goes on to explore the paradox of our fear and yet longing for obliteration, but I shall leave his thoughts in this area at this point. My take is that the dominant Anglo-Australian culture in Australia does have a fear of the unconscious and that stepping into the world of deep affect and emotions could put us in dangerous waters (water as the common symbol for feelings). It is much safer to stay on dry land and use tools that will keep us in the conscious and cognitive world of positivist 'hard' science and social science.

LB: Aboriginal cultures are the oldest living cultures in the world, and embedded in our knowledge systems is the deep knowledge of this continent and the traditional ecological knowledge of how to care for Country (land and water). As Bruce Pascoe (2015) points out, if non-Aboriginal people 'alter [their] perspective by as little as 15 degrees toward an Aboriginal point of view' (p. 166), they will understand Aboriginal people's interconnected relationship with water. These peoples will also have, as Patty suggests, 'an embodied and experiential response' to Country (land and water), and in so doing, change the way in which they view, engage and make policy about water.

If non-Aboriginal people changed their view of the Brewarrina fish traps as a sightseeing novelty to an Aboriginal frame of reference, they will see evidence of Aboriginal people's ecologically sustainable practices and engineering ingenuity of Biaime's Ngunnhu. However, as Pascoe (2015) also points out, 'they have had very little research interest' (p. 167), other than the original research on the 'stone bedding technique', which dates the Ngunnhu to 40,000 years old and much older than Stonehenge (pp. 167–168). Biaime's Ngunnhu, known as the Brewarrina fish traps at Brewarrina in northwest New South Wales, is 'thought be the oldest human structure in the world' (p. 167). The Ngemba, the custodians of the Ngunnhu, have always known that this intricate locking system was created during the Dreaming and Biaime, the great creator's time on earth. After he had completed his epic and creative journey, he laid down the law and built the Ngunnhu, which became his stepping stones back into the sky.

JW: Listening to Lorina's story has confirmed for me that other approaches are of great importance. I have found eco-psychology really helpful in assisting with the inner journey of reconnection to nature and importantly to ancestral Earth-based knowledge systems. Eco-psychology developed in the 1990s as a means to examine the foundation of human dysfunctional relationships with nature (predominantly in the western world) and assist with reconnecting with nature and the creation of more sustainable cultures. From a white Australian perspective, eco-psychology offers insights of how the western world's disconnection with nature began from Judeo-Christian belief systems that repressed the connections with nature, animals, creative fantasy and the primitive side of humans. The 'Green Man' (a spirit of nature personified as man), spanning thousands of years and found throughout European architecture, archaeology and folklore (Raglan, 1939), is evidence of the west's long forgotten TEKS. Eco-psychology attempts to unite modern humans with the 'two million year old man within', an often forgotten ancestral lineage with western Indigenous systems of knowledge disregarded or dismissed through Christian-based anthropocentric theology domination (Merritt, 2012). In very generalized terms, it is here that the disconnect from nature and each other began, as western Indigenous culture practices were originally grounded in the Earth, the cosmos and in close relationships with people, as other Indigenous cultural practices around the world (Merritt, 2012; Scott, 2014). Healing on an individual and collective level through connection to nature is now considered fundamental for healthy human mental health (Bratman et al., 2012; Mayer et al., 2009; Nisbet et al., 2011). We are returned again to the question: how might we apply these insights into recovering and supporting TEKS?

PP: As we are busy scratching our heads trying to figure out what is it that we do not know, or do not know how to know, relating to land and water issues, I think one of many solutions lies in the application of depth psychology practices and principles to the issues and those working on the issues. Much of this ground has already been well developed and applied under eco-psychology and as Jacqueline points out, provides us with a means to examine the foundation of human dysfunctional relationships with nature (predominantly in the western world) and assist with reconnecting with nature and the creation of more sustainable cultures (Merritt, 2012). Within the eco-psychology framework, Elizabeth Bragg (1996) has provided us with a deep insight into the concept of 'ecological self', a term first coined by deep ecologist Arne Naess (1973). Implicit in deep ecology is 'indigenous ecology', which looks at underlying value systems and worldviews that can be linked to environmental issues and conflicts. At its heart is a belief in the intrinsic value of all life rather than its instrumental value for humans. It is primarily a phenomenological psychological approach rather than a moral or prescriptive approach (Fox, 1990). It embraces an expansive sense of self 'as [the] sense of self that extends beyond one's egoic, biographic, or personal sense of self' (Maslow, 1968, p. iv) hence the reference to the 'ecological self'. The ecological self of a person reacts to the interests of another—be it human, animal, plant, water or land—as if they were their own, and these reactions can include emotions, perceptions and cognitions, spiritual experiences and physical behaviours (Bragg, 1996). Empathy and 'feeling with the world' are strong expressions of the ecological self. The argument goes that the experiences of an ecological self are crucial steps towards changing our individual and collective relationships with the natural world and in so doing, learning to, and behaving in an environmentally responsible way. Spontaneous ecological behaviour, defending or nurturing aspects of the natural world, follow naturally from a developing or developed ecological self.

To get a better understanding of ecological self, the following is a summary of three different self states with their possible boundaries of self: 1) A 'narcissistic self' is where the boundaries and relationship interests of the self do not go much beyond the individual; 2) a more functional, 'interconnected self' is one that is the focus of interest of most western psychology where human relationships (and often those that are close to us and similar to us) are central; and 3) is the 'ecological self' where the boundaries of self extend to the natural world and to individuals and communities that may have very different perspectives to us—covering human and natural ecologies. In terms of TEKS and water and land issues, it would be very beneficial to have people working on this issue who embody the ecological self and can assist others to gain experiences of the complexity of the place and the issues from an expanded self-perspective. The overall aim would be to assist stakeholders to move beyond their self-interest or 'familiar world interest' to take a more expanded holistic view and work together to find beneficial solutions for all involved, including those with TEK—which is at the heart of a solution to many of our current land and water challenges.

Many of these ideas have been woven together into a framework that can be applied to land and water issues. Renee Lertzman (2014), working on one of the more complex issues of our time—climate change—has provided us with a framework that can assist in clarifying the complexities associated with a more holistic and engaged approach to water resource management and TEKS. In the climate change arena there is a need to respond to highly complex emotional and ideological forces, combined with powerful and complex psychological, cultural and social processes. A similar description could be applied to land and water issues. At the heart of her framework is her assessment that 'psychosocial' thinking and research represent a significant 'skills gap' that could potentially inform engagement and communications to assist in this challenge, echoing Syme and Hatfield-Dodds (2007) and Hussey and Dovers (2007). Renee sees these skills as 'under-represented and under-utilized'. She offers an integrative four-quadrant engagement model that includes the behavioural, socio-cultural, systems and psychosocial/emotional experience (see Table 8.1).

It is the top right hand quadrant, the psychosocial/emotional experience quadrant, which is highlighted as the 'missing piece' with respect to climate change psychology challenges. Psychosocial research recognizes both conscious and non- (un- or sub-) conscious processes, experiential and 'affective'/ emotional dimensions, and the interplay of the individual with social forces. At the heart of this framework is the conceptualization of the human subject as dynamic, conflicted, multiple and capable of profound ambivalence. The focus is more on dilemmas and conflicts than on what is reported as unified or constant. This model brings in the need for skills that involve listening, conversation and dialogue. The aim is to breakdown resistance and defences and support more coherent and collaborative efforts to address the issue. In terms of engagement strategies and applications, psychosocial work is in the early stages of development (Lertzman, 2014). Integration across the four engagement quadrants can provide a powerful vehicle for deeper engagement, understanding and transformation.

The water sector mirrors the climate change sector in many ways. Behavioural, socio-cultural and systems research is often undertaken these days. But I agree with Renee in that it is the application of a psychosocial approach that is commonly missing and undervalued.

Table 8.1 The engagement quadrant

Socio-Cultural	Emotional Experience
• worldviews; values; ethics and moral; ideologies; beliefs; political affiliations; messaging; framing; targeted campaigns; values-based engagement	• emotions; experience; conflicts and dilemmas; construction of meaning; affect; defence mechanisms (denial, etc.); psychodynamic; empathy; narrative
acting-on	*acting-with*
• faith-based programs; the 'Six Americas', surveying and polling public opinions • contexts: political messaging, marketing, policy segmentation, opinion polling	• dialogue; listening; conversation groups; motivational interviewing; qualitative research; support • contexts: awareness-raising (group work), conversation-based platforms, immersion experiences, workshops, leadership development, arts
Behavioural	*Systems*
• motivations; reasoning; probabilities; drivers; levers; cognitive processes; rationality; triggers; shift; switch; nudge incentives; rewards; gamification; rebates	• collaboration; design science; social practices; systems thinking; resilience; infrastructure, interdisciplinary, empathy, solutions focus
Behavioural	*Systems*
acting-on	*acting-on and with*
• social marketing; champions; penalties; quantitative research • contexts: behaviour change approaches, energy efficiency, utilities transportation (incentives/taxes), employee engagement	• social innovation projects; public-private partnerships; community-based projects; participatory design; piloting; prototyping • contexts: resource issues (developing regions), complex regional-based issues, sparking innovation, new markets/technologies, field-based education, living laboratory models

Source: After Lertzman, 2013, p. 7

One of the issues is that our western culture has a long history of separating body and mind (Damasio, 1994). Engaging at a deeper, affective level requires us to move away from this way of thinking and see ourselves living in a truly 'embodied mind'. Associated with such a shift is the need for a move from the dominance of 'masculine' principles—analysis, logic, faster pace, past and future time frames—being given priority to an approach where both 'masculine' and 'feminine' principles (Jung, 1971) are given equal standing.

The incorporation of feminine principles would include holism, intuition, slower pace, meandering spiral patterns and a present centred time frame (de Castillejo, 1973; Woodman, 1982). Inclusion of narrative, relational and affective and emotional approaches fit within feminine principles. Lertzman (2014) also highlights the need for capacity building of core psychosocial concepts and skills which include psychoanalytic and psychodynamic concepts most germane to complex environmental issues: namely how people respond and handle anxieties, fears and related psychodynamic processes such as defences and denial. At a very fundamental level, capacity building in these areas involves people developing their emotional and social intelligence (Goleman 1995; 2006; Salovey & Mayer, 1990).

LB: If people listen to Aboriginal oral histories they will learn the 'deep history' and view the Australian landscape through different lens (McGrath, 2015). For example, by listening to the oral histories and attending the Festival of the Ngunnhu they will learn about the traditional ecological knowledge and sustainability practices of the Ngemba and neighbouring Nations, that are at odds with the legislation that was set up to protect and preserve the Ngunnhu as a state heritage place of significance. But protect it from whom? The Ngemba, who have not been able to perform ceremonies or carry out cultural and ecological sustainability practices for over 80 years? By changing their perspective, decolonizing and centring themself, they will learn how the Ngemba and their neighbours worry for Country (land and water) and all living beings in the natural and spirit worlds, because they are unable to harvest the reeds to make baskets and fishnets, and to use as thatching to construct shelters and wind breaks. Lilly Shearer, a Ngemba woman, explained how the Ngemba and their neighbours are denied the cultural practices of carrying out their responsibilities and obligations—to clean up the fish traps and as a consequence, the Barwon River is sick and so are the people (Shearer, personal communication, March 23, 2017). The health of the waterways and the impacts on people is reiterated by Matt Rigney, a Ngarrindjeri man, and his concern for the Murray and Darling rivers:

> We are of these waters … and all of its estuaries are the veins within our body. You want to plug one up, we become sick. And we are getting sick … because our waterways are not clean. So it is not sustaining us as it was meant to by the creators of our world. (Weir, 2008, p. 161)

By listening to the recording of Nyarri Nyarri Morgan, a Martu Elder from the western desert talking about Martu country, the effects of mining and his experience of making the award winning virtual reality film *Collisions* (Wallworth et al., 2016), we will gain a deeper understanding of connection to Country. You will hear Nyarri and his family explain the painting Kalyu (water), which was created by nine Martu Elders in 2014. The painting focuses on Kalyu (water) and the deep knowledge of the underground water systems beneath the Western Australia deserts. The painting also depicts Martu Country and how they care for Country through their land management practices and ecological systems. The painting is also a form of protest against mining on their land that expresses their deep knowledge and concern for water and the wider effects of mining, as all life depends on Kalyu.

In watching the award winning documentary film *Putuparri and the Rainmakers: A remarkable story of national significance* (Ma, 2015), viewers will go on a journey through Wangkajunga Country that will evoke a 'embodied and experiential response' and understanding of the importance of *jila* (springs) *jamu* (soaks) to the Wangkajunga people. The Wangkajunga 'inter-relations of responsibility and reciprocity' to Country (land and water) is skilfully and carefully woven throughout the film. You see this when Putuparri and his Elders take a six-day journey in search of *Kurtal* (Helena Springs), the sacred waterhole in the Western Australia's Great Sandy Desert and the Elder's worry for Kurtal and the water snake, that he may abandon, die or leave the site if they did not visit and perform ceremonies at the site. By the end of the film, people will come to realize they have witnessed the transference of cultural knowledge from the Elders to Putuparri in his preparation as a cultural leader and Rainmaker. They will recall these poignant moments as the Elders call out, run and burn the grass so the smoke will inform *Kurtal* and the water snake that they are coming; as the Elders touch, drink, immerse themselves in and swim in the water; and as they talk, sing, worry and cry for *Kurtal*, people will come to understand Aboriginal people's relationship with water—that it is not merely a body of water, but a living entity, with its own spirit and is 'central to life in the desert'.

JW: The Anthropocene creates challenges and opportunities for humanity to change from a dominant individualistic, unconscious self-interested cultural complex into a more conscious, collective focused environmental stewardship psychological state, as Scharmer and Kaufer (2013) highlight, moving from ego-system to eco-system economies. Whilst neo-liberalism dominates our human institutions, those whose cultural foundation is one of stewardship and ethical moral obligation to nature are the most marginalized in the world, particularly First Nations Peoples and increasingly peasant/agrarian communities (Corntassel & Bryce, 2012; International Assessment of Agricultural Knowledge, Science and Technology for Development (IAASTD), 2009; La Via Campesina, 2017). How can the power dynamics be changed so that dominant society is one that fosters stewardship based on inter-dependency with nature and reviving and recognizing the important role of TEKS to achieve sustainability, nurturing food systems and restoring ecosystems?

I have come to think that the Anthropocene could be considered the result of a human unconscious collective 'midlife crisis', aligning with the baby boomer cohort (also spiritually in midlife transition) who dominate the neoliberal regime governed by archetypes playing out and projecting from a dream-like state. These earth system changes and challenges for our very survival have a parallel with Jung's theory of the 'meeting with the collective humanity shadow self'. Jungian perspectives offer another lens to view the psychological and spiritual effects of consumerism and the current global crisis to facilitate and support the human behavioural changes, individually and collectively. I think it is crucial that we learn to focus on 'wholeness' instead of 'goodness' and individual and collective 'individuation' as a means to foster environmental stewardship and healthy societies. Jung's 'meeting of the shadow self' offers a process through the various stages of denial: projection, integration and transformation as a means to achieve wholeness, individually and collectively. Searle's (2005) theory on institutions and deontic power offers a different lens to view the current lack of recognition (and subsequent disempowerment) of TEKS. Eco-psychology and decolonizing methodologies utilized within an understanding of the current human narcissist epidemic and the Anthropocene also offer opportunities to assist in decolonizing western governance systems through behaviour change at the individual and collective level. New governance systems are required to shift from neoliberalism to a more earth centric stewardship governance system through decolonizing western governance and changing the behaviour of the system individually and collectively to empower and revive TEKS. Deborah Bird Rose (1996) wrote of the 'nourishing' terrains from Aboriginal views of Australian landscape and wilderness. It is an approach that could encourage a new view of Australian landscapes as mosaics of cultural ecosystems, both Aboriginal and European pastoral, that nourish nature and humans through a variety of TEKS that heal the Country and all its Peoples, forming the foundation of a new stewardship form of governance. To commence this process requires reconciling assumptions held by European settlers with new frames for integration across law and science through the involvement of intercultural 'knowledge bridgers' (Bohensky & Maru, 2011; von der Porten & de Loë, R, 2013).

> Our story is in the land. . . it is written in those sacred places, that's the law. Dreaming place . . . you can't change it, no matter who you are.
>   The white man's law is always changing, but Aboriginal Law never changes, and is valid for all people. (Neidjie et al., 1986)

As a white Australian on a decolonizing journey, these two statements by Gagadju Elder Bill Neidjie 'Kakadu Man' remind me of how my scholarly disciplinary practice, western upbringing and education become barriers to reconciling assumptions about, and connections to, TEKS. From my perspective the journey for white Australians requires reclaiming one's own ancestral TEKS to reconnect to a stewardship culture within, at the same time learning, acknowledging and respecting the oldest living culture, the deep history and knowledge systems of Aboriginal Australians.

PP: Writers and practitioners actively working to promote the embodiment of the 'ecological self' and eco-psychology have provided us with some useful approaches that could expand the boundaries of self in individuals and groups so that people bring a more empathic and inclusive approach into their work on water and land-based issues. In Australia, those taking an eco-psychological approach utilize workshops in nature, wilderness experiences, gestalt therapy with the Earth, pastoral counselling and ecological self therapy to increase the feelings of connection with nature and enhance the experiences of 'ecological self' (Bragg, 1995). My own experience of expanding my boundaries of self to be more inclusive of the natural world was accelerated during my years of training to be a geologist, where I spent a lot of time immersed in nature, at one level focussing on specific aspects of the natural environment as a scientist does, but at another level, engaging deeply with the wonders of our natural geological and water resources world. More recently, working on human dimensions of pest animal issues, I have found myself engaging more deeply with the animal aspects of our world and the relationships that we humans have with animals. I have definitely experienced my boundaries of self-expanding through all these experiences as they have enhanced my receptivity to natural phenomena and provided me with a deeper sense of well-being.

As mentioned my recent research position involves the application of more conventional behavioural science and behaviour change approaches to prompt communities to become more engaged on pest animal issues. Whilst this work is not specifically about water resources, there are many practices and principles we use that could be applied to engagement on water resources issues. By combining the use of behaviour change frameworks like Community-Based Social Marketing (McKenzie-Mohr, 2011) and the Behaviour Change Wheel (Michie et al., 2014) (see Chapter 10, this volume) with a deeper psychological approach (Please & Hine, 2014), more effective engagement on water issues could be achieved. Tools from the behaviour change frameworks could be used to prompt and obtain commitments from workshop participants for them to undertake actions to promote and maintain behaviours that assist in facilitating an expanded ecological self. In this way, work that was initiated in a workshop can be continued long after it is over. For some it may be as easy as taking time to be in nature and appreciate what is there; for others it could be about making an effort to connect and relate to people with different perspectives. As well, we could look into framing our conversations and communications around deeper shared values to do with caring for the environment, connection to native wildlife and working together to obtain the benefits of a diverse community with many different voices.

There will always be those who maintain a defensive posture towards the notion of trying to increase their connection with the natural environment and the people who hold traditional ecological knowledge. With that sector of people, perhaps the simplest exercise would be to have multiple gatherings of interested parties at sites of disagreement or conflict about a water resource issue. The inclusion of the natural environment in a meeting can be powerful in itself.

In Australia we may have a head start with feelings of connection to land and water. Research utilizing measures to examine felt connectedness to environment indicated that 'nature looms large in the psyche of Australians' and that we have a 'strong identification with the natural environment' (Bragg & Reser, 2012, p. 4). Many of us would actually like to conserve, sustain and protect our natural resources. My experience of working in many different agencies is that the individual will is there to engage and embrace these environmental challenges. The evidence seems to point to our governance systems making this a very difficult task for individuals to take up. Our systems do not often facilitate capacity building for skills in engagement, empathic relationships and expanding our sense of self to include a deeper connection to nature. These skills are often considered as 'soft skills', skills associated with 'feminine' principles and world views, and are not given the same prominence and value in the water management environment as approaches that reflect the masculine principles, e.g. analysis, logic, focussed delivery.

LB: The earlier examples of Brewarrina fish traps, Aboriginal oral histories and the documentary *Putuparri and the Rainmakers* demonstrate the urgency with which many Elders are conveying Aboriginal water knowledge to government agencies and the wider public. By inviting creative artists into their homes, communities and Countries to film, to share stories of their artworks and to explore and engage with new immersive technologies, like virtual reality. There are many complex reasons why our Elders are sharing these water stories and have chosen multimedia as the vehicle for that transmission; these include the concerns about mining on their land and the impact of the Native Title legislation on people, communities and Country. But the more immediate concern for our Elders is ensuring the cultural, spiritual and social and emotional wellbeing of people, place and the environment. Another is climate change, which, as Tony Birch (2016) explains, will have future impacts that will further increase the pressure on already vulnerable communities (p. 93). However, it is important to point out, that while some cultural knowledge is being shared, the sacred and secret knowledge meant for members of Aboriginal communities is never shared with outsiders.

The following free verse excerpt is an assemblage of oral histories highlighting people's disbelief at the 2006 drought that caused the Darling River to dry-up. It also highlights people's anger at the over use of water irrigation and lack of consultation with the Aboriginal communities in the Bourke region:

Back in the nineties
The people remember the time
When the great mighty Darling River
Almost died
Drained by irrigated pipes
Infected with Blue-green algae
Made worse by the drought

The people remember
When the mighty Darling River ran dry
We walked up and down her banks
Along the riverbed
We wandered
Sad and bewildered at what we had caused
People gathered old fishhooks and sinkers
Some drove cars
Others collected white sand
To spread in their gardens
The mighty Darling River
Many people depend
We have an obligation to protect and conserve
What is left of our beautiful river
If there is no Darling river
Then there will be
No you and me

(Excerpt from L.L. Barker poem,
The Darling River (Bourke), 2006).

The following photograph (Figure 8.1) is part of a collection of photographs that captures a group of family members at the river in Bourke, but there is no water in which it is safe to swim. This photograph taken at the weir illustrates the obvious devastating environmental impacts; but what about the human impact? The family's pain and shock is expressed in their body language, and their questions, 'How did this happen?', 'Why don't they listen?', were left unanswered.

Aboriginal cultures have cared for this Country for millennia, and our ecological knowledge systems and land management practices have helped to sustain humans, the Country (land and water) and all living things. Evidence that Aboriginal cultures have much knowledge to share, but how do you get people to listen? What does 'real' Aboriginal consultation look like? How do you integrate Aboriginal knowledge systems with western science? How can water policy be done differently? That is the challenge.

The ground-breaking research by Patrick Nunn and Nicholas Reid (2015), titled *Aboriginal Memories of Inundation of the Australian Coast Dating from More Than 7000 Years Ago* is a start. Nunn and Reid (2015, p. 3), used empirical analysis to interpret and test the documented oral histories of the sea-level rises from 21 places on the coast of Australia. They found that the oral histories were 'empirically informed' and 'they say essentially the same thing': that these oral histories are

Figure 8.1 The Darling River in drought.
Author's own.

evidence that Aboriginal people witnessed the last glacial Ice Age when the sea levels rose and covered parts of the Australian coastline around 8,000 to 10,000 years ago. This confirms what Aboriginal people have always known, that these stories are real and indeed did happen and so represent reliable forms of historical knowledge (Klapproth, 2004). In fact this study demonstrates how western science is only starting to catch up with Aboriginal knowledge systems, and that some scientists are beginning to listen and recognize the value and richness of Aboriginal knowledge systems. As a result, partnerships are being developed between key stakeholders: scientists, Aboriginal people and wider community (Birch, 2016). There are projects that aim to develop synergies between western and Aboriginal knowledge systems and where they converge, and some are well underway (Anderson, 2000; Stasiuk, 2015). However, for the most part Aboriginal knowledge is being ignored and undervalued in water policy reforms. In many instances, Aboriginal people feel powerless to do anything and those that are active are often viewed with scepticism or ignored, resulting in Aboriginal knowledge systems being dismissed as 'myths'.

Note

1 Ground tanks is a term used in northwest NSW to describe the man-made excavated pits built to capture and store water for station stock. In other parts of eastern Australia they are referred to as dams. See Goodall (2008).

References

Anderson, D.G. (2000). *Identity and ecology in Artic Siberia: The number one reindeer brigade*. Oxford, Oxford University Press.

Aseron, J., Williams, J., & Greymorning, N. (2015). Inclusive practices, innovative collaboration, governance and recognising cultural capital: Environmental law through a cultural lens. In P. Martin, S. Z. Bigdeli, T. Daya-Winterbottom, W. du Plessis, & Kennedy, A. (Eds.), *The Search for Environmental Justice* (pp. 333–356). Cheltenham, UK: Edward Elgar.

Australian Broadcasting Corporation (ABC) (2016). *Investigation into Aboriginal heritage on proposed Shenhua mine site launched*. Retrieved from http://www.abc.net.au/news/2016-09-09/invesigation-launched-aboriginal-heritage-shenhua-mine-site/7829648

Barker, L. L. (2006). The Darling River. In Barker, L. L. (2014). *Ngarraka Yaan: A Murdi history of Weilmoringle*. PhD thesis, University of New England Armidale, pp. 358–61.

Barker, L. L. (2014). *Ngarraka Yaan: A Murdi history of Weilmoringle*. PhD thesis, University of New England Armidale.

Birch, T. (2016). Climate change, mining and traditional Indigenous knowledge in Australia. *Social Inclusion*, 4(1), 92–101.

Boer, B. (2000). The globalization of environmental law. *Australian Law Reform Commission Journal*, Reform Issue 76, 33–39.

Bohensky, E. L., & Maru, Y. (2011). Indigenous knowledge, science, and resilience: What have we learned from a decade of international literature on 'integration'? *Ecology and Society*, 16(4), 6. doi:10.5751/ES-04342-160406

Bragg, E. A. (1995). *Towards ecological self: Individual and shared understandings of the relationship between 'self' and the natural environment*. Townsville, Australia: James Cook University.

Bragg, E. A. (1996). Towards ecological self: Deep ecology meets constructionist self-theory. *Journal of Environmental Psychology*, 16(2), 93–108. doi:10.1006/jevp.1996.0008.

Bragg, E., & Reser, J. (2012). Ecopsychology in the Antipodes: Perspectives from Australia and New Zealand. *Ecopsychology*, 4(4), 253–265. doi: 10.1089/eco.2012.0085.

Bratman, G. N., Hamilton, J. P., & Daily, G. C. (2012). The impacts of nature experience on human cognitive function and mental health. *Annals of the New York Academy of Sciences*, 1249, 118–136. doi:10.1111/j.1749-6632.2011.06400.x

Choudry, A. (2003). *New Wave/Old wave; Aotearoa/New Zealand's colonial continuum*. Retrieved from http://www.coloursofresistance.org/397/new-wave-old-wave-aotearoa-new-zealands-colonial-continuum/

Corntassel, J., & Bryce, C. (2012). Practicing sustainable self determination: Indigenous approaches to cultural restoration and revitalisation. *The Brown Journal of World Affairs*, 8(11), 151–162.

Damasio, A. (1994). *Descartes' error – Emotion, reason and the human brain*. New York, New York: Putnam Berkley Group, Inc.

de Castillejo, I. C. (1973). *Knowing woman: The feminine psychology*. New York, New York: Harper and Row.

Fischer, J., Gardner, T. A., Bennett, E. M., Balvanera, P., Biggs, R., Carpenter, S., Daw, T., Folke, C., Hill, R., & Hughes, T. P. (2015). Advancing sustainability through mainstreaming a social–eco-logical systems perspective. *Current Opinion in Environmental Sustainability*, 14, 144–149.

Fox, W. (1990). *Toward a transpersonal ecology. Developing new foundations for environmentalism*. Boston, Massachusetts: Shambhala Publications Inc.

Goleman, D. (1995). *Emotional intelligence: Why it can matter more than IQ*. London, UK: Bloomsbury Publishing.

Goleman, D. (2006). *Social intelligence*. New York, New York: Bantam Dell Publishers.

Gómez-Baggethun, E., Corbera, E., & Reyes-García, V. (2013). Traditional ecological knowledge and global environmental change: Research findings and policy implications. *Ecology and Society*, 18(4) 72. doi:10.5751/ES-06288-180472

Goodall, H. (2008). Digging deeper: Ground tanks and the elusive Indian Achipelago. In A. Mayne (Ed.), *Beyond the Black Stump: Histories of Outback Australia* (p. 134). Kent Town: Wakefield Press.

Hussey, K., & Dovers, S. (2007). International perspectives on water policy and management – Emerging principles, common challenges. In K. Hussey & S. Dovers (Eds.), *Managing water for Australia: The social and institutional challenges* (pp. 141–154). Collingwood, VIC, Australia: CSIRO Publishing.

International Assessment of Agricultural Knowledge, Science and Technology for Development (IAASTD) (2009). *Agriculture at a Crossroads Global Report*. Washington: Island Press.

Jung, C. G. (1971). Aion: Phenomenology of the self. In J. Campbell (Ed.), *The Portable Jung* (pp. 139–162). New York, New York: Penguin Publishers (Original work published 1951).

Klapproth, D. M. (2004). *Narrative as social practice: Anglo-Western and Australian Aboriginal oral traditions*. Berlin, New York: Mouton de Gruyter (p. 381).

Kohut, H. (1984). *How does analysis cure?* Chicago, Illinois: University of Chicago Press.

La Via Campesina (2017). Retrieved from http://viacampesina.org/en/

Lertzman, R. (2013). *Engaging with climate change. How we think about engagement*. PhD for Skoll Global Threats Fund, The Skoll Foundation. Retrieved from http://reneelertzman.com/wp-content/uploads/2016/12/Engaging-With-Climate_Landscape-Report-June-2013.pdf

Lertzman, R. (2014). *Psychosocial contributions to climate sciences communications research and practice*. London: University College London. Retrieved from http://www.ucl.ac.uk/public-policy/for-policy-professionals/commissions/psychosocial.pdf

Ma, N. (2015). *Putuparri and the Rainmakers*. DVD Film, Sensible Films, Victoria, Australia. Retrieved from http://sensiblefilms.com/portfolio/putuparri-and-the-rainmakers/

Mackay, H. (2017, January). *The state of the nation starts in your street*. Annual Gandhi Oration, University of New South Wales, Sydney, Australia.

Maslow, A. (1968). *Towards a psychology of being, 2nd edition*. New York: Van Nostrand.

Matthews, J. (1977). *The two worlds of Jimmie Barker: The life of an Australian Aboriginal, 1900–1972*. Canberra: Australian Institute of Aboriginal Studies.

Mayer, F. S., Frantz, C. M., Bruehlman-Senecal, E., & Dolliver, K. (2009). Why is nature beneficial? The role of connectedness to nature. *Environment and Behaviour*, 41(5), 607–643.

McGrath, A. (2015). Deep histories in time, or crossing the great divide? In A. McGrath & M. Jebb (Eds.), *Long History, Deep Time: Deepening Histories of Place* (pp. 163–170). Canberra, Australia: ANU Press.

McKenzie-Mohr, D. (2011). *Fostering sustainable behaviour: An introduction to community-based social marketing*. Canada: New Society Publishers.

Merritt, D. L. (2012). *Jung and ecopsychology: The dairy farmer's guide to the universe.* Volume 1. Carmel, California: Fisher King Press.

Michie, S., Atkins, L., & West, R. (2014). *The Behaviour change wheel: A guide to designing interventions.* United Kingdom: Silverback Publishing.

Mollon, P. (2001). *Releasing the self: The healing legacy of Heinz Kohut.* London: Whirr Publishers.

Moore, K., Kleinman, D. L., Hess, D., & Frickel, S. (2011) Science and neoliberal globalization: A political sociological approach. *Theory and Society*, 40(5), 505–532.

Naess, A. (1973). The shallow and the deep, long range ecology movements: A summary. *Inquiry*, 16(1), 95–100.

Neidjie, B., Davis, D., & Fox, A. (1986). *Australia's Kakadu Man Bill Neidjie.* Australia: HarperCollins Publishers Pty Ltd.

Nisbet, E. K., Zelenski, J. M., & Murphy, S. A. (2011). Happiness is in our nature: Exploring nature relatedness as a contributor to subjective well-being. *Journal of Happiness Studies*, 12(2), 303–322.

Nunn, P. D., & Reid. N. J. (2015). Aboriginal memories of inundation of the Australian coast dating from more than 7000 years ago. *Australian Geographer*, 47(1), 11–47.

Pascoe, B. (2015). Panara. In A. McGrath & M. Jebb (Eds.), *Long History, Deep Time: Deepening Histories of Place* (pp. 163–170). Canberra, Australia: ANU Press.

Please, P., & Hine, D. (2014, May). Deep framing versus social marketing: Eliciting long-term, meaningful behaviour change in community-engaged invasive animal management, *16th Australasian Vertebrate Pest Conference*, Brisbane, Australia.

Please, P. M. (2011). The feeling of salt, land, and water. In C. San Roque, A. Dowd, & D. Tacey (Eds.), *Placing Psyche* (pp. 95–119). New Orleans, USA: Spring Journal, Inc.

Preston, B. J. (2011). *Internalising ecocentricism in environmental law.* Paper presented at 3rd Wild Law Conference: Earth Jurisprudence – Building Theory and Practice, Brisbane, Queensland.

Raglan, Lady (1939, March). The green man in church architecture. *Folklore*, 50(90990), 45–57.

Robertson, J. (2016). *Adani Carmichael mine opponents join Indigenous climate change project.* Retrieved from https://www.theguardian.com/business/2016/nov/13/adani-carmichael-mine-opponents-join-indigenous-climate-change-project

Rose, D. B. (1992). *Dingo makes us human: Life and land in an Aboriginal Australian culture.* Cambridge, England: Cambridge University Press.

Rose, D. B. (1996). *Nourishing terrains: Australian Aboriginal views of landscape and wilderness.* Canberra Australia: Australian Heritage Commission.

Salovey, P., & Mayer, J. (1990). Emotional intelligence. *Imagination, Cognition and Personality*, 9(3), 185–211.

San Roque, C., Dowd, A., & Tacey, D. (2011). *Placing psyche – Exploring cultural complexes in Australia.* New Orleans: Spring Journal Books.

Scharmer, O., & Kaufer, K. (2013). *Leading from the emerging future: From ego-system to eco-system economies.* San Francisco, California: Berrett-Koehler Publishers Inc.

Scott, E. (2014). The visionary psyche: Jung's analytical psychology and its impact on theories of shamanic imagery. *Anthropology of Consciousness*, 25(1), 91–115.

Searle, J. R. (2005). What is an institution? *Journal of Institutional Economics*, 1(1), 1–22.

Singer, T., & Kimbles, S. L. (2004). *The cultural complex – Contemporary Jungian perspectives on psyche and society.* United Kingdom: Routledge Publishers.

Smith, L. T. (2012). *Decolonizing methodologies: Research and indigenous peoples.* London: Zed Books.

Stasiuk, G. (2015). *Synergies: Walking together – Belonging to Country.* DVD Film, BlackRussian Productions, Australia. Retrieved from https://www.youtube.com/watch?v=aeGqTpLDYjQ

Syme, G. J., & Hatfield-Dodds, S. (2007). The role of communication and attitudes research in the evolution of effective resource management arrangements. In K. Hussey & S. Dovers (Eds.), *Managing Water for Australia – The Social and Institutional Challenges* (pp. 11–22). Australia: CSIRO Publishing.

Tacey, D. (2014). The Australian resistance to individuation: Patrick White's knotted mandala. In C. San Roque, A. Dowd, & D. Tacey (Eds.), *Placing Psyche* (pp. 283–303). New Orleans, USA: Spring Journal, Inc.

Tang, R., & Gavin, M.C. (2016). A classification of threats to traditional ecological knowledge and conservation responses. *Conservation Society,* 14(1), 57–70.

Venn, D. (2009). *Confronting the monoculture of scientific knowledge.* Institute for Knowledge Mobilization. Retrieved from http://www.knowledgemobilization.net/essay-by-david-venn-confronting-the-monoculture-of-scientific-knowledge/

Verschuuren, J. M. (2012). Global environmental law. *Tilburg Law Review, Journal on International and Comparative Law,* 17, 222–227.

Von der Porten, S., & de Loë, R. (2013). Water governance and Indigenous governance: Towards a synthesis. *Indigenous Policy Journal,* 23(4), 1–12.

Wallworth, L., Morgan, N. N., & Taylor, C. (2016). In conversation, 'Nyarri Nyarri Morgan: Virtual Reality, History and Indigenous Experience'. ACMIOnline, *Collisions.* Retrieved from https://guides.acmi.net.au/collisions/

Weir, J. K. (2008). Connectivity. *Australian Humanities Review,* 45, 53–64.

Wikaire, R. K. L., & Newman, J. I. (2014). Neoliberalism as neocolonialism?: Considerations on the marketisation of Waka Ama in Aotearoa/New Zealand. In C. Hallinan & B. Judd (Eds.), *Native Games: Indigenous Peoples and Sports in the Post-Colonial World* (Research in the Sociology of Sport, Volume 7) (pp. 59–83). Emerald Group Publishing Limited.

Woodman, M. (1982). *Addiction of perfection: The still unravished bride.* Toronto, Canada: Inner City Books.

**Water policy for resilient
agri-environmental landscapes**

Lessons from the Australian experience

Richard Stayner and Melissa Parsons

Introduction

The chequered history of the development of Australia's water resources for agriculture since European settlement owes much to the pursuit of successive nation-building visions in a complex and only partially understood natural environment. It is also a story of water management policies and institutions that were similarly based on a partial understanding of the economic and social environments. The nation-building visions imagined a stable future that has never eventuated. Instead, complexity and surprise in dynamic biophysical, economic and social systems continue to characterize the water policy domain. Australian policy makers have, however, recently attempted to apply the lessons of history and emerging ecological, economic and social knowledge, and to focus on the 'resilience' of the system within which water is used.

Past mistakes had high economic, social and ecological costs (Hillman, 2008; Musgrave, 2008; Tisdell, Ward & Grudzinski, 2002; Ward, 2009). The further development of Australia's water resources must therefore be wary of fanciful visions that imagine a future of certainty and stability. The comprehensive reform of water policy and governance embedded in Australia's Intergovernmental Agreement on a National Water Initiative (Council of Australian Governments, 2004) was a significant break from the past, but its implementation remains both incomplete and vulnerable to political opportunism. A recent Australian Government policy paper (Australian Government, 2015a) on the development of northern Australia reprises a nation-building rhetoric that envisages another round of water development, but risks giving insufficient attention to the complexity of the region's natural, social and economic systems, and the continuing likelihood of surprise.

In this chapter we suggest that the adoption of an approach that sees water for agriculture as a complex social-ecological system would avoid the high costs of repeating past mistakes. We first describe some of the unforeseen consequences of past water development. We then explain the concept of resilience and how it has been applied in the design and application of water reforms over the past twenty years in Australia. The literature on the resilience of social-ecological systems provides an insight into past mistakes and can provide guidance on policy for the future use of water for agriculture, especially in northern Australia. We conclude with a reflection on how a resilience focus changes the way in which the future of water use is imagined. The Australian experience constitutes a case study in the recent international literature examining the relevance of resilience concepts to water resource management (see, for example, Namoi Catchment Management Authority, 2013; Pahl-Wostl et al., 2007) and illuminates some of the obstacles to implementing such a framework both in Australia and elsewhere.

Nation-building visions

The perceived colonial-era imperative to occupy and settle the Australian continent (ignoring the fact that it was already occupied by its Indigenous inhabitants) took a crucial turn in the mid-19th century when the end of gold rushes in the eastern States presented colonial administrations with the problem of how to engage large numbers of former miners in a nation-building task (Davidson, 1997, pp. 30–31). A key response was the closer settlement schemes of the mid- to late 19th century, under the spell of a heroic vision of yeoman farmers transforming what were deemed empty wastelands into bountiful landscapes (Davidson, 1981).

Flaws in this vision soon became apparent. Many farms proved too small to provide settlers with a secure living in the face of volatile commodity prices and an unfamiliar climate. The failing early vision of an egalitarian agrarian society was revived, most notably in Victoria, by the idea that adding irrigation to the system would at least solve the problem of rainfall variability. Again, this vision imagined a bountiful rural hinterland occupied by hard-working citizens, dotted with growing rural communities. The vision was again undermined, however, by successive droughts and economic cycles (Musgrave, 2008).

The gathering momentum towards Federation in 1901 was fertile ground for a resurgence of nation-building rhetoric and grand visions. This is perhaps epitomised in George Essex Evans's poem *The Nation Builders* (1906) in which he envisaged '[a] handful of heroes scattered to conquer a continent' and a future golden age with 'the victory won' and 'the work done'. The heroic pioneer-settler was thus cast in a battle to conquer nature, while the future was imagined as a stable state in which pioneering hardship would cease and prosperity descend, with nature now tamed and beneficent.

The late 19th and early 20th centuries also saw large state-funded investments in water storage and distribution infrastructure, notably on the riverine plains of northern Victoria and southern New South Wales, giving farms without river frontage access to irrigation. This 'march of irrigation' (Musgrave, 2008, pp. 35–41) continued apace until the early 1960s throughout the lower Murray Darling Basin (MDB). Soldier-settler schemes after both World Wars pursued both social and economic objectives by combining small-scale farming with irrigation. Again, poor planning and inadequate knowledge led to much hardship and many failed farms, and generated a 'small farm problem' in which farmers found it difficult to adapt to changing economic conditions, with many enduring chronic poverty (Malcolm, Davidson & Vandenberg, 2000, pp. 2–3).

This expansion phase of water development saw a surge in the construction of water storage and delivery infrastructure, epitomised by the Snowy Mountains Scheme (a hydro-electric complex in south-eastern Australia), based on the worthy objective of post-World War Two nation building. It again envisaged irrigated agricultural landscapes as the growth engine for rural economic development, complemented by secondary industries supported by protectionist economic policies (Musgrave, 2008).

The dream of 'drought-proofing' agriculture by harnessing and controlling water resources was a recurring theme in these nation-building visions. In reality, storing and diverting water from river systems to farms has not removed the effects of drought, even in many irrigation regions (Williams, 2003). Instead, as the Millennium drought of 1996–2010 demonstrated, intensive irrigation development has resulted in many water-dependent industries remaining vulnerable to the kinds of decadal droughts that are a recurrent feature of Australia's climate.

There have also been several iterations, dating from 1912, of development visions specifically for northern Australia (Cook, 2009). One plank of their rationale has been the perceived national security benefit of bringing settled agriculture to an under-populated region seen as vulnerable to incursion from more densely populated countries to Australia's north. Irrigated agriculture was integral to these northern visions, but reality again exposed a limited understanding of the specific climatic, agronomic and economic contexts of northern Australia, and has delivered little to repay their investments and expectations. The history of the Ord River irrigation schemes (Hassall & Associates, 1993; Cook, 2009) exemplifies the gap between these frontier visions and the reality.

Complexity in the water-in-agriculture system

The history of irrigated agriculture in Australia and the policies developed to deal with it are well documented (Crase, 2008; Musgrave, 2008; Ward, 2009). The MDB is the exemplar of this history (Connell & Grafton, 2011; Quiggin, Mallawaarachchi & Chambers, 2012) and highlights the complexity of managing water in an unpredictable climate in the face of multiple competing uses. The biophysical and socio-economic elements of the system are dynamic and interact in complex ways. The system also includes the institutional setting: arrangements introduced by governments and other agencies in order to achieve objectives as perceived and pursued at different times. Design flaws in these institutions have also generated costly unintended consequences. The following examples illustrate some of the complexity inherent in the water-for-agriculture system in Australia and the adaptation issues this has raised.

The riparian doctrine

Colonial administrations had little knowledge of the extent and behaviour of local water resources. In an unfamiliar climate with high seasonal and year-to-year variability of rainfall and stream flow the inherited British common law doctrine of riparianism, where rights to water accrued to landholders bordering the river, soon proved inadequate, not only to the needs of agriculture but also of urban development and industries such as mining (Musgrave, 2008).

An early example of institutional failure

The early experience of irrigation development, especially in Victoria, exemplifies institutional failure. Locally run water trusts suffered financial collapse and were taken over by the State. This led to the recognition that clarifying the rights to water required public administration (Powell, 1989; Tisdell et al., 2002).

Salinity in the southern MDB

The clearing of deep-rooted perennial native vegetation for dryland and irrigated farms in the upper catchments and riverine plains in northern Victoria and southern New South Wales allowed rainfall and irrigation water to infiltrate into saline soil strata and water tables. Over time, salt rose close to, and in places into contact with, the root zone of irrigated crops and pastures. This imposed severe constraints on the management of irrigated farms, and eventually led to the abandonment of some irrigation districts, at considerable cost to governments and socio-economic stress to irrigators and rural communities (Wilson, 2004). This is an example of a complex system generating 'surprise': the biophysical system generated little feedback until an unforeseen 'tipping point' was reached, threatening the viability of agricultural production (Walker, Abel, Anderies & Ryan, 2009).

Managing sub-systems; overlooking externalities

Complex systems have multiple outputs and users. In contrast, system managers typically pursue limited objectives for specific uses and sub-systems. As demand for water in its various uses in the MDB grew it became clear that existing allocation and governance mechanisms were inadequate. The Constitutional authority for water, being vested in the States, encouraged States to focus on developing water resources within their own borders. Consequently they paid little attention to downstream effects. For example in the late 1980s, anticipating a Basin-wide 'cap' on water used in agriculture, the Queensland Government allowed rapid irrigation development in the northern MDB, using the flows in south-flowing rivers. This led to reduced cross-border flows, including overland flood flows previously relied on by downstream landholders for the periodic rejuvenation of pasture, and reduced flows to ecosystems including protected wetlands (Thoms, 2003). Similar downstream effects have been the source of ongoing tension over water management between the State of South Australia at the mouth of the system and upstream States (Connell, 2007).

Path dependence and irreversibility

Irrigated agriculture based on large-scale storage and delivery infrastructure inevitably creates highly modified natural environments. These changes have often proved to be effectively irreversible. This makes it difficult to specify desired future environmental states, to negotiate the volume of water required to achieve environmental outcomes and to specify the steps needed to achieve them (Cummins & Watson, 2012). Key socio-economic elements of the system also display path dependence (see Chapter 1 in this volume) and resistance to reversal, as the difficulty of negotiating and implementing reduced diversions for irrigated agriculture in the Murray-Darling Basin Plan (Australian Government, 2016) have demonstrated (Marshall & Alexandra, 2016). The legacy impacts of past policy and water infrastructure decisions also create path dependence. Cummins and Watson (2012) highlighted the role of past policy shortcomings in limiting the effectiveness of recent more soundly based policy, and imposing constraints on adaptation. Fixed infrastructure can constrain adaptation, and make it costly to modify layouts. There is a continuing recourse to new infrastructure to shield Basin irrigators from more desirable adjustment. Reversing the impacts of ill-advised or obsolescent investment in fixed infrastructure is economically, socially and politically costly.

Under-pricing of water for agriculture

Consistent with nation-building objectives, the construction costs of storage and distribution infrastructure during the 'march of irrigation' were largely funded by governments. Davidson (1969) highlighted the dubious economic benefits of large public investments in dam building, given competing national development needs. As well, water charges to irrigators were slow to cover even maintenance and delivery costs. To promote economic development, water licences were distributed freely, leading to over-allocation relative to the capacity of the highly variable hydrological system (Williams, 2011). This inevitably led to over-use of the resource. The subsidy also encouraged irrigators to over-invest in on-farm development, increasing the difficulty of subsequent agricultural adjustment and reinforcing the strong psycho-social attachments farmers develop to the land on which they have lived and laboured. These factors raise the socio-economic costs of industry adaptation to market conditions.

Transferring water between uses and regions

Irrigation development was based on the vision of creating permanently settled rural communities. Water licences were originally attached to land titles, ruling out their transfer between farms and regions in response to changing economic and climatic conditions. Until recent institutional changes allowing water trading, this reduced the capacity of irrigation farms and districts to adapt by trading seasonal water allocations or permanent entitlements. While water trading now enables adaptation for both buyer and seller, it can also result in socio-economic stress in local communities reliant on input-supply or output-processing businesses. In the latter years of the Millennium drought many rice growers in the southern MDB sold water allocations downstream to irrigators of permanent plantings (tree and vine crops) at a price sufficient to compensate for their lack of an annual rice crop (National Water Commission, 2012, pp. 42–43), but this resulted in socio-economic stress on the rice processing sector in those communities.

The complexity of groundwater

The connection between groundwater and surface water can be highly complex and requires sophisticated methods to monitor quality and quantity. In one groundwater zone of the upper Namoi catchment in the northern MDB, irrigators adapted to reduced extraction limits by using the latest technology, sharing information and trading water (National Water Commission, 2011). The benefits of this co-operative management of the shared common-pool resources are at risk, however, from the potential exercise of market power by a single buyer, namely a nearby coal mine. In a nearby zone in the same catchment, groundwater irrigators have concerns over the potential impacts on groundwater of another proposed coal mine (Australian Broadcasting Corporation, 2015). Similar concerns are held by irrigators in a nearby region where intensive exploration for coal-seam gas is occurring (Australian Broadcasting Corporation, 2017). In these instances, an unregulated water market would not protect agriculture from the power of a dominant market player.

A summary of the lessons

These examples demonstrate some of the complexity and unforeseen consequences of past water-resource development policies (see also Chapter 11, this volume). The hydrology of the MDB is highly variable over time and space, and is subject to both cyclical and trend shifts associated with climate change (Letcher & Powell, 2008; Linton & Budds, 2014). Path dependence can effectively render ecosystem modification irreparable; thus, the conversion of land to irrigated agriculture is largely irreversible. Infrastructure investments have persistent legacy effects that severely constrain adaptation and future policy options, whilst multiple values and cross-border impacts require a governance structure that allows whole-of-system trade-offs to be negotiated. Policy errors also have legacy effects and can constrain necessary future adaptation, and the socio-economic environment is inherently volatile and subject to surprise, requiring resources, including people, to move. While social systems may be inherently more adaptable than ecological systems over time, once established, vested interests within governments and communities are capable of exerting direct pressure on policy institutions in order to resist changes that would be in the overall system's interest (Marshall & Alexandra, 2016).

A case for a resilience approach to water policy

These lessons from the past call for fundamental changes to the way the relationships between water and the socio-economic system of agriculture are imagined. The future is not a stable state and certainty is not possible. A resilience-based approach to natural resource management offers a means of addressing the uncertainty of the future by requiring consideration of the interacting social and ecological components as a complex system. Resilience-based natural resource management advocates ecosystem stewardship, where emphasis is placed on building the adaptive capacity of social and ecological systems to prevent transformations to undesirable states (Chapin, Kofinas & Folke, 2009) and the loss of ecosystem services (Biggs, Schluter & Schoon, 2015). A resilience approach can help re-imagine the behaviour of the water-in-agriculture system as a sustainable interaction between people and their environment, and offer guidance on system management (Parsons, Thoms & Flotemersch, 2017).

Resilience concepts and principles

The interrelationships between human societies and their natural environment can be viewed as social-ecological systems (SES). In a SES, people depend on the resources and services provided by ecosystems (e.g. agriculture, forestry, fisheries) and at the same time, ecosystem resources and services are influenced by human activities (Chapin et al., 2009). The type and balance of feedbacks between the social and ecological components of a SES determine the mechanisms and capacities available to respond and adapt to external shocks (Meadows, 2008). Ultimately, a sustainable SES needs to be able to supply ecosystem resources into the future against a backdrop of spatial and temporal variability arising from social and environmental sources (Chapin et al., 2009).

Resilience thinking is an approach to understanding and managing sustainability in complex social-ecological systems characterized by variability and unpredictability (Marshall, 2010). Resilience is often defined as the amount of disturbance a system can withstand while remaining in the same state with essentially the same function, structure and feedbacks (Walker & Salt, 2006). A resilient social-ecological system has the capacity to buffer or adapt to shocks or surprises. The human actors in a resilient social-ecological system must be able to adapt and transform to influence sustainability.

Resilience assumes that systems are not stable, but rather cycle through an adaptive loop with phases of accumulation, collapse and reorganisation. This change cycle presents opportunities for production, renewal and improvement of systems; however, past some threshold a system may cross into an alternative state with different structure, function and feedbacks (Gunderson & Holling, 2002). This alternative state is often undesirable in the sense of resulting in the reduced provision of resources. A resilient SES persists and maintains ecosystem structure and function and associated human use of ecosystem services (Biggs et al., 2015).

There are seven general principles that enhance resilience in a social-ecological system (Biggs et al., 2015). Attention to these principles in managing a SES allows the system to retain structure and function and supply ecosystem services, while being able to absorb shock and constantly change, adapt and renew. These characteristics are:

1   Maintain diversity and redundancy because diversity provides options for responding to change and disturbance;
2   Manage connectivity of the ways in which parts of a SES interact with each other;
3   Manage 'slow' variables and feedbacks because changes in controlling slow variables can cause a regime shift, often rapidly and unexpectedly;
4   Foster complex adaptive systems thinking because this relates to the inter-connectedness, non-linear change, uncertainty and surprise inherent in SES;
5   Encourage learning and foster the conditions for effective learning in SES;
6   Broaden participation and actively engage relevant stakeholders in management and governance process; and
7   Promote polycentric governance, with multiple interacting governing bodies (Biggs et al., 2015).

The application of these principles to managing ecosystem services assumes that decisions have already been made about which services are desirable in the SES and that the trade-offs involved in making those decisions have been resolved (Schoon et al., 2015).

The emergence of resilience in Australian water policy

Resilience is a sound umbrella concept for managing water-in-agriculture in Australia because it explicitly addresses the social-ecological system in which agriculture occurs, and the inevitability of surprise. After several decades of theoretical and empirical refinement, the concept of resilience is beginning to be more widely applied to natural resource management policy and programs (Chapin et al., 2009; Namoi Catchment Management Authority, 2013). However, a potentially major constraint to these innovations is that they are generally implemented within existing institutional structures and legislation. These structures emphasise a linear command and control paradigm, which may also divide social and ecological resilience into separate components with different departmental responsibility (Benson & Garmestani, 2011; Cork, 2010). Most existing governance arrangements are not conducive to the tools of resilience: adaptive management; adaptive monitoring; co-generation of knowledge; treatment of complexity; communication of uncertainty; and social-learning (Berkes & Folke, 1998; McLoughlin & Thoms, 2015; Parsons et al., 2017; Rogers et al., 2013). Substantial shifts in governance and legislation will be required to embed the resilience approach into water management practice, because resilience is not a blanket that can be overlain onto existing structures (Benson & Garmestani, 2011; Dovers & Hussey, 2013). Despite these challenges, some progress has occurred in Australia in designing policy that is based on concepts of water resources as complex social-ecological systems.

By the early 1990s it had become clear that there was a critical need for fundamental reform of water institutions and policy in Australia (Crase, 2008; McKay, 2008). The experience of the previous hundred years had been costly, and the system was failing to adapt to contemporary expectations. Water for agriculture had ceased to be the free gift of a munificent nature that could be stored, diverted and used for agriculture without regard to its environmental and social consequences. The complexity of the socio-ecological system in which agriculture and water interact had been brought into sharp relief. Water had become a scarce commodity, and fundamental policy reform was required to manage that scarcity.

In response, the policy innovations of the past two decades have attempted to take into account the growing understanding of the complexity and the multiple interests involved in the water-for-agriculture system. Key events include the Agreement on Water Resources (Water Reform Framework) by the Council of Australian Governments (CoAG, 1994), which recognized that addressing policy reform was the shared responsibility of all Australian governments (McKay, 2008). Policy reforms first focused on facilitating the movement of water between users and uses in response to changing economic conditions and environmental values (Crase, 2008). Ten years later CoAG (2004) announced the National Water Initiative (NWI), a set of principles that 'aims to achieve national compatibility in the markets, regulatory and planning schemes to achieve sustainable management of surface and groundwater' (McKay, 2008, p. 52). Under the NWI, governments made commitments whose full implementation aimed to achieve the economically efficient use of water and related investment that maximizes the economic, social and environmental value of Australia's water resources.

The articulation and implementation of the NWI principles required a comprehensive program of research, monitoring and reporting. The National Water Commission (NWC) was created in 2004 for that purpose, and served as the independent watchdog on the progress and performance of State and Territory jurisdictions in implementing the NWI. There were numerous references to the concept of resilience in the work of the NWC. Matthews (2010) identified several resilience challenges in water reform, particularly in relation to climate change, and noted that '[i]n effect, the need for resilience to deal with climate change has been driving Australian water reform for ten years already' (p. 10). The NWC's commitment to a resilience approach was also indicated by commissioning research that applied resilience thinking to water management issues (see Capon, Parsons & Thoms, 2009; Parsons et al., 2009). In its final triennial assessment, the Commission stated: 'At its heart, the NWI sets out the principles by which surface water and groundwater resources are to be shared to support resilient and viable communities, healthy ecosystems and economic development' (National Water Commission, 2014, p. 15).

The Australian Government wound up the NWC in 2014, raising doubts about how faithfully NWI objectives and principles would be adhered to in the future. In its final triennial assessment report, the Commission signalled its concerns in that regard, stating that '[g]overnments should not backtrack on water reform' and '[g]overnments should not "mark their own scorecards" on water reform' (National Water Commission, 2014, pp. 4–5).

Resilience concepts have also been applied to the performance of water governance institutions. Marshall and Alexandra (2016, p. 684) consider that:

> institutional path dependence is of particular concern to scholars interested in the resilience of social-ecological systems because it may constrain such systems from achieving the institutional adaptations required for their robustness in the face of unforeseen events (e.g. arising from climate change). Concerns have been raised that institutional path dependence may threaten the robustness of the MDB's social-ecological systems (e.g. Walker et al., 2009), with Loch et al. (2016, p. 3) observing that 'under increasing uncertainty surrounding future Australian (and global) climate contexts, it becomes important [in MDB water policy choices] to create flexible pathways that avoid lock-in actions or irreversible investment'. (p. 684)

A resilience approach also allows for future sources of stress and surprise and their possible impacts, and to build capacity to adapt to them. Shifts in weather patterns associated with climate change have already affected agricultural production and exports in the southern MDB (CSIRO, 2012), requiring adaptation of the irrigation sector. More generally, climate change will impact the behaviour of food-energy-water systems in all parts of Australia in ways that cannot be fully anticipated (Hughes, Steffen, Rice & Pearce, 2015). The future economic environment of Australian agriculture is likely to be no less volatile than in the past, given the uncertainties inherent in the world economy and in emerging Asian economies. Increased foreign involvement in the markets for Australian land and water by countries seen as future customers for food exports is another source of uncertainty. In the context of recent trade treaties such as the China-Australia Free Trade Agreement (Australian Government 2015b), Australian governments may be constrained in their ability to respond to emerging surprises in the natural and socio-economic environments. Turbulence in the mining industry is another source of risk that needs to be integrated into water planning (Fermio & Hamstead, 2012), given its evident competition with agriculture for both land and water. From a resilience perspective, the potentially irreversible impacts of mining on agricultural land and water, and the inherent power of mining interests in water markets, are matters that require substantial policy and legislative attention.

An opportunity for resilience thinking: Future water development in Northern Australia

A key test of how well the policy process has absorbed the lessons of the past and applies contemporary paradigms of resilience will be how it deals with what some see as the next frontier for the development of water resources for agriculture:

northern Australia. The Northern Australia Land and Water Taskforce (NALWT) produced comprehensive reviews of the state of the science and the prospects for further development of the region's resources (Northern Australia Land and Water Taskforce, 2009). These reviews stressed the limited opportunities for such development, given the knowledge gaps relating to the hydrological system, the changing climate, the lessons from the history of Australian agricultural and water development and the fate of similar past visions.

The report, entitled Sustainable Development in Northern Australia, noted that:

> The north is not a vacant land. It needs to be actively managed for resilience and sustainability, based on a contemporary and informed understanding of the complexities of the landscape and its people. Contrary to popular belief, water resources in the north are neither unlimited, nor wasted. Equally, the potential for northern Australia to become a 'food bowl' is not supported by evidence. (p. iii)

Six years later, however, in *Our North Our Future* (Australian Government, 2015a), a policy 'White Paper' on developing northern Australia, there were echoes of the familiar nation-building rhetoric. In its Foreword, the responsible Ministers asserted that '[w]e will fix the roads and telecommunications, build the dams and deliver the certainty that landholders and water users need' (Australian Government, 2015a, p. iv). The opening statement went on to trumpet the 'untapped promise, abundant resources' of the north and assert that 'a strong north means a strong nation' (p. 1). Those familiar with the adaptation problems that resulted from fixed water infrastructure in the south would be wary of the emphasis on dam building. Similarly, the promise to deliver certainty obscures the need to adapt to the inevitability of future surprises, a key focus of a resilience paradigm, and devalues scientists' warnings regarding the incomplete understanding of northern Australia's water resources (CSIRO, 2009; CSIRO, 2010). Indeed, the word 'certainty' appears 31 times in the document, while 'uncertainty' appears eleven times, and then only in the context of actions that purport to remove it.

Beneath its surface rhetoric, however, the White Paper takes a more guarded tone. For example, there is a relatively modest commitment to fund new water infrastructure. There is also a recognition of the incomplete knowledge base in its intention to fund further research on the capacity of northern water resources. This suggests that the feasible and likely scale of further water development in the north would be modest compared with that in the MDB, and does not represent a major source of economic growth or export opportunity in Northern Australia. There are also repeated references to the leading role that private sector investors will take in determining the extent and distribution of infrastructure, and assertions that business is 'far better placed to understand the risks and rewards from northern development' (Australian Government, 2015a, p. 3). This ignores the fact that while businesses make investment decisions based only on their (private) benefits and costs, the complexity of the water system, with its multiple uses and stakeholders, and externalities over time and space, requires a whole-of-system analysis of total community benefits and costs to ensure that the resulting investments are in the public interest. The stated government role to create successful business environments and to reduce 'regulatory risk' downplays the necessary complementary roles of regulation and comprehensive planning, which are two key planks of the NWI.

Northern Australia has seen several iterations of grand visions for its future, only to experience their failure to materialize. The region would be better served by programs and investments that acknowledge, indeed highlight, the complexity of its social, economic and ecological environments; in other words, programmes and investments built on a platform of resilience. The resilience-aligned focus of the NWI may have been too late to repair the damage of past mistakes in the south, but its core principles would, if appropriately revised (Connell, 2009), provide a good foundation for water management in northern Australia. Reimagining the development of water resources in Northern Australia as a social-ecological system acknowledges the feedbacks between people and their environment and the surprises and uncertainties in supply, use and condition of water resources that these feedbacks can generate.

Conclusion: What price visions?

Since European settlement, Australia's development visions have implicitly imagined the future as a stable state. These visions were imposed on a natural environment whose complexity was poorly understood, and in the context of institutions that often proved resistant to adaptation (Marshall & Alexandra, 2016). Farmers often have a similar tendency to form stable expectations of the future; in stressful times many call themselves optimists and await the return of 'normal' seasons and market conditions, probably knowing that such seasons are atypically favourable. Visions, without the sobering detail of the trials of the journey and the consequences of potential failure, might be a useful spur for personal development. Both classical literature and Australia's frontier mythology contain tales of heroic quests: explorers and settlers venturing into an unknown interior. This was also how collective nation-building enterprises were promoted in colonial times, when the territory and its hardships were mostly unknown, and personal doubts might be stilled by the conviction that the pioneer was contributing to a great national purpose. The risks and costs, however, were never borne only by the individual, but were spread across time and space to subsequent generations, governments and natural systems.

The frontier rhetoric of a settler society is now surely out-dated. It is irresponsible for modern governments, cognizant of the lessons of history and possessing more, but still incomplete, scientific knowledge, to promote further development by declaring that certainty in a stable water-for-agriculture system is possible. Rather, a resilience framework offers a way to imagine the future as a complex system of interacting natural, socio-economic and institutional elements in which surprise is inevitable, and for which adaptations to such surprise must be designed into the system.

Acknowledgements

We thank Graham Marshall (UNE) and Jason Alexandra (Alexandra and Associates) for helpful comments on the manuscript.

References

Australian Broadcasting Corporation. (2015) *At the coalface*, Landline 4 August 2015. Retrieved from http://www.abc.net.au/landline/content/2015/s4285159.htm

Australian Broadcasting Corporation. (2017) *Pilliga coal seam gas project at crossroads as NSW Government considers environmental impact*, ABC News, 6 April 2017. Retrieved from http://www.abc.net.au/news/2017-04-06/nsw-csg-project-sparks-fierce-debate-over-energy-future/8418102

Australian Government. (2015a). *Our north, our future: White Paper on developing northern Australia*. Canberra, Australia: Australian Government.

Australian Government. (2015b). *China-Australia free trade agreement*. Canberra: Department of Foreign Affairs and Trade.

Australian Government. (2016). *Water Act, 2007. Part 2, Management of basin water resources*. Retrieved from https://www.legislation.gov.au/Details/C2016C00469

Benson, M. H. & Garmestani, A. S. (2011). Can we manage for resilience? The integration of resilience thinking into natural resource management in the United States. *Environmental Management*, 48(3), 392–399.

Berkes, F. & Folke, C. (1998). Linking social and ecological systems for resilience and sustainability. In Berkes, F., Folke, C. & Colding, J. (Eds.), *Linking social and ecological systems: Management practices and social mechanisms for building resilience* (pp. 1–29). Cambridge, UK: Cambridge University Press.

Biggs, R., Schluter, M. & Schoon, M. L. (2015). *Principles for building resilience: Sustaining ecosystem services in social-ecological systems*. Cambridge, UK: Cambridge University Press.

Capon, T., Parsons, M. & Thoms, M. (2009). Floodplain ecosystems: Resilience, value of ecosystem services and principles for diverting water from floodplains. *Waterlines Report Series*, Number 22, October 2009. Canberra, Australia: National Water Commission.

Chapin, F. S., Kofinas, G. P. & Folke, C. (2009). *Principles of ecosystem stewardship: Resilience-based natural resource management in a changing world*. New York, New York: Springer.

Commonwealth Scientific and Industrial Research Organization (CSIRO). (2009). *Water in northern Australia. Summary of reports to the Australian government from the CSIRO Northern Australia Sustainable Yields Project*. Australia: CSIRO.

Commonwealth Scientific and Industrial Research Organization (CSIRO). (2010). *Northern Australia land and water science review. A report to the Northern Australia Land and Water Taskforce*. Australia: CSIRO.

Commonwealth Scientific and Industrial Research Organization (CSIRO). (2012). *Climate and water availability in south-eastern Australia: A synthesis of findings from Phase 2 of the South Eastern Australian Climate Initiative*. Australia: CSIRO.

Connell, D. (2007). *Water politics in the Murray-Darling Basin*. Annandale, NSW, Australia: Federation Press.

Connell, D. (2009). The case for a revised National Water Initiative for Northern Australia. *Northern Australia Land and Water Science Review*. (Ch. 23). Retrieved from http://northernaustralia.gov.au/sites/prod.office-northern-australia.gov.au/files/files/Chapter_23-Case_for_revised_NWI_for_NA.pdf

Connell, D. & Grafton, R. Q. (2011). *Basin futures: Water reform in the Murray-Darling Basin.* Canberra, Australia: ANU ePress.

Cook, G. (2009). Historical perspectives on land use development in northern Australia: With emphasis on the Northern Territory. In Australian Government (Ed), *Northern Australia Land and Water Science Review* (Ch. 6). Canberra, Australia: CSIRO.

Cork, S. (2010). Resilience of social-ecological systems. In Cork, S. (Ed), *Resilience and Transformation: Preparing Australia for an Uncertain Future* (pp. 131–142). Melbourne, Australia: CSIRO Publishing.

Council of Australian Governments (CoAG). (1994). Extracts from council of Australian governments: Water reform framework communiqué (Hobart 25 February, 1994). Canberra, Australia: Environment Australia, Marine and Water Division.

Council of Australian Governments (CoAG). (2004). *Intergovernmental Agreement on a National Water Initiative.* Retrieved from http://www.agriculture.gov.au/water/policy/nwi

Crase, L. (2008). An introduction to Australian water policy. In Crase, L. (Ed), *Water Policy in Australia: The Impact of Change and Uncertainty* (pp. 1–16). Washington, DC: Resources for the Future.

Cummins, T. & Watson, A. (2012). A hundred-year policy experiment: The Murray-Darling Basin in Australia. In Quiggin, J., Mallawaarachchi, T. and Chambers, S. (Eds), *Water Policy Reform: Lessons in Sustainability from the Murray-Darling Basin* (pp. 9–36). Cheltenham, UK: Edward Elgar.

Davidson, B. R. (1969). *Australia wet or dry: The physical and economic limits to the expansion of irrigation.* Melbourne, Australia: Melbourne University Press.

Davidson, B. R. (1981). *European farming in Australia: An economic history of Australian farming.* Amsterdam, The Netherlands: Elsevier.

Davidson, B. (1997). An historical perspective of agricultural land ownership in Australia. In Lees, J. (Ed), *A Legacy Under Threat? Family Farming in Australia* (pp. 15-47). Armidale, New South Wales: University of New England Press.

Dovers, S. & Hussey, K. (2013). *Environment and sustainability: A policy handbook.* Sydney, Australia: The Federation Press.

Fermio, S. & Hamstead, M. (2012). Integrating the mining sector into water planning and entitlements regimes. *Waterlines Report Series*, Number 77. Canberra, Australia: National Water Commission.

Gunderson, L. H. & Holling, C. S. (2002). *Panarchy: Understanding transformations in human and natural systems.* Washington, DC: Island Press.

Hassall & Associates. (1993). *The Ord River irrigation project, past, present and future: An economic evaluation.* Western Australia, Australia: Kimberley Water Resources Development Advisory Board.

Hillman, T. (2008). Ecological requirements: Creating a working river in the Murray-Darling Basin. In Crase, L. (Ed), *Water Policy in Australia: The Impact of Change and Uncertainty* (pp. 124–142). Washington, D.C: Resources for the Future.

Hughes, L., Steffen, W., Rice, M. & Pearce, A. (2015). *Feeding a hungry nation: Climate change, food and farming in Australia.* Australia: Climate Council of Australia.

Letcher, R. & Powell, S. (2008). The hydrological setting. In Crase, L. (Ed), *Water Policy in Australia: The Impact of Change and Uncertainty* (pp. 17–27). Washington, DC: Resources for the Future.

Linton, J. & Budds, J. (2014). The hydrosocial cycle: Defining and mobilizing a relational-dialectical approach to water. *Geoforum*, 57, 170–180.

Loch, A., Boxall, P. & Wheeler, S. A. (2016). Using proportional modelling to evaluate irrigator preferences for market-based water reallocation. *Agricultural Economics*, 47(4), 387–398.

Malcolm, B., Davidson, B. & Vandenberg, E. (2000). *The rural adjustment scheme: Its role, operation and effectiveness*. RIRDC Publication No. 99/19. Barton, Australian Capital Territory: Rural Industries Research and Development Corporation. Retrieved from https://rirdc.infoservices.com.au/downloads/99-019

Marshall, G. R. (2010). Governance for a surprising world. In Cork, S. (Ed), *Resilience and Transformation: Preparing Australia for Uncertain Futures*. Melbourne, Australia: CSIRO Publishing.

Marshall, G. R. & Alexandra, J. (2016). Institutional path dependence and environmental water recovery in Australia's Murray-Darling Basin. *Water Alternatives*, 9(3), 679–703.

Matthews, K. (2010 February). *Building resilience through water reform*. Presentation to Australia21 Conference, Canberra, Australia.

McKay, J. (2008). The legal frameworks of Australian water: Progression from common law rights to sustainable shares. In Crase, L. (Ed), *Water Policy in Australia: The Impact of Change and Uncertainty* (pp. 44–60). Washington, DC: Resources for the Future.

McLoughlin, C. A. & Thoms, M. C. (2015). Integrative learning for practicing adaptive resource management. *Ecology and Society*, 20(1), 34. Retrieved from http://dx.doi.org/10.5751/ES-07303-200134

Meadows, D. (2008). *Thinking in systems*. White River Junction, Vermont USA: Chelsea Green Publishing Company.

Musgrave, W. (2008). Historical development of water resources in Australia: Irrigation policy in the Murray-Darling Basin. In Crase, L. (Ed), *Water Policy in Australia: The Impact of Change and Uncertainty* (pp. 28–43). Washington DC: Resources for the Future.

Namoi Catchment Management Authority. (2013). *Namoi Catchment Action Plan 2010–2020, 2013 update*. Retrieved from http://archive.lls.nsw.gov.au/__data/assets/pdf_file/0005/496364/archive-namoi-catchment-action-plan-2010-2020-2013-update.pdf

National Water Commission (NWC). (2011). Water trading: An irrigator's perspective. The Warnock family – Narrabri, New South Wales. *Irrigator Case Study Series*. Canberra, Australia: National Water Commission.

National Water Commission (NWC). (2012). *Impacts of water trading in the southern Murray–Darling Basin between 2006–07 and 2010–11*. Canberra, Australia: National Water Commission.

National Water Commission (NWC). (2014). *Australia's water blueprint: National reform assessment 2014*. Canberra, Australia: National Water Commission.

Northern Australia Land and Water Taskforce (NALWT). (2009). *Sustainable development in Northern Australia. A report to government from the Northern Australia Land and Water Taskforce*. Canberra, Australia: Department of Infrastructure, Transport, Regional Development and Local Government.

Pahl-Wostl, C., Craps, M., Dewulf, A., Mostert, E., Tabara, D. & Taillieu. T. (2007). Social learning and water resource management. *Ecology and Society*, 12(2), 5. Retrieved at http://www.ecologyandsociety.org/vol12/iss2/art5/

Parsons, M., Thoms, M., Capon, T., Capon, S. & Reid, M. (2009). Resilience and thresholds in river ecosystems. *Waterlines Report Series*, Number 21, September 2009. Canberra, Australia: National Water Commission.

Parsons, M., Thoms, M. C. & Flotemersch, J. (2017). Eight river principles for navigating the science-policy interface. *Marine and Freshwater Research*, 68(3), 401–410.

Powell, J. M. (1989). *Watering the garden state: Water, land and community in Victoria.* Sydney, Australia: Allen & Unwin.

Quiggin, J., Mallawaarachchi, T. & Chambers, S. (2012). *Water policy reform: Lessons in sustainability from the Murray-Darling Basin.* Cheltenham, UK: Edward Elgar.

Rogers, K. H., Luton, R., Biggs, H., Biggs, R., Blignaut, S., Choles, A. G., Palmer, C. G. & Tangwe, P. (2013). Fostering complexity thinking in action research for change in social-ecological systems. *Ecology and Society*, 18(2), 31. Retrieved from http://dx.doi. org/10.5751/ES-05330-180231

Schoon, M. L., Robards, M. D., Brown, K., Engle, N., Meek, C. L. & Biggs, R. (2015). Politics and the resilience of ecosystem services. In Biggs, R., Schluter, M. & Schoon, M. L. (Eds), *Principles for Building Resilience: Sustaining Ecosystem Services in Social-Ecological Systems* (pp. 32–49). Cambridge, UK: Cambridge University Press.

Thoms, M. C. (2003). Floodplain-river ecosystems: Lateral connections and the implications of human interference. *Geomorphology*, 56(3), 335–349.

Tisdell, J., Ward, J. & Grudzinski, T. (2002). *The development of water reform in Australia.* Technical Report 02/5. Melbourne, Australia: Cooperative Research Centre for Catchment Hydrology.

Walker, B. & Salt, D. (2006). *Resilience thinking: Sustaining ecosystems and people in a changing world.* Washington DC: Island Press.

Walker, B. H., Abel, N., Anderies, J. M. & Ryan, P. (2009). Resilience, adaptability, and transformability in the Goulburn-Broken Catchment, Australia. *Ecology and Society*, 14(1), 12. Retrieved from http://www.ecologyandsociety.org/vol14/iss1/art12/

Ward, J. (2009). Palisades and pathways: Historical lessons from Australian water reform. In Australian Government (Ed), *Northern Australia Land and Water Science Review* (Ch. 14). Canberra, Australia: CSIRO.

Williams, J. (2003). Can we myth-proof Australia? *Australian Science*, 24(11), 40–42.

Williams, J. (2011). Understanding the Basin and its dynamics. In Connell, D. & Grafton, R. Q. (Eds), *Basin futures: Water reform in the Murray-Darling Basin* (pp. 1–38). Canberra, Australia: ANU ePress.

Wilson, S. M. (2004). *Dryland and urban salinity costs across the Murray-Darling Basin. An overview and guidelines for identifying and valuing the impacts.* Canberra, Australia: Murray-Darling Basin Commission.

10 Waterworks

Developing behaviourally effective policies to manage household water use

*Donald W. Hine, Lynette McLeod and
Aaron B. Driver*

Introduction

Water scarcity is becoming a significant problem in many countries and is linked to a complex mix of demographic, environmental, economic, and social issues. Vörösmarty, Green, Salisbury, and Lammers (2000) identified human population growth and economic development as the primary drivers of scarcity. However, recent modelling suggests that climate change will exacerbate the problem in many vulnerable regions, with a projected 40 per cent increase in the number of people worldwide who will face 'absolute water scarcity' – less than 500 cubic metres per year (Schiemeier, 2014). In this chapter, we argue that the effective water management will require policy responses that are informed by theory and methods from the behavioural sciences. Examples are drawn primarily from Australia and other Western countries, but the general principles of behaviour change introduced in the chapter generalize to other contexts, including developing nations.

Australia is the driest inhabited continent. Rainfall patterns are highly variable, and extended droughts are common. With global climate change, Australian weather events are projected to become even more extreme (Whetton, 2015). Each year, Australians use approximately 20,000GL of water (Australian Bureau of Statistics, 2014). Of this, close to 9 per cent is used by households, considerably less than the agricultural sector (65%), but more than the mining and manufacturing industries combined, which account for about 3 per cent each (Australian Bureau of Statistics, 2014).

From 2002, state and local governments regularly have introduced mandatory water restrictions on households in drought-affected parts of the country (Australian Bureau of Statistics, 2013). There also have been numerous education and behaviour change programs aimed at reducing domestic water use (Lehane, 2014; Syme, Nancarrow, & Seligman, 2000). For example, a national water efficiency labelling scheme was introduced in 2006, providing consumers with information about the expected water consumption of a range of products such as shower heads, dishwashers, and clothes washing machines. The scheme also banned the sale of unregistered products (Chong, Kazagalis, & Giurco, 2008). Incentive campaigns have been popular with local councils, which offer cash rebates to consumers to cover a proportion of the purchase price of water-saving products (Department of Environment, 1996). In Queensland, the Urban Water Security Research Alliance recently ran a program comparing the impact on water consumption of general education information, feedback where water was being used within a household, and normative feedback about other households' water use (Fielding et al., 2012).

These policies and programs have helped drive changes in water conservation practices, with substantial increases in the installation of water tanks, grey water systems, dual-flush toilets, and low-flow shower heads (Australian Bureau of Statistics, 2010). Many Australians also report taking steps to reduce household water use by using washing machines only when fully loaded, taking shorter showers, turning off taps when cleaning teeth and shaving, and using the half flush button on dual-flush toilets. Outside the home, many Australians report water-saving activities such as using mulch to retain moisture in gardens, only watering when necessary using a trigger hose, and car washing less often or not at all (Australian Bureau of Statistics, 2013). Despite these encouraging trends, Australia continues to have one of the highest rates of per capita water usage in the world (Lehane, 2014). Furthermore, with the easing of drought conditions and the removal of restrictions, water use is once again increasing in many parts of the country (Australian Bureau of Statistics, 2014).

This troubling lack of progress has prompted growing awareness that the successful delivery of water policy requires a more sophisticated understanding of the drivers of human behaviour, and how behaviour change is best accomplished (Halpern, Bates, Beales, & Heathfield, 2004). In a recent review of the literature, Michie and her colleagues (Michie, West, Campbell, Brown, & Gainforth, 2014) identified 83 theories relevant to behaviour change. In this section, we review five of these theories that are particularly relevant for household water conservation. We then introduce the Behaviour Change Wheel (Michie, Atkins, & West, 2014; Michie, van Stralen, & West, 2011), a tool for integrating these theories into a single, practical framework in order to: (1) identify and understand the causal factors that lead individuals to engage and fail to engage in water conservation practices in the home; (2) link these causal factors to specific behaviour change interventions and policies to reduce water consumption.

Behavioural theories

Behavioural theories describe factors that exert a causal influence on behaviour, as well as the nature of this influence. These models are useful for identifying the main motivational drivers of behaviour, and the internal and external barriers that sometimes prevent behaviour change. Many behavioural theories are based on the assumption that behaviour arises from a deliberate decision making process involving a systematic evaluation of potential costs and benefits associated with a range of behavioural options. These consequentialist models view conscious expectancies about future outcomes as the key driver of decision making.

The theory of planned behaviour

The theory of planned behaviour (TPB; Ajzen, 1991) is perhaps the most broadly applied consequentialist theory of human behaviour. According to TPB, the primary determinant of specific water conservation behaviours is an individual's conscious decision – or reasoned intention – to engage in one or more of these behaviours. In turn, TPB proposes that intentions are determined by three main psychological factors: (1) attitudes (the extent to which we feel positive or negative towards the behaviour), (2) subjective norms (the extent to which important others in our lives think that engaging in the behaviour is a good idea), and (3) perceived behavioural control (the extent to which we believe we can successfully engage in the behaviour). Thus, according to the theory, if people have positive feelings about water conservation, expect they will receive social approval for conserving water, and believe they have knowledge, skills, and resources to complete relevant behaviours, then they will be more likely to develop intentions to conserve water, and initiate action.

Importantly, TPB suggests that unless a person has adequate levels of perceived behavioural control, positive attitudes and normative pressure may not be enough to change behaviours. This helps to explain why attitudes and norms are inconsistent predictors of intentions and behaviour. It also highlights the need to identify internal (e.g., lack of knowledge about how to conserve water) and external (e.g., unavailability of water-saving technology) barriers to intentions and behaviours. A range of studies has shown one or more TPB variables to be important predictors of intentions and/or behaviours related to using less water and installing water-saving devices (Corral-Verdugo, Bechtel, & Fraijo-Sing, 2003; Kantola, Syme, & Campbell, 1982; Lam, 1999, 2006; Syme, Shao, Po, & Campbell, 2004; Trumbo & O'Keefe, 2005).

Focus theory of normative conduct

Social norms are the accepted standards of behaviour within social groups. Psychological research has shown that these norms can be a powerful force to either increase or decrease the probability of a broad range of behaviours relevant to environmental sustainability. Focus Theory of Normative Conduct (Cialdini, Reno, & Kallgren, 1990) differentiates between two kinds of social norms: (1) injunctive norms – behaviours that are perceived to be approved by other people – that is, beliefs about what ought to be done, and (2) descriptive norms – perceptions of how other people are actually behaving – that is, what is actually being done. Injunctive social norms reflect the moral rules and guidelines of the social group, and tend to motivate or constrain behaviours by promising social rewards or threatening sanctions. Descriptive social norms play an adaptive role in human behaviour, functioning as a kind of mental shortcut for guiding behaviour when individuals are unsure of how to act in social situations (Jackson, 2005).

The practical implications of Cialdini's theory, with respect to water conservation, become apparent when campaigns try to persuade an audience to behave in a desired manner. In a study investigating hotel towel reuse, Goldstein, Cialdini, and Griskevicius (2008) found that persuasive messages containing descriptive norms were more effective in eliciting this water conservation behaviour in guests than the standard environmental protection message. In addition, they discovered that normative appeals were most effective when they described the behaviour of similar others (i.e., other hotel guests who had stayed in the same room).

Another study by Aronson and O'Leary (1982–1983), conducted in public showers, demonstrated the strong influence of peer behaviour. They employed a confederate to model water conservation by turning off taps while soaping, a behaviour requested by prominent, nearby signage. The presence of the water-conserving confederate elicited 49 per cent compliance for the desired behaviour, compared to only 6 per cent compliance in the control condition for which a sign was present requesting that shower users conserve water. Adding a second confederate increased compliance to 67 per cent. For better or worse, humans are very similar to herd animals. As the number of people engaging in a behaviour increases, the probability that others will follow also increases.

Another important finding from the social norms literature is that to maximize effectiveness, descriptive and injunctive normative messages must be aligned, thus prompting behaviour in the same direction (Cialdini et al., 2006). For example, interventions can fail if an injunctive normative message ('people should take shorter showers to conserve water') is undercut by a descriptive normative message indicating that most people are doing the opposite ('most people are enjoying longer showers').

Affect heuristic

Many behavioural models assume that behaviour is the result of conscious reflection, often involving the evaluation of costs and benefits. In practice, however, many human behaviours stem from a combination of controlled cognitive processes and automatic responses driven by emotion and/or habit. Research has shown that affect can play an important role in guiding judgments and decisions (Bhullar et al., 2014; Finucane, Alhakami, Slovic, & Johnson, 2000; Loewenstein, Weber, Hsee, & Welch, 2001; Mellers & Schwartz, 1997; Peters & Slovic, 2000; Shiv & Fedorikhin, 1999). People are not only influenced by what they think about a situation, but also by how they feel about it.

The affect heuristic is based on the premise that human decision processes are guided by two distinct information processing systems: (1) an analytic system that is intentional, effortful, and logic-based, and (2) an experiential system that is passive, effortless, rapid, and closely tied to intuition and affect. The analytic system is under the conscious control of the individuals, whereas the experiential system operates automatically with conscious input from the decision maker. According to Finucane et al. (2000), thoughts and images stored in memory are tagged with affective markers that vary in terms of valance and strength. Stimuli in the decision context activate relevant thoughts and images, which in turn spread activation to their associated affective markers. The activated markers combine to generate an 'affect pool' – in this context, a general feeling of goodness or badness about a water conservation policy or activity. In turn, this feeling guides subsequent judgments and decisions. The model predicts that positive affective responses will lead people to perceive more benefits and fewer costs associated with water conservation, whereas negative affective responses will lead them to perceive fewer benefits and more costs.

An important implication of this model is that when affect is the primary driver of cost-benefit judgments, perceptions of costs and benefits will be inversely correlated with each other. That is, individuals who have strong negative affective associations with water conservation will perceive conservation as more costly and less beneficial. Conversely, individuals who have strong positive affective associations with water conservation will view conservation behaviours as more beneficial and less costly. This indicates that so-called 'rational cost-benefit assessments' are often predetermined by our initial emotional responses. The model also highlights potential opportunities for using emotion-based appeals to encourage household water conservation. Persuasive messages need not always appeal to reason to be effective.

Theory of interpersonal behaviour

Similar to the affect heuristic model, Triandis's theory of interpersonal behaviour (TIB) asserts that behaviour is determined by both automatic and controlled cognitive processes (Triandis, 1977). Like TPB, TIB proposes that we often consciously deliberate and develop intentions to engage or not engage in specific behaviours, and that these intentions are influenced by attitudes, norms, and other facilitating conditions – factors in the external environment that make it easier or harder to engage in water conservation practices. However, TIB also recognizes that not all behaviours are driven by conscious consequentialist decision making. It notes that some behaviours are driven primarily by habit, established patterns of past behaviours. This is particularly relevant to highly repetitive behaviours, like showering, which are generally done in essentially the same manner day after day, month after month. Research suggests that habits act as an important boundary condition. When habit is strong the attitude-intention-behaviour relation is weak, because an individual's 'habitual mind-set' makes them less attentive to new information and courses of action. Strong habits undermine people's best intentions to change by reinforcing short-term rewards rather than long-term benefits. But when habit is weak (low frequency of repetition, not well learned, unstable context, some awareness and/or control) conscious decision making becomes more prominent, and attitudes and intentions stronger predictors of behaviour (Aarts, Verplanken, & van Knippenberg, 1998; Verplanken & Aarts, 1999; Verplanken, Aarts, van Knippenberg, & Moonen, 1998). This highlights the importance of understanding the nature of the behaviour that one wants to change. Habitual behaviours like showering will require very different types of interventions than non-habitual behaviours such as purchasing a water-efficient washing machine.

The behaviour change wheel

Many behaviour interventions are based on the ISLAGIATT principle – 'it seemed like a good idea at the time' (Michie, Atkins, & West, 2014). In a similar vein, Martin and Verbeek (2006, p. 5) note that many policy frameworks to support sustainability are 'irrational, poorly designed, and inefficiently administered', particularly when compared to similar frameworks that have been implemented to support wealth generation. Strategies and policies related to public goods are often developed without first systematically assessing what behaviours to target, the main drivers and barriers for these behaviours, and the specific behaviour change techniques and policies that maximize the chances of success.

To address this general *ad hoc* approach, social scientists have developed a range of frameworks that provide practitioners with step-by-step guides for developing, delivering, and evaluating behaviour change interventions. In their recent review of the literature, Michie et al. (2011) identified nineteen such frameworks. However, they noted that most failed to make explicit connections between the underlying causes of behaviour, behaviour change intervention tools, and public policy, often leaving practitioners unclear about which intervention and policy tools are most appropriate for specific contexts and populations.

In response to this problem, Michie et al. (2011; Michie, Atkins, & West, 2014) developed the *Behaviour Change Wheel* (BCW) that links the behavioural factors to interventions and policy (see Figure 10.1). The BCW enables policy makers to understand the mechanisms underlying problematic behaviours, such

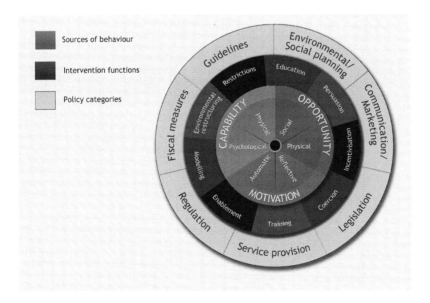

Figure 10.1 The Behaviour Change Wheel, an integrated framework for guiding behavior change interventions.

Reproduced from Michie et al. (2011), by permission of BioMed Central.

as excessive water consumption, and select appropriate interventions and policies to invoke behaviour change. In this section of the chapter, we describe the main elements of behaviour change projects, and how the BCW can help improve the quality of interventions and policies related to domestic water use.

Selecting the 'right behaviours' to target

Many attempts to increase environmentally sustainable behaviours disappoint because they target the wrong behaviours. In an influential paper, Gardner and Stern (2008) produced a short-list that ranked technology upgrades and behavioural changes based on their effectiveness in reducing household energy use. The short list provided households with a practical guide for prioritising behaviours to maximize energy savings.

Following this tradition, Inskeep and Attari (2014) extended the short-list concept to household water conservation. Using US data, they demonstrated that installing a water efficient toilet saved substantially more water (reducing indoor water use by 18.6%) than flushing 25 per cent less frequently (7.3% reduction). Reducing showers from 8.2 minutes to 5 minutes was much more effective in reducing water use (8.2% reduction) than installing a low-flow showerhead (1.9% reduction). In terms of outdoor water use, lawn and plant watering using water collected from a rain harvesting system (i.e., a water tank that catches run-off from the roof) was much more water efficient (up to 100% reduction in outdoor water use) than watering lawns with a hose (33% reduction) or using a programmable irrigation system (30% reduction).

Taking impact into account when choosing which behaviours to target is obviously important. But impact is not the only criterion worth considering. In his book on Community Based Social Marketing, McKenzie-Mohr (2011) proposes a simple framework for prioritising behaviour based on: (1) the impact of the behaviour on tangible ecological and economic outcomes, similar to the short-list approach outlined above, (2) the probability that the behaviour will actually be adopted, and (3) the proportion of the target population already engaged in the behaviour (penetration). In most instances, interventions should aim to influence a small number of high-impact behaviours that have a high probability of being adopted, and currently have low penetration rates within the target community. In this context, practitioners and policy makers should avoid spending time, energy, and money promoting activities that will have little impact on overall water use. Nor should they allocate resources to encouraging behaviours that are unlikely to be adopted or that most people are already performing.

COM-B system for understanding behaviour

The COM-B (Capability, Opportunity, Motivation – Behaviour) system is an overarching, integrative model of behaviour that lies at the hub of the BCW (Michie et al., 2011; Michie, Atkins, & West, 2014). COM-B can help water policy makers and behaviour change practitioners understand behaviour in context by identifying the main causes of problematic behaviours, such as the failure to adopt water-saving practices. COM-B also helps identify what exactly needs to change to increase the probability that desirable behaviours will occur. According to COM-B, behaviour is determined by three main factors:

(1) Capability – an individual's capacity to engage in the behaviour of interest. COM-B distinguishes between two types of capability. *Physical capability* refers to the extent to which an individual can engage in the behaviour. For example, does the householder have the financial resources, equipment, and/or physical ability to install a low-flow showerhead or rainwater tank? *Psychological capability* refers to the capacity to engage in the necessary mental activities (risk assessments, mental simulation of possible outcomes, decision making etc.) to select appropriate options and actions. Installing a water tank may seem like a straightforward exercise requiring few cognitive demands. However, there are many different types to choose from (e.g., polyethylene, concrete, steel, bladder, above ground, below ground, etc.), all with various strengths and limitations. In addition, most states and municipalities have guidelines about how and where tanks must be installed, which can further complicate the process. Depending on the accessibility and complexity of relevant information, psychological capability can be easily stretched, even for tasks that initially appear to be quite easy. Of course, health and mental health issues such as dementia and depression can also have negative impacts on psychological capacity to successfully engage in behaviour change across a range of domains.
(2) Opportunity – factors external to the individual that prompt or enable the behaviour to occur. COM-B distinguishes between two types of opportunity. *Physical opportunity* refers to situational factors such as having relevant water-saving products and/or installation equipment readily available. It is difficult to install a water tank if the type best suited to climate and personal circumstances cannot be purchased locally. *Social opportunity* refers to cultural or community values and norms that may make engaging in recommended best practices more or less likely. For example, if most households within a community are complying with water restrictions, this creates a descriptive social norm that increases the likelihood that others in the region will also engage in this practice.
(3) Motivation – factors internal to the individual that energise or direct behaviour. There are two main types of motivating factors: reflective and automatic. *Reflective motivation* consists of conscious deliberation and reasoning, and often involves evaluating threats, planning, goal setting, and mentally simulating possible outcomes associated with various types of actions. For example, prior to purchasing a water-efficient appliance, an individual may make a list of the costs and benefits of purchasing or not purchasing the appliance, and select the option that he or she believes is most likely to produce the most positive outcome. *Automatic motivation* refers to mental processes that operate largely outside conscious control

of the individual, including habits, impulses, and emotionally driven behaviour. For example, an individual may initially take long, hot showers because of the pleasure and emotional satisfaction this behaviour affords. However, over time this behaviour may become automatised, and become driven primarily by habit.

According to the COM-B model, capability, opportunity, and motivation both influence and are influenced by behaviour. For example, individuals who perceive many benefits and few costs associated with installing a water-saving appliance or backyard water storage tank (high motivation), have the relevant knowledge and skills to conduct the installation (high capability), and live in communities where these tanks and appliances are readily available and commonly used (high opportunity), are more likely to purchase and install them. In turn, engaging in water-saving behaviours can have a reciprocal reinforcing effect, increasing capability, opportunity, and motivation. Successfully installing a water-efficient showerhead can build competence and self-efficacy, increasing the likelihood that other water-saving projects will be attempted. Purchasing water-saving appliances can help build local markets for these products, increasing purchasing opportunities for future like-minded customers. And, of course, using less water produces a financial benefit in the form of reduced water bills, an important motivator that can encourage further savings.

Although money is an important driver of behaviour change, it is not the only driver; other potent motivators are unrelated to financial outcomes. For example, a recent study by Taufik, Boderdijk, and Steg (2015) found that simply engaging in environmentally friendly behaviours can be psychologically rewarding – eliciting positive feelings and a literal 'warm glow' in the form of reliable increases in perceived temperature. And in the context of developing effective sustainability strategies, Martin and Verbeek (2006) highlight the potential benefits of other non-financial incentives such as public recognition, social rewards, and opportunity.

It is also worth noting that the pattern of drivers and barriers influencing a given behaviour may vary somewhat across individuals within a target community. Not everyone views water conservation in the same way. Distinct segments of the community may have very different driver barrier profiles, reflecting their values, beliefs, and current behaviours. Thus, a policy maker or behaviour change practitioner may not be dealing with a single target community, but rather several. The number and nature of these communities should be understood prior to designing and implementing relevant policies and interventions (Hine et al., 2013; Hine et al., 2014; Slater, 1996).

Linking behavioural theories to COM-B

Importantly, all of the individual components of the behavioural theories reviewed in the first part of this chapter can be classified into the COM-B system (see Table 10.1). From an applied perspective, we consider this to be an extremely important advance.

Michie, West, et al. (2014) compiled a compendium of 83 theories of behaviour and behaviour change, a number substantial enough to overwhelm even the most competent of policy makers. The COM-B system provides a straightforward approach for integrating a highly disparate behavioural science literature into a single manageable framework that will enable practitioners to identify behavioural drivers and barriers that are most relevant for the water usage problem they wish to solve. The COM-B system, as part of the behaviour change wheel, also enables practitioners to explicitly link drivers and barriers to specific behaviour change strategies, our next topic.

Table 10.1 Translating behavioural theories into COM-B system

COM-B Categories	Definition	Model Factors
CAPABILITY – Physical	Capacity to physically engage in the behaviour.	TPB - Perceived behavioural control TIB – Self-efficacy
CAPABILITY – Psychological	Capacity to engage in the thought processes (comprehension, reasoning, etc.) that underlie the behaviour.	TPB – Perceived behavioural control TIB – Self-efficacy
OPPORTUNITY – Physical	Features of the physical environment prompting or making possible a behaviour.	TIB – Facilitating conditions
OPPORTUNITY – Social	Features of the socio-cultural environment prompting or making possible a behaviour.	TIB – Facilitating condition
MOTIVATION – Reflective	Conscious brain processes that energise and guide the behaviour (e.g., evaluations and plans).	TPB – Attitudes; Normative beliefs; Subjective Norms; Outcome beliefs; Outcome evaluations; Motivation to comply; Intention FTNC – Injunctive Norms; Descriptive Norms TIB – Behavioural intention; Social normative beliefs; Personal normative beliefs; Perceived consequences; Role beliefs; Normative beliefs; Personal norms; Professional norms
MOTIVATION – Automatic	Automatic brain processes that energise and guide behaviour (e.g., emotions, impulses, etc.).	TPB – Subjective Norms; Normative beliefs FTNC – Injunctive Norms; Descriptive Norms TIB – Affect; Habit; Affective attitudinal beliefs AH – Affect heuristic

Note: TPB (Theory of Planned Behaviour), FTNC (Focus theory of Normative Conduct), TIB (Theory of Interpersonal Behaviour), AH (Affect Heuristic)

Linking drivers and barriers to interventions

The middle ring of the BCW consists of nine intervention functions for changing behaviour: education, persuasion, training, incentivisation, restriction, environmental restructuring, modelling, and enablement. Definitions and examples related to domestic water use are provided in Table 10.2. An important strength

Table 10.2 Definitions and examples of interventions

Intervention Functions	Definitions	Domestic Water Use Examples
Education	Increasing knowledge and understanding	Providing written factsheets, technical manuals and videos, or practical courses to disseminate information and demonstrate household water conservation practices.
Persuasion	Using communication to induce positive or negative feelings or stimulate action	Providing information about descriptive norms (what people are doing) and injunctive norms (what people should be doing) to encourage people to engage in water conservation behaviours.
Incentivisation	Creating expectation of reward	Providing rebates for purchasing water storage tanks and water efficient appliances.
Coercion	Creating expectation of punishment or cost	Introducing legislation makes water conservation practices (e.g., restrictions on watering lawns and gardens) mandatory, with fines for non-compliance.
Training	Imparting skills	Running courses to train homeowners to install water-efficient shower heads, and create and maintain drought resistant gardens.
Restriction	Using rules to influence the engagement in the target behaviour	Introducing rules about days and times when outside watering can occur.
Environmental restructuring	Changing the physical or social context to encourage desired behaviours	Passing legislation that bans the sale of plumbing products and white goods if they do not meet the standards of the water efficiency labelling scheme.
Modelling	Providing an example for people to aspire to or imitate	Setting up a 'demonstration site' on a local property to display best-practice water conservation methods.
Enablement	Increasing means/ reducing barriers to increase capability or opportunity	Developing new technologies such as more water-efficient appliances. Providing householders with smart meters that provide users with online feedback of current and daily water use, with comparisons to neighbourhood or town/city norms.

Source: Based on Michie et al. (2011)

of the BCW is that it links identified causes of behaviour (from COM-B analysis) to specific intervention types (in the middle ring).

To succeed, practitioners must be aware of the wide range of behaviour change interventions available to them, and understand that different interventions may be required depending on the specific drivers and barriers in a given context. For example, in a large study of Australian households, Dolnicar and Hurlimann (2010) found a number of important barriers to adopting water-conserving appliances such as front-loading washing machines and low-flow showerheads: high purchase costs (which undermine physical capability), the perceived impracticality (the perceived added burden associated with using these appliances), and lack of knowledge about how much water the appliance would actually save (two factors that undermine reflective motivation). Given this set of barriers the BCW suggests an optimal intervention could include enablement (e.g., providing rebates on water-saving appliances to enhance physical capability), and education and persuasion (e.g., providing information that emphasizes the appliances in question are effective and easy to use).

The BCW does not provide specific guidance about how to design behaviour change interventions, but it is a systematic, empirically grounded approach for identifying the general types of interventions that should be most effective for specific behaviours in specific contexts. By understanding the mechanisms that drive and prevent target behaviours, practitioners will be in a much stronger position to develop appropriate strategies. Table 10.3 summarises how the components of the COM-B model link to the nine intervention functions.

Linking interventions to policy

Australia's water resources are managed through a co-operative water reform framework implementing reforms through a variety of instruments, often with perverse results (Martin & Williams, 2014), further reinforcing the need for a greater focus on behaviourally effective strategies.

The outermost ring of the BCW focuses on policies – plans of action and strategies to help governments and organisations to achieve their goals. The BCW distinguishes between seven policy types: communication/marketing, guidelines, legislation, regulation, fiscal, environmental/social planning, and service provision.

Just as the BCW links behavioural causes to intervention types, it also bridges the gap between interventions and policy tools in this outermost ring. The BCW provides a common framework for practitioners and policy makers to jointly identify which policy tools are most likely to benefit behaviour change initiatives. For example, interventions aimed at persuading households about the benefits of installing water tanks or other conservation technologies would be best supported by a policy mix involving one or more of the following: communication/marketing, guidelines, regulation, legislation, and service provision. On the other hand, interventions aimed at increasing household water conservation behaviours through modelling would be best supported by policies related to

Table 10.3 Links between COM-B components and intervention functions

COM-B	Education	Persuasion	Incentives	Coercion	Training	Restriction	Environmental Restructuring	Modelling	Enablement
Capability									
Physical	✓				✓				✓
Psychological					✓				✓
Opportunity									
Physical						✓	✓		✓
Social						✓	✓		✓
Motivation									
Reflective	✓	✓	✓	✓					
Automatic		✓	✓	✓			✓	✓	✓

Source: Based on Michie et al. (2011)

Table 10.4 Links between intervention functions and policy tools

Intervention Function

Policy Tool	Education	Persuasion	Incentives	Coercion	Training	Restriction	Environmental Restructuring	Modelling	Enablement
Communication/Marketing	☑	☑	☑	☑				☑	
Guidelines	☑	☑	☑	☑	☑	☑	☑		☑
Fiscal			☑	☑	☑		☑		☑
Regulation	☑	☑	☑	☑	☑	☑	☑		☑
Legislation	☑	☑	☑	☑	☑	☑	☑		☑
Environmental/Social Planning									
Service Provision	☑	☑	☑	☑	☑			☑	☑

Source: Based on Michie et al. (2011)

communication/marketing and service provision. The types of policy tools that are best matched to intervention types are summarised in Table 10.4.

Again it is worth highlighting that the BCW was not designed to provide advice about how to construct or what to include in a specific policy or intervention related to behaviour change. Rather, the BCW provides a general framework for understanding the main drivers and barriers for a given behaviour, and then, based on that behavioural analysis, identifying which general types of behaviour change interventions and policy delivery systems are best suited to address the behavioural problem under investigation. For practitioners who are interested in step-by-step guides about how the BCW and related behaviour change frameworks can be applied in field settings, several excellent resources are available (McKenzie-Mohr, 2011; Michie, Atkins, & West, 2014).

Conclusions

Water scarcity is a growing problem in many countries across the world. Pressures associated with global population growth, economic development, and climate change are projected to make matters worse. The social sciences have produced a large number of behavioural theories relevant to managing domestic water use, and important to overall water consumption. These theories can help policy makers and practitioners understand the main causes of problematic water use behaviours, and identify the most appropriate intervention strategies for changing behaviours. Michie et al.'s (2011; Michie, Atkins, & West, 2014) BCW provides an integrated framework for understanding behavioural causes in context, and explicitly linking these causes to interventions and policy tools. The framework provides a common language and mental model for policy makers and practitioners to conceptualise and communicate about behaviour change. It provides a sound foundation for launching a systematic program of analysis and action to address the behaviours that lie at the heart of many water conservation problems.

References

Aarts, H., Verplanken, B., & van Knippenberg, A. (1998). Predicting behavior from actions in the past: Repeated decision making or a matter of habit? *Journal of Applied Social Psychology, 28*(15), 1355–1374. doi:10.1111/j.1559-1816.1998.tb01681.x

Ajzen, I. (1991). The theory of planned behavior. *Organizational Behavior and Human Decision Processes, 50*(2), 179–211. doi:10.1016/0749-5978(91)90020-T

Aronson, E., & O'Leary, M. (1982–83). The relative effectiveness of models and prompts on energy conservation: A field experiment in a shower room. *Journal of Environmental Systems, 12*(3), 219–224.

Australian Bureau of Statistics. (2010). Australia's environment: Issues and trends, January 2010. Retrieved from http://www.abs.gov.au/AUSSTATS/abs@.nsf/Lookup/4613.0Chapter75Jan+2010

Australian Bureau of Statistics. (2013). Environmental issues: Water use and conservation, March 2013. Retrieved from http://www.abs.gov.au/ausstats/abs@.nsf/Lookup/4602.0.55.003main+features3Mar 2013

Australian Bureau of Statistics. (2014). Water account, Australia, 2012–2013. Retrieved from http://www.abs.gov.au/ausstats/abs@.nsf/Latestproducts/4610.0Main Features2 2012–13?opendocument&tabname=Summary&prodno=4610.0&issue=2012–13&num=&view=

Bhullar, N., Hine, D. W., Marks, A., Davies, C., Scott, J. G., & Phillips, W. (2014). The affect heuristic and public support for three types of wood mitigation strategies. *Air Quality, Atmosphere and Health, 7*(3), 347–356. doi:10.1007/s11869-014-0243-1

Chong, J., Kazagalis, A., & Giurco, D. (2008). *Cost effectiveness analysis of WELS: Final report.* Retrieved from Sydney, NSW: http://www.waterrating.gov.au/sitecollectionimages/publications/2012/10/110/cost-effectiveness-wels.pdf

Cialdini, R. B., Demaine, L. J., Sagarin, B. J., Barrett, D. W., Rhoads, K., & Winter, P. L. (2006). Managing social norms for persuasive impact. *Social Influence, 1*(1), 3–15. doi:10.1080/15534510500181459

Cialdini, R. B., Reno, R. R., & Kallgren, C. A. (1990). A focus theory of normative conduct: Recycling the concept of norms to reduce littering in public places. *Journal of personality and social psychology, 58*(6), 1015.

Corral-Verdugo, V., Bechtel, R. B., & Fraijo-Sing, B. (2003). Environmental beliefs and water conservation: An empirical study. *Journal of Environmental Psychology, 23*(3), 247–257. doi:10.1016/S0272–4944(02)00086–5

Deni Greene Consulting Services & Australian Consumers' Association & National Key Centre for Design (Australia) & Australia. Dept. of the Environment, Sport and Territories (1996). *More with less: Initiatives to promote sustainable consumption.* Canberra, A.C.T: Dept. of the Environment, Sport and Territories.

Dolnicar, S., & Hurlimann, A. (2010). Australians' water conservation behaviours and attitudes. *Australian Journal of Water Resources, 14*(1), 43–53.

Fielding, K. S., Spinks, A., Russell, S., Mankad, A., McCrea, R., & Gardner, J. (2012). *Water demand management: Interventions to reduce household water use.* Retrieved from http://www.urbanwateralliance.org.au/publications/UWSRA-tr94.pdf

Finucane, M. L., Alhakami, A., Slovic, P., & Johnson, S. M. (2000). The affect heuristic in judgments of risks and benefits. *Journal of Behavioral Decision Making, 13*(1), 1–17. doi:10.1002/(SICI)1099–0771(200001/03)13:1<1::AID-BDM333>3.0.CO;2-S

Gardner, G. T., & Stern, P. C. (2008). The short list: The most effective actions US households can take to curb climate change. *Environment: Science and policy for sustainable development, 50*(5), 12–25.

Goldstein, N. J., Cialdini, R. B., & Griskevicius, V. (2008). A room with a viewpoint: Using social norms to motivate environmental conservation in hotels. *Journal of Consumer Research, 35*(3), 472–482.

Halpern, D., Bates, C., Beales, G., & Heathfield, A. (2004). *Personal responsibility and changing behaviour: The state of knowledge and its implications for public policy.* Retrieved from London, UK: http://217.35.77.12/archive/england/papers/welfare/pdfs/pr.pdf

Hine, D. W., Phillips, W. J., Reser, J. P., Cooksey, R., Marks, A. D. G., Nunn, P., & Ellul, M. (2013). *Enhancing climate change communication: Strategies for profiling and targeting Australian interpretive communities.* Gold Coast, QLD: National Climate Change Adaptation Research Facility.

Hine, D. W., Reser, J. P., Morrison, M., Phillips, W. J., Nunn, P., & Cooksey, R. (2014). Audience segmentation and climate change communication: Conceptual and methodological considerations. *Wiley Interdisciplinary Reviews-Climate Change, 5*(4), 441–459. doi:10.1002/wcc.279

Inskeep, B. D., & Attari, S. Z. (2014). The water short list: The most effective actions US households can take to curb water use. *Environment: Science and policy for sustainable development, 56*(4), 4–15.

Jackson, T. (2005). *Motivating sustainable consumption: A review of evidence on consumer behaviour and behavioural change.* Retrieved from London: http://www.sustainablelife-styles.ac.uk/sites/default/files/motivating_sc_final.pdf

Kantola, S. J., Syme, G. J., & Campbell, N. A. (1982). The role of individual differences and external variables in a test of the sufficiency of Fishbein's model to explain behavioral intentions to conserve water. *Journal of Applied Social Psychology, 12*(1), 70–83. doi:10.1111/j.1559-1816.1982.tb00850.x

Lam, S.-P. (1999). Predicting intentions to conserve water from the Theory of Planned Behavior, perceived moral obligation, and perceived water right. *Journal of Applied Social Psychology, 29*(5), 1058–1071. doi:10.1111/j.1559-1816.1999.tb00140.x

Lam, S.-P. (2006). Predicting intention to save water: Theory of Planned Behavior, response efficacy, vulnerability, and perceived efficiency of alternative solutions. *Journal of Applied Social Psychology, 36*(11), 2803–2824. doi:10.1111/j.0021-9029.2006.00129.x

Lehane, S. (2014). *Australia's water security Part 2: Water use.* Retrieved from Dalkeith, WA: http://www.futuredirections.org.au/publication/australia-s-water-security-part-2-water-use

Loewenstein, G. F., Weber, E. U., Hsee, C. K., & Welch, N. (2001). Risk as feelings. *Psychological Bulletin, 127*(2), 267–286. doi:10.1037/0033-2909.127.2.267

Martin, P., & Verbeek, M. (2006). *Sustainability strategy.* Sydney, NSW: Freedom Press.

Martin, P., & Williams, J. (2014). Science hubris and insufficient legal safeguards. *Environmental Planning and Law Journal, 31*(4), 311–326.

McKenzie-Mohr, D. (2011). *Fostering sustainable behavior: An introduction to community based social marketing.* Gabriola Island, Canada: New Society.

Mellers, B. A., & Schwartz, A. (1997). Decision affect theory: Emotional reactions to the outcomes of risky options. *Psychological Science, 8*(6), 423–429.

Michie, S., van Stralen, M. M., & West, R. (2011). The behaviour change wheel: A new method for characterising and designing behaviour change interventions. *Implementation Science, 6*(1), 1–11.

Michie, S., Atkins, L., & West, R. (2014). *The behaviour change wheel: A guide to designing interventions.* Great Britain: Silverback.

Michie, S., West, R., Campbell, R., Brown, J., & Gainforth, H. (2014). *ABC of behaviour change theories.* Great Britain: Silverback.

Peters, E., & Slovic, P. (2000). The springs of action: Affective and analytical information processing in choice. *Personality and Social Psychology Bulletin, 26*(12), 1465–1475. doi:10.1177/01461672002612002

Schiemeier, Q. (2014). Water risk as world warms: First comprehensive globab-impact project shows that water scarcity is a major worry. *Nature, 505*(7481), 10.

Shiv, B., & Fedorikhin, A. (1999). Heart and mind in conflict: The interplay of affect and cognition in consumer decision making. *Journal of Consumer Research, 26*(3), 278–292. doi:10.1086/209563.

Slater, M. D. (1996). Theory and method in health audience segmentation. *Journal of Health Communication: International Perspectives, 1*(3), 267–284. doi:10.1080/108 107396128059

Syme, G. J., Nancarrow, B. E., & Seligman, C. (2000). The evaluation of information campaigns to promote volutnary household water conservation. *Evaluation Review, 24,* 539–578.

Syme, G. J., Shao, Q., Po, M., & Campbell, E. (2004). Predicting and understanding home garden water use. *Landscape and Urban Planning, 68*(1), 121–128. doi:10.1016/j. landurbplan.2003.08.002

Taufik, D., Bolderdijk, J. W., & Steg, L. (2015). Acting green elicits a literal warm glow. *Nature Climate Change, 5*(1), 37–40.

Triandis, H. (1977). *Interpersonal behaviour.* Monterey, CA: Brooks / Cole.

Trumbo, C. W., & O'Keefe, G. J. (2005). Intention to conserve water: Environmental values, reasoned action, and information effects across time. *Society & Natural Resources, 18*(6), 573–585. doi:10.1080/08941920590948002

Verplanken, B., & Aarts, H. (1999). Habit, attitude, and planned behaviour: Is habit an empty construct or an interesting case of goal-directed automaticity? *European Review of Social Psychology, 10*(1), 101–134. doi:10.1080/14792779943000035

Verplanken, B., Aarts, H., van Knippenberg, A., & Moonen, A. (1998). Habit versus planned behaviour: A field experiment. *British Journal of Social Psychology, 37*(1), 111–128. doi:10.1111/j.2044-8309.1998.tb01160.x

Vörösmarty, C. J., Green, P., Salisbury, J., & Lammers, R. B. (2000). Global water resources: Vulnerability from climate change and population growth. *Science, 289*(5477), 284–288. doi:10.1126/science.289.5477.284

Whetton, P. (2015). *Climate change in Australia: Projections for Australia's NRM regions.* Retrieved from Canberra: https://publications.csiro.au/rpr/download?pid=csiro:EP154 327&dsid=DS2

11 Quixotic water policy and the prudence of place-based voices

Robyn Bartel, Louise Noble and Wendy Beck

Introduction

Mainland Australia is a round island: there is a large area of land in the interior, far from the ocean. Australia is also an exceptionally old and weathered place, and is therefore very flat, with low likelihood of orographic rainfall. It is the driest inhabited continent on Earth (see for example Wahlqvist, 2008). Surface freshwater is often an intermittent, rather than permanent, feature of the landscape, and indeed is scarce for significant periods of the year (see for example Chartres & Williams, 2006; Wahlqvist, 2008). And rivers – imagined, mapped and worked as reliable features – are often in reality 'chain-o-ponds', long knotted strings of deeper waterholes separated by shallow or dried-out reaches (see for example Eyles, 1977; Selby, 1981). On first seeing the Namoi River in New South Wales, Eric Rolls's wife, Joan, is mightily disappointed: 'Is that the river?... It just looks like a muddy waterhole' (Rolls, 1974, p. 4). The river*bed* may be a permanent feature, although also often indistinct, and flowing water may be transient. Rainfall patterns are unpredictable but evaporation rates and flow regimes reliably extreme. Open plains may transform into vast lakes overnight, while almost equally quickly a shining water-body may disappear into sand and salt. Such conditions challenged settlers and policy-makers alike. For example, one absurd situation involved an unlicensed dam being built to capture water from a creek, although – as the watercourse was unable to be included within the legislative definition of 'a river', as it could not be described as experiencing permanent flow – such action could not be found to be illegal (*Latta v Klinberg* [1977], cited in Tan, 2002, p. 450). This interpretation has been affirmed as recently as 1998 in the case of *Mitchell v Vella*, in which Justice Sheahan of the New South Wales Land and Environment Court considered a waterbody must 'exhibit features of continuity, permanence and unity' (para 93). Similarly incongruent with natural conditions is the fact that Australians are very high per capita users of water – using more than all other countries of the Organisation for Economic Co–operation and Development (OECD) except New Zealand, and the United States (OECD, 2008). The nation is also a net water exporter (Hoekstra & Mekonnen, 2012). How did such a situation arise? And how did such a mammoth failure of water policy occur?

The antecedents of the 'modern' water management practices and policies employed in Australia were originally designed for European landscapes, where rivers may be believed to be permanent features and rainfall patterns more predictable. According to Linton (2010), the 'Western imaginary had been . . . pervaded by a measure of ignorance of – or a contempt for – aridity' (p. 123). This 'northern temperate bias' (Linton, 2010, p. 106) towards a 'well watered Earth' (Tuan, 1968, p. 144) with 'copious volumes of liquid' (Linton, 2010, p. 123) has created a marked disjuncture with the material reality of Australian water systems (see for example Botterill, 2003; Lamaro et al., 2007). However, even 'European rivers' do not behave like 'European rivers'. Northern river systems were heavily modified before their behaviour was studied, and anthropogenic disturbance continues to be confused with 'natural' operation (see Maybeck, 2003). Nonetheless, artificial constructions of rivers were exported along with the convicts; and beyond the transportation of hopeful but hopeless biophysical water models, Australia was also forced to bear the burden of a much larger and perhaps even more problematic cultural inheritance. This was the often crushing weight of a progressive and resourcist ethic, and a humanist as well as modernist paradigm, that considered people both capable and entitled to comprehend and conquer all, bolstered by a population and economic agenda devoted to growth, and political ideologies and scientific ontologies suffused with human arrogance and avarice. Encouraged by government subsidies, over-allocation and an over-ambitious programme for 'making the desert bloom', failed irrigation schemes provide one of the more potent examples of what can happen when biophysical reality meets human hope and fallibility (see for example Strang, 2004; Wahlqvist, 2008; Weir, 2009).

The consequences of such quixotic[1] water policies are social as well as biophysical. To date, deteriorating river health and declining water quality and quantity caused by poor management practices have impeded human wellbeing as well as the operation of ecological systems. This chapter will analyse these issues further through undertaking a case study of a northern New South Wales catchment of the Murray-Darling Basin, identifying management fashions and deficiencies and then, in Chapter 12 (this volume), build on this analysis to develop recommendations for the transformation of water policy.

Background: The Murray Darling Basin and the Namoi catchment, NSW, Australia

The Murray-Darling Basin

Water management issues in the Murray-Darling Basin (MDB) dominate global and national water policy discussions (WEF, 2015). Approaches applied in the Basin are sought out for lessons applicable elsewhere, precisely due to their perceived rigour and robustness in having been designed to meet the extreme challenges of attempting to support water security, environmental health and human livelihoods, including agriculture and diverse industries, in a location of already heavily circumscribed water resources. With an area of just over

1 million sq km, the MDB is the continent's largest catchment. It is home to 70 per cent of its irrigated agriculture, 40 per cent of Australia's farms, just over 2 million people and supports a third of the nation's food supply (MDBA, 2014). It is also home to sixteen Ramsar sites, over 30,000 wetlands and the Willandra Lakes World Heritage Site (MDBA, 2014). It furthermore bears the consequences of poor land and water management, including both dryland and irrigation-induced salinity, excessive land clearing, inappropriate tillage practices and overstocking, which have contributed to severe erosion. Water quality has deteriorated through increased turbidity and chemical pollutants, and water quantity has declined through over-allocation and misuse. River ecology has further suffered through water regulation, diversion and drainage, as well as the introduction of invasive species. Native species have been rendered extinct and ecosystems decimated. These are just some of the major environmental issues (see for example Chartres & Williams, 2006; Kingsford & Thomas, 1995; Selby, 1981; Wahlqvist, 2008). Human populations and economies reliant on Basin systems have also suffered as a result.

The Namoi catchment of New South Wales, Australia

The Namoi catchment sits within the MDB in northern New South Wales (NSW) (see Figure 11.1). The current population is 100,000, largely centred in the urban areas of Tamworth, Gunnedah, Narrabri and Walgett (MDBA, 2010; Namoi CMA, 2013). Around 42,000 sq kms in area, it is traditionally the land of the Kamilaroi peoples. It exhibits a varied biophysical profile west to east, with the rainfall gradient 400–1300 mm/yr and evaporation 2200–1000 mm/yr (see Green et al., 2011). It possesses two unique environmental features that suggest that water management approaches should be able to be made workable here. The first feature is the soil. The area contains arable land uncommon in Australia. In contrast to the geological senescence of the remainder of the continent, it has experienced relatively recent volcanic activity – and therefore boasts productive, heavy black and grey clay soils – which remain in largely good condition (Namoi CMA, 2013; NSWSoC, 2010). The second feature is the water. In addition to existing storages of the Chaffey, Split Rock and Keepit Dams (Figure 11.1), and the extensive but relatively lower quality Great Artesian Basin, it has access to rare stores of fresh groundwater from the Upper and Lower Namoi Alluvium layers. As a result, it has one of the highest extractions of groundwater and greatest development of this resource nationally (Namoi CMA, 2013; NSWSoE, 2015). The area uses 2.6 per cent of the total surface water diverted for irrigation in the MDB and 15.2 per cent of the groundwater resource (CSIRO, 2007). The economic value of the area's agricultural production is $AUD 1.6 billion (Namoi CMA, 2013). Major primary industries are grazing (i.e. ranching or pastoralism), cropping (including cotton, grain [mainly wheat], hay and some small crop horticulture), poultry and forestry. Cropping occurs mainly on alluvial floodplains and in some areas is in conflict with mining exploration and extraction (see for example Hannam, 2015; and also NSWLC, 2012; RATRC, 2011; SCUGM, 2016;

SWS, 2012). Coal and coal seam gas mining are especially contentious, particularly for potential impacts on the groundwater of the Upper Namoi Alluvium upstream of Narrabri (SWS, 2012). Cotton has also been controversial for its use of water, and it is the second largest (after dairying) agricultural user of water nationwide (ABS, 2016).

Even with its relative biophysical advantages, the Namoi shares the poor ecosystem health of most other catchments in the MDB (Davies et al., 2013; Jackson et al., 2017). There have been many attempts to address the situation. In addition to the MDB Plan introduced in 2010–2012 (Table 11.1), state-level interventions include a Water Sharing Plan (2004, replaced in 2016), a Groundwater Sharing Plan (2006, due for replacement or extension in 2017), a Great Artesian Basin Sharing Plan (2008) and a separate Water Sharing Plan for the Peel River (2010). There is also a Namoi Catchment Action Plan (upgraded in 2010), while the regionally governing Catchment Management Authorities were 'transitioned' to Local Land Services in 2014. Market instruments have also been used: a State Water Trading programme was introduced in 1989 and separation of licenses from land titles in 2004 (earlier trading also occurred through transfers since 1983, see Tan, 2003). Basin-wide water trading followed in 2014. Many of the more unpopular reforms have been driven by the Commonwealth and relate to broader MDB issues (see Table 11.1). Given this history, it is not surprising that the Namoi populace may be suffering from 'reform fatigue' (NRC, 2013, p. 15; and see further Green et al., 2011; House Standing Committee on Regional Australia, 2011; see also Sharp & Curtis, 2012). However, and perhaps because of, the plethora of policy activity and dearth of desirable outcomes, water remains the environmental issue of most concern to residents (IPSOS, 2007).

The discussion below reports on semi-structured interviews of one to two hours in length conducted with a purposive sample of ten key stakeholders in the Namoi in the wake of the Draft Basin Guide Community Consultations in late 2010. All participants are long-term residents of the catchment, the cohort is of mixed age and gender and all are occupied directly in agricultural production, including irrigated agriculture, and/or natural resource management, including in government positions and the private sector. The chapter is structured in part to reflect the interview themes, derived through a grounded theory approach (Glaser & Strauss 1967; Corbin & Strauss 2007). Selected quotes from the interviews appear below, identified by generic appellations to preserve anonymity.

Discussion

Demand vs supply: Advancing economic growth and environmental decline

Water management in Australia has been, and remains, focused on extraction, provision and production – on accessing abundance and accelerating consumption (see for example Barr & Cary, 1992; Connell, 2007; O'Gorman, 2012b). Water supply plans have been ambitious and over-optimistic, fuelled by hopes

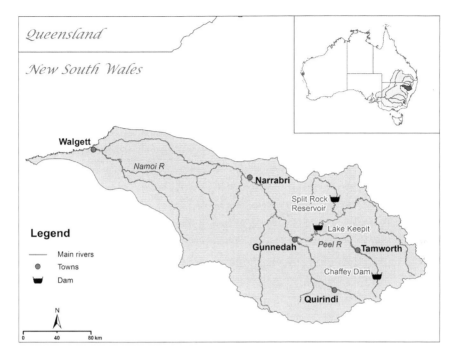

Figure 11.1 Location of Murray-Darling Basin (inset) and Namoi catchment, NSW, Australia.

Table 11.1 Abridged timeline of Murray-Darling Basin floods, droughts and management 1895–2016

Federation Drought 1895–1902 Australia as a Federation was brought into being in 1901

1914–1915 Drought

1915 – River Murray Waters Agreement signed between Australian government and governments of Victoria, South Australia (SA) and NSW.
1917 – River Murray Commission established.
1917 – Flooding
1931 – Flooding
1930–40s – Barrages built in SA to prevent seawater incursions into lower Murray during times of low flow.

1944–1945 Drought

1956 – Flooding

1967 – Drought

1981 – Murray mouth at Goolwa, SA closes due to lack of water.
1982 – River Murray Waters Agreement amended.

(continued)

Table 11.1 (continued)

1983–1984 Drought

1985 – Murray Darling Basin Ministerial Council established including the Commonwealth, Victoria, SA, NSW, Queensland (Qld) and the Australian Capital Territory (ACT).
1987 – Murray Darling Basin Agreement signed by relevant states (and later by ACT).

1991–1995 Drought

1991–1992 – 1,000km of the Darling River affected by the largest blue-green algal (cyanobacteria) bloom in the world.
1992 – New Murray Darling Basin agreement signed (later by ACT and Qld). Murray Darling Basin Commission replaces River Murray Commission.
1993 – Murray Darling Basin Act.
1994 – Council of Australian Governments (COAG) Water Reform agenda.
1995 – Cap imposed to limit extraction to 1993–1994 levels. National Competition Agreement includes water trading.
1996 – National Heritage Trust established.
1997 – Permanent cap on diversions from SA, NSW and Victoria.
2000 – Salinity package introduced.
2001 – Integrated Catchment Management policy introduced.

2002–2006 Millennium Drought (from 1996 and into 2015 in some areas)

2002 – Murray mouth at Goolwa, SA remains open only via dredging.
2002 – Independent expert reference panel reports that return of 1,900 GL needed to secure a 'moderate' probability of achieving a healthy working river and 4,000 GL needed to secure a 'high' probability of achieving a healthy working river.
2002 – Living Murray programme introduced. 500GL to be returned to environment.
2004 – National Water Initiative and National Water Commission established.
2005 – Return of 500GL planned between 2004–2009.
2007 – National Plan for Water Security, Commonwealth Water Act passed (amended in 2008) and Murray Darling Basin Authority (MDBA) established.
2008 – Water for the Future plan announced, including focus on buy-backs of water licenses and allocations (also pursued further in Restoring the Future program in 2011).
2010 – Draft Basin Guide released in October, to considerable controversy. Includes scenarios for the return of 3,000, 3,500 and 4,000 GL.
2010 – Chair of the MDBA resigns in December 2010 (effective January 2011).
2010–2011 Flooding
2011 – Inquiry into the impact of the Murray-Darling Basin Plan in Regional Australia.
2011 – Draft Basin Plan released in November.
2012 – Revised Draft Basin Plan released in May.
2012 – Altered Proposed Basin plan released in August.
2012 – Basin Plan released in November, provides for return of 2,750 GL, less than the original amount advised, distributed variably across 23 Catchment areas through Sustainable Diversion Limits (SDLs). Groundwater SDLs increased from the original. All State water resource plans to be consistent with the MDB Plan. Targets to be achieved by 2019, through the use of buy-backs and infrastructure improvements. Northern Basin limit for consumptive use set at 3,468 GL.
2014 – National Water Commission dismantled. Final report is highly critical of government. Independent review of the Water Act 2007.

2015 – Buy-backs capped at 1,500 GL.

2016 – Select Committee Report on the Murray Darling Basin Plan recommends scaling-back of water buy-backs and re-assessment of economic and social impacts. Northern Basin Review commenced and recommends reduction in water recovery target from 390 GL to 320 GL (effectively returning northern basin extraction to the 3,858 GL level when the Basin Plan was introduced).

Note that Table 1 is an abridged timeline of the history across the Basin and does not include state-based initiatives that occurred independently and simultaneously, for example the NSW Water Sharing Plans. In addition to State Water Acts and policies there are also national and state drought policies, as well as national and state Environmental Impact Assessment (EIA) legislation. The National EIA legislation does not usually oversee water, but does include Ramsar Wetlands, World Heritage areas and some mining proposals that impact on water resources. State Water Acts and mining policies may also assess the latter. See also Chapter 9, this volume.

of 'progress' and abundance, rather than by attempting to 'live within Earth's limits' (Postel Last Oasis Part II cited in Linton, 2010, p. 53). Such is the case perhaps most obviously in irrigation, with vast impoundments engineered to not only meet but grow demand. From the Murray Irrigation Area to the Ord River scheme, irrigated agriculture encompasses less than 1 per cent of the land area of the nation but uses about half of the water consumed annually and is responsible for about a third of Australia's agricultural production and half of its profits (NPSI, n.d). About half of this is in the MDB, where, in addition to food, fibre (mainly cotton), fodder crops and wine are also grown. Regulation, in terms of policy rather than impoundments, has largely resulted in more of the same: an over-allocation of licenses, poor (or absent) monitoring and enforcement and yet more plans for water supply. And not just surface water, but groundwater as well, as one of our participants noted:

> in that period from 1982 to 1984 they (State Government Department) knowingly over-allocated the resource by over 400%. (Namoi 2)

Rather than managing demand to meet natural supply (see Brooks, 2002; Falkenmark, 2003; Wallace et al., 2003; also known as 'soft path' approaches, see Brooks & Brandes, 2011; Gleick, 2002), a supply focus remains primary, even overriding attempts to save water through reducing losses from existing infrastructure (such as replacing open channels with closed pipes and using trickle instead of flood irrigation). Government decisions can be contrasted with those of landholders in the Namoi, who were concerned about over-allocation as early as 1966 (see Pigram, 1986, p. 188):

> They [local irrigator group] unanimously voted against the policy to hand out more water licenses. (Namoi 2)

This approach, informed by local, place-based prudence, was referred to by the interviewee as a 'sustainable policy' in comparison with the 'mining policy' of over-allocation perpetrated by the government. The Namoi region hosted the first trial of volumetric allocation in 1967–1968 (Sturgess & Wright, 1993); however, these systems were not always hydrologically – or administratively – reliable, nor successful in limiting water usage, and embargoes on the granting of new licenses were introduced in 1976 (Tan, 2002; 2003). Local landholder groups in the Namoi also instituted the State's first voluntary floodplain management scheme (Tan, 2002, p. 448).

Over-extraction of groundwater now mirrors overuse of surface water. However, groundwater is still often viewed as akin to the promised inland sea that was sought but never found (Courtney, 2015). Plans to increase groundwater utilisation proliferate, despite its status as the great 'known unknown' of the Australian environment (Murray et al., 2003). Dreams of abundance continue to prove resilient to water scarcity, even in the face of the mounting evidence of the harm occasioned by overuse:

> *we've seen the decline of the rivers, they've been overallocated, the ecosystems are in crisis.* (Namoi 5)

Current incarnations of the dream include 'moving' Australian agriculture northwards (Hart, 2004; Australian Government, 2015), despite the unsuitability of the tropics for intensive agriculture (as was, indeed, the south, see for example O'Gorman, 2013). Pipelines and irrigation tools continue to be proposed to solve water access issues (see for example Australian Government, 2015). And the dream includes further impoundments (Australian Government, 2015); the belief remains that they are a secure option to capture water in times of plenty and so to provide during lean times (SMH, 2013). Their allure appears undiminished even in the face of overwhelming evidence as to their shortcomings – including abbreviated life-spans, high evaporation and siltation rates, low replenishment, the loss of land as well as interference in natural flow regimes and therefore also biotic life cycles (see for example Baxter, 1977; Goldsmith & Hildyard, 1984; Vörösmarty & Sahagian, 2000).

Although the agency of water in the landscape has thus been potent, it is also poorly recognized, and rarely given explicit mention or weight in policy (see Latour, 2014). However, there is a marked cyclicity to the policy activity in the MDB over time (see Table 11.1). Agreements and interventions appear to have been largely reactive – arising in response to the inevitable droughts; however, the impacts of these are apparently quickly forgotten, and the policies watered down, when the rains return and the floods arrive. This *ad hoc* pattern of 'punctuated evolution' (Cummins & Watson, 2012, p. 22) or, somewhat less generously, 'haphazard, drought-driven spasms . . . of policy development' (Hussey & Dovers, 2006, p. 38; and see also Tan, 2003, p. 187), has continued to the present day. For example, the Millennium Drought was followed by the wettest two-year period (2010–2011) on record (Jackson et al., 2017, p. 28). In a radio interview (ABC, 2017) Professor Lin Crase has observed that 'if it had stayed dry . . . all of the elements of the Basin Plan would have continued largely untouched. The fact that it did rain provided wiggle room for the politicians to backslide'.

Both the National Water Commission in 2014, and the Australia State of the Environment Report in 2016, identified that the impetus for water reform had evaporated, with much yet to be achieved (Argent, 2017; NWC, 2014). Katherine Boo (2015) has observed that, for those living in the slum communities of Annawadi in India, 'forgetting catastrophe is a large part of the art of perseverance' (p. 308). However, 'forgetting' may also be a type of catastrophe. The National Water Commission delivered this damning assessment, in what was to be the final report following the Commission's abolition in 2014 (NWC, 2014):

> At present water is not a priority for most governments across Australia, partly because of the substantial gains made through the reform process and partly because the major urban centres and irrigation districts are no longer in drought. Submissions to the 2014 assessment noted the potential for water to drop off government agendas, with the National Farmers' Federation stating 'the cessation of the Millennium Drought has meant that water reform is no longer "front of mind". This combined with the apparent budgetary constraints of most jurisdictions will challenge any future reform effort'. Water reform principles have not been fully embedded in government processes. Realising all the benefits of the efforts made so far is therefore at risk. (p. 3)

As one of our Namoi participants also identified:

> *governments for a hundred years or more in Australia have been promoting people to get out there and slash and burn. And then dealing belatedly with some of the consequences without saying 'stop slashing and burning'.* (Namoi 5)

The same pattern has been repeated worldwide. 'Wet growth' has been proposed as a water-focused version of 'smart growth', in which all land use decisions are to be informed primarily by impacts on water quantity and quality, which requires much more detailed consideration of local constraints and interactions between human activities and water management. Arnold (2005, pp. 13, 17) has observed that site, location and watershed specific land use regulations offer much promise; however, current political boundaries across space, and administrative boundaries between public policy areas, often prove significant barriers to such reforms (see also Cosens, 2010).

Jurisdiction vs catchment: Exerting power and creating distance

One of the most enduring paradigms of natural resource policy has been the management of resources individually and in isolation from everything else (see for example Arnold, 2005; Farrier, 2002; and see also Schoemen et al., 2014), and at scales that reflect political and social considerations rather than environmental factors. Jurisdictional boundaries (see inset map in Figure 11.1) are far too coarse for environmentally sympathetic policies. Catchment (also known as watershed, see Chapter 7, this volume) scale planning has been introduced in an attempt to provide more biophysically defined and appropriate regions for management, as well as to facilitate local governance, ownership and decision-making (and/or appear more democratic) (Namoi CMA, 2013; Tan, 2003). In NSW, catchment scale-planning, including in the Namoi, has been accompanied by some limited moves towards integrated natural resource management, also known as 'total' catchment management, which ideally would include land use change, biodiversity, climate and mining (Hussey & Dovers, 2006, p. 38; Linton, 2010, p. 54, p. 216ff; Tan, 2002, p. 457; see also Daniell & Barrateau, 2014; Gupta et al., 2013). However, catchment-level natural resource management is not without significant challenges, including the requirement for subsidiarity and supportive policy settings at all scales (see for example Bartel, 2014; Cash et al., 2006; Curtis & Farrelly, 2005; Curtis et al., 2014; Guerrero et al., 2013; Curtis & Lockwood, 2000; Lockwood et al., 2009; Vörösmarty et al., 2015). This reform attempt appears to have produced little success so far (Crase & Cooper, 2015) and exhibits contradictory elements, including a concerning concentration of power (Nheu, 2002) and favouring of certain interests, particularly of irrigators (Tan, 2003). In a particularly regressive administrative reorganization, the Catchment Management Authorities established in 2004 were disbanded in 2014 and replaced by Local Land Services – ostensibly focused on integrated natural resource management but based on local government rather than catchment boundaries.

The MDB itself is a catchment, one that spans four States and one Territory, all with divergent jurisdictional-based views regarding water use and economic interests as well as heterogeneous perspectives within their boundaries. There have been major problems in developing an effective cross-border management framework. Increasing Commonwealth activity and the dilution of State power since Federation has facilitated several Federal Government programmes in addition to the MDB Plan, including the National Water Initiative and National Action Plans. 'Federalizing' water, however, risks taking control out of the hands of the regions (Australian Government, 2014) and creates social distance (see Bartel, 2014; Scott, 1998). As one interviewee observed:

> The Commonwealth [government] is very bad at relating to the public. ... I think the further you remove, physically and temporally, not [just] in time but in the amount of time you are there on the ground. (Namoi 5)

Social distance is often also geographic, and is problematic for regulatory success given the associated disconnections and distrust, poor compliance and resistance, crowding out of local initiatives and perverse and suboptimal on-ground outcomes (Bartel & Barclay, 2011; Faysee et al., 2014; Hulme, 2010; Marshall, 2013; Ostrom, 2000; Sendzimir et al., 2007). Although the MDB Plan may appear to be a catchment-based framework, any intended biophysical sympathy is at once undermined by the size and heterogeneity within the Basin, and the overriding power held by the Commonwealth under the governing piece of legislation, the

Water Act 2007 (Table 11.1). In many respects control has become re-centralized, and therefore increasingly separated from catchment-based community concerns (see Connell et al., 2005; Cummins & Watson, 2012; Daniell, 2011; Hussey & Dovers, 2006; Strang, 2009, p. 280; Wheeler et al., 2017). Furthermore, as Crase and Cooper (2015, p. 390–391) make clear, it appears that a commitment to achieve environmental outcomes co-operatively with a range of stakeholders – perhaps the most critical ingredient – has been absent.

Government vs governance: Let the public or the market decide?

At Federation, Australians put their faith in government rather than the market (see for example Connell, 2007, p. 59–60). More recently, however, 'government' has become unfashionable, due to the influence of two very different political and ideological movements: firstly, economic neoliberalism, which promotes smaller government and greater market involvement, and secondly, liberal democratic, or progressive and radical, positions supportive of an expansion of the decision-making franchise in order to improve the feasibility and legitimacy of government action as well as to advance equity and justice (Bartel et al., 2014; Berger, 2003; Curtis et al., 2014; Dryzek, 1997). The small 'g' government that is forefront in many policy-makers' minds is not the government that makes way for governance, but the government that makes way for the market, particularly for economic efficiency and competitiveness (see for example Connell et al., 2005; Hussey & Dovers, 2006, p. 38; Jackson, 2017), which may undermine other objectives, as in the example of water trading.

Water trading is being actively promoted through a disconnection between water and land, however artificial and inconsistent this may be with traditional notions of property (see Chapter 3, this volume). According to economic theory this (artificial, created) water should 'flow' to higher value uses, with greater returns per unit water input (see for example Chartres & Williams, 2006; MacDonald & Young, 2001). The introduction of the environment as a 'trader' or 'customer' in this system (such as the Commonwealth Environmental Water Holder), through the idea of environmental water, flows and water buy-backs (see Table 11.1), only partially addresses market failure (see Bennett, 2003). Water privatization and commensuration (i.e. valuation in monetary terms) may attempt to address the so-called global commons 'problem', as well as attempting to 'make-good' negative externalities, but does not address other more fundamental market failures, not least the incomparability of natural and economic capital, and could crowd out more innovative governance options (Cox et al., 2010; Gruber, 2010; Markulev & Long, 2013; Ostrom, 2007; Williams & McNeill, 2005).

Furthermore, a commoditized view of non-human nature perpetuates modernist, humanist and resourcist perspectives, and diminishes the potential for more vital water relationships to develop (Strang, 2004; Tarlock, 2000; Weir, 2009; Linton, 2010; Wilcock et al., 2013; Tuck & McKenzie, 2015; Wright, 2015; see also Chapter 5, this volume). Additionally, the market failure that originally demanded regulatory intervention remains unaddressed:

You know 'the marketplace will solve everything'? Well… we've just lived through that experience [the Global Financial Crisis] where the marketplace hasn't solved the market's problems. (Namoi 5)

An alternative approach is collaborative governance (Gunningham, 2009; Holley et al., 2011; Kallis et al., 2009; Lockwood & Davidson, 2010) that draws on the views and knowledges of the peoples and communities who live in the landscape (see Cranney & Tan, 2011; Tan et al., 2012; Woodward & McTaggart, 2015; and see also Daniell & Barrateau, 2014; Ryder this volume). Local knowledge, also known as vernacular knowledge, is the knowledge people have of their places, including their own unique water stories; it reflects the co-constitution of people and place (Bartel, 2014). It is known by a variety of related terms, including folklore, folk or customary law, local ecological knowledge or traditional environmental knowledge (see also Williams et al., this volume). It is often over-looked and under-utilized despite the growing evidence of its value, particularly in ensuring better policy development and regulatory success (see Bartel, 2014, among many others[2]). It has particular potential for adaptation to global change in the Anthropocene (Crutzen, 2002; Crutzen & Stoermer, 2000; Vörösmarty et al., 2000). Our research participants understood this also:

> *You gotta change your system to match your environment. You're gonna have wet years, you're gonna have dry years… it's only one mob seems to be doing the climate change [adaptation] and that is the farmers. Everyone else is talking about it, but they're the ones actually living with it.* (Namoi 8)

Even when community participation is a stated policy goal or process, engagement methods are often cursory and ill-suited to the task (see for example Arnstein, 1969; Crase et al., 2005; Curtis et al., 2014; Daniell & Barrateau, 2014; Innes & Booher, 2004; Smith & McDonough, 2001). As is being increasingly recognized, to ignore vernacular knowledge is to guarantee policy failure: most obviously through resistance, such as the spectacular burning of the Draft Basin Guide (see Table 11.1) in the NSW towns of Griffith and Deniliquin in 2010, and more recently in Barham, NSW (Dunlevy, 2010; Neales, 2015; Long & Pepper, 2015). The 2010 protests occurred in response to the Guide to the Draft MDB Plan, whose belated "community consultation" programme consisted of a travelling roadshow comprising laborious presentations including interminable powerpoint slides and an inherent 'deficiency model' (Bartel, 2014, p. 904) of the community, with insufficient time for any two-way flow of information or engagement, let alone co-creation or collaboration.[3] These 'public participation' sessions were used by the Murray-Darling Basin Authority as decision-delivery platforms, and therefore were met with a high degree of resistance and anger (Loch et al., 2014; see also Daniell, 2011). According to Quiggin (2012) the central problem arose due to 'poor communication by the Authority combined with an exceptionally narrow and bureaucratic interpretation of its role in the process' (p. 55). And as one Namoi informant said, somewhat in resignation:

The other thing is that we will get more of the same… Not one point [in the Guide]
actually says it will engage with communities one-on-one. (Namoi 1)

Increasingly, good governance is described as requiring collaboration, and so public participation is considered necessary, although much of this remains rhetoric rather than realized, particularly where responsibility but not power is devolved (Strang, 2009, pp. 61, 246). It is very concerning, particularly given the wealth of evidence of its importance, that deeper consultation was not a requirement for the MDB Plan. The Water Act (see Table 11.1) also establishes a top-down approach almost devoid of public participation, and, most critically, lacking the social science and humanities expertise that could have drawn attention to the inevitable conflict – and solutions to it (Boully & Maywald, 2011; Wahlqvist, 2011; Wheeler et al., 2017; and see also Ross et al., 2012). And despite the apparent reliance on techno-cratic approaches, there was also insufficient scientific research conducted into the changes necessary in order to achieve longer-term sustainability, both for the MDB Plan (Cosier et al., 2012; Cummins & Watson, 2012, p. 30; Horne, 2014), as well as for the broader National Water Initiative (Hussey & Dovers, 2006).

In response to increasing concerns about the Plan's social and economic (as distinct from environmental) impacts, a review of its objectives in the Northern Basin (an area including the Namoi and towns of Gunnedah, Narrabri and Walgett, and all tributaries of the Barwon-Darling upstream of the Menindee Lakes (see the upper north-eastern reaches in the inset of Figure 11.1)), was insti-gated, recommending cuts to water recovery targets (see Table 11.1). A 2016 review recommended than no further reductions in water entitlements occur until the Northern Basin Review had been fully completed (SSC, 2016; see also House Standing Committee on Regional Australia, 2011). Tellingly, the 2016 review also recommended that the MDBA needed to 'review its communication methods. . . and improve its ability to incorporate the views of communities and landholders into decisions and reports' (SSC, 2016, p. xvi). The NSW Natural Resource Commission, in its review of State Groundwater Sharing Plans, rec-ommended that local knowledge and perspectives be included, and that the engagement of Aboriginal people should be prioritized – and specifically for 'their cultural knowledge and groundwater dependent cultural values' – to be incorporated (NRC, 2016). The Independent Review of the Water Act 2007 also recommended that Indigenous values and uses be prioritized (Australian Government, 2015, p. xiv) and further came to the conclusion that 'decision-makers – governments, their agencies and water managers – need to more transparently demonstrate how economic, social and environmental considera-tions are taken into account in decision-making under the Act and the Basin Plan' (Australian Government, 2015, p. xix).

Substantive incorporation of place-based appreciations is now key, as is greater consideration and transparency regarding the process of inclusion. Such a trans-formation would build better policy in terms of substantive content – as it would incorporate the vital vernacular knowledges of those with direct experience of water in the landscape – as well as achieve more just outcomes through participation.

Perhaps more pragmatically, at least in political terms, wider – and more meaningful – participation would also assist in securing the necessary support of those who must abide by and implement these same policies, thereby further increasing the likelihood that sustainable policy outcomes will be achieved.

Conclusion

> … *the whole world's over-allocated their water and people are seeing how Australia's going to sort it out.* (Namoi 3)

Australia has a history of quixotic water policies, forged by dreams of abundance and facilitated by technological and regulatory mechanisms, including impoundments, irrigation schemes and over-allocation of both surface and artesian water. When droughts hit, as they invariably do, they may precipitate reactive political and legislative responses. These, however, are largely short-lived, washed away by the subsequent, and again inevitable, floods. Periods of drought and water variability appear to have enforced rather than moderated colonial and settler dreams of 'making the deserts bloom'. Even now, these dreams – 'supply-oriented paradigms' (Linton, 2010, p. 227) – are being extended to the imagined great, untapped tropical north, and delusions of abundance, through dam-building and groundwater extraction, continue to infuse policy. Reliance on inherently unsustainable approaches, as well as thinking, has become sclerotic (Harvey, 2000, p. 185), normalised (Linton, 2010), sedimented (see Nash, 2000), and therefore appears entrenched. Governments remain fervently wedded to deeply held beliefs in human capacity, competence, progress and growth, even in the face of their proven incapacity to 'overcome' the biophysical reality of ephemeral water and an arid and ancient landscape. The imagined river-as-reliable-resource has remained the dominant frame (see Linton, 2010; see also King, 2015), and it will take a shift into deeper policy learning, incorporating vernacular knowledge and other appreciations, for more fundamental reform to be engaged and to bear fruit. The lessons of people who have already adapted to the Australian landscape are evermore vital: Indigenous owners and all those who have survived not only the vagaries of the weather, climate and landscape but also the shifting priorities and fortunes of policies administered at all levels: State, Commonwealth and catchment (see also Chapter 2, this volume). Basin dwellers have surely been resilient, as the biophysical environment has had to be (see discussion in Greenhill et al., 2009; see also Cosens et al., 2014). However, we need to do more than simply 'bounce back' from floods and droughts – we need to transform water policy.

Acknowledgements

Thanks to Catherine MacGregor for cartographic assistance. Thanks to Arts New England, our interview participants and the Namoi catchment for supporting our fieldwork. The research was conducted following Australian Human Research Ethics protocols, University of New England approval no. HE10/193.

Notes

1 The reference is to Cervante's *Don Quixote*; the analogy is to tilting at water that simply isn't there (rather than absent windmill-monsters).
2 See Ayre & Mackenzie, 2012; Bäckstrand, 2003; Blaikie et al., 1997; Callon, 1999; Cohen, 2009; Cranney & Tan, 2011; Evans, 2006; Evans & Pratchett, 2013; Fazey et al., 2010; Fischer, 2000; Gibson-Graham, 2003; Greiw, 2010; Holling & Meffe, 1996; Irwin, 1995; Kemmis, 2002; McNamara & Westoby, 2011; O'Gorman, 2012a, b; Robertson et al., 2000; van Tol Smit et al., 2015; Williams, 2013; Woodward & McTaggart, 2015, among others.
3 The authors attended the Community Consultation in Narrabri, NSW, which followed meetings in several other towns including Griffith and Deniliquin, where protests ultimately led to revisions of the delivery programme (see House Standing Committee on Regional Australia, 2011).

References

Argent, R. M. (2017). *Australia State of the Environment 2016: Inland water*, independent report to the Australian Government Minister for the Environment and Energy, Department of the Environment and Energy, Canberra: Australian Government.

Arnold, C. A. (2005). Introduction. Integrating water controls and land use controls: New ideas and old obstacles. In C. A. Arnold (Ed.), *Wet Growth: Should Water Law Control Land Use* (pp. 1–56). Washington, D. C.: Environmental Law Institute.

Arnstein, S. R. (1969). A ladder of citizen participation. *Journal of the American Planning Association*, 35(4), 216–224.

Australian Broadcasting Corporation (ABC). (2017). Up the creek: ess water for the Murray-Darling Basin (Jo Chandler), *Background Briefing*, Sunday 19 February, Radio National.

Australian Bureau of Statistics (ABS). (2016). *Water Account Australia*, November 2016 Summary. Canberra: Australian Government.

Australian Government. (2014). *Report of the Independent Review of the Water Act 2007*. Canberra: Australian Government.

Australian Government. (2015). *White Paper on Developing Northern Australia*. Canberra: Australian Government. Online at http://northernaustralia.infrastructure.gov.au/

Ayre, M., and Mackenzie, J. (2012). 'Unwritten, unsaid, just known': The role of indigenous knowledge(s) in water planning in Australia. *Local Environment: The International Journal of Justice and Sustainability*, 18(7), 753–768.

Bäckstrand, K. (2003). Civic science for sustainability: Reframing the role of experts, policy-makers and citizens in environmental governance. *Global Environmental Politics*, 3(4), 24–41.

Barr, N., and Cary, J. (1992). *Greening a Brown Land: The Australian Search for Sustainable Land Use*. Melbourne: Macmillan.

Bartel, R. (2014). Vernacular knowledge and environmental law: Cause and cure for regulatory failure. *Local Environment: The International Journal of Justice and Sustainability*, 19(8), 891–914.

Bartel, R., and Barclay, E. (2011). Motivational postures and compliance with environmental law in Australian agriculture. *Journal of Rural Studies*, 27, 153–170.

Bartel, R., McFarland, P., and Hearfield, C. (2014). Taking a de-binarized envirosocial approach to reconciling the environment vs economy debate: Lessons from climate change litigation for planning in NSW, Australia. *Town Planning Review*, 85(1), 67–96.

Baxter, R. M. (1977). Environmental effects of dams and impoundments. *Annual review of Ecology and Systematics*, 8, 255–283.

Bennett, J. (2003). Environmental values and water policy. *Australian Geographical Studies*, 41(3), 237–250.

Berger, G. (2003). Reflections on governance: Power relations and policy making in regional sustainable development. *Journal of Environmental Policy and Planning*, 5(3), 219–234.

Blaikie, P., Brown, K., Stocking, M., Tang, L., Dixon, P. B., and Sillitoe, P. (1997). Knowledge in action: Local knowledge as a development resource and barriers to its incorporation in natural resource research and development. *Agricultural Systems*, 55(2), 217–237.

Boo, K. (2015). Annawadi, India: Another way of seeing. *Granta*, 130, 207–226.

Botterill, L. C. (2003). Uncertain climate: The recent history of drought policy in Australia. *Australian Journal of Politics and History*, 49(1), 61–74.

Boully, L., and Maywald, K. (2011). Basin bookends, the community perspective. In D. Connell and R. Q. Grafton (Eds.), *Basin Futures: Water Reform in the Murray-Darling Basin*, ANU E-press. Online at http://press.anu.edu.au//apps/bookworm/view/Basin+Futures+Water+reform+in+the+Murray-Darling+Basin/5971/upfront.xhtml

Brooks, D. B. (2002). *Water: Local Level Management*. Ottawa: International Development Research Institute.

Brooks, D. B., and Brandes, O. M. (2011). Why a water soft path, why now and what then? *Water Resources Development*, 27(2), 315–344.

Callon, M. (1999). The role of lay people in the production and dissemination of scientific knowledge. *Science Technology and Human Values*, 4(1), 81–94.

Cash, D. W., Adger, W., Berkes, F., Garden, P., Lebel, L., Olsson, P., Pritchard, L., and Young, O. (2006). Scale and cross-scale dynamics: Governance and information in a multilevel world. *Ecology and Society*, 11(2), 8. Online at http://www.ecologyandsociety.org/vol11/iss2/art8/

Chartres, C., and Williams, J. (2006). Can Australia overcome its water scarcity problems? *Journal of Developments in Sustainable Agriculture*, 1(1), 17–24.

Cohen, B. R. (2009). The once and future georgic: Agricultural practice, environmental knowledge, and the place for an ethic of experience. *Agriculture and Human Values*, 26(3), 153–165.

Connell, D. (2007). *Water Politics in the Murray Darling Basin*. Canberra: Federation Press.

Connell, D., Dovers, S., and Grafton, R. (2005). A critical analysis of the National Water Initiative. *The Australasian Journal of Natural Resources Law and Policy*, 10(1), 81–107.

Corbin, J., and Strauss, A. (2007). *Basics of Qualitative Research: Techniques and Procedures for Developing Grounded Theory*. 3rd ed. Thousand Oaks: Sage Publications.

Cosens, B. (2010). Transboundary river governance in the face of uncertainty: Resilience theory and the Columbia River Treaty. *Journal of Land, Resources and Environmental Law*, 30(2), 229–265.

Cosens, B. A., Gunderson, L., and Chaffin, B. C. (2014). The adaptive water governance project: Assessing law, resilience and governance in regional socio-ecological water systems facing a changing climate. *Idaho Law Review, Natural Resources and Environmental Law Edition*, 51(1), 1–27.

Cosier, P., Davis, R., Flannery, T., Harding, R., Hughes, L., Karoly, D., Possingham, H., Purves, R., Saunders, D., Thom, B., Williams, J., and Young, M. (2012). *Once Again the Taxpayer Pays, Individuals Benefit and the River Loses*, Wentworth Group evaluation of proposed basin plan, August 2012.

Courtney, P. (2015). Andrew Forrest calls for harvesting of rivers, underground aquifers to droughtproof agricultural areas, *ABC News* 9 March. Online at http://www.abc.net.au/news/2015-03-08/twiggy-forrest-plan-for-droughtproofing-australia/6287690

Cox, M., Arnold, G., and Villamayor Tomás, S. (2010). A review of design principles for community-based natural resource management. *Ecology and Society*, 15(4), 38. Online at http://www.ecologyandsociety.org/vol15/iss4/art38/

Cranney, K., and Tan, P. (2011). Old knowledge in freshwater: Traditional ecological knowledge in the Richmond catchment. *Australasian Journal of Natural Resources Law and Policy*, 14(2), 71–113.

Crase, L., and Cooper, B. (2015). Politics, socio-economics and water allocations: A note on the limits of Integrated Water Resources Management. *Australian Journal of Environmental Management*, 22(4), 388–399.

Crase, L., Dollery, B., and Wallis, J. (2005). Community consultation in public policy: The case of the Murray-Darling Basin of Australia. *Australian Journal of Political Science*, 40(2), 221–237.

Crutzen, P. J. (2002). The geology of mankind. *Nature*, 415(6867), 23.

Crutzen, P. J., and Stoermer, E.F. (2000). The Anthropocene. *IGBP Newsletter*, 41(17), 17–18.

CSIRO. (2007). *Water Availability in the Namoi*. Summary of a report to the Australian Government from the CSIRO Murray-Darling Basin Sustainable Yields Project. Canberra: CSIRO.

Cummins, T., and Watson, A. (2012). A hundred year policy experiment: The Murray-Darling Basin in Australia. In J. Quiggan, T. Mallawaarachchi, and S. Chambers (Eds.), *Water Policy Reform* (pp. 9–36). Cheltenham, UK: Edward Elgar Publishing.

Curtis, A., and Lockwood, M. (2000). Landcare and catchment management in Australia: Lessons for state sponsored community participation. *Society and Natural Resources*, 13(1), 61–73.

Curtis, A., Ross, H., Marshall, G. R., Baldwin, C., Cavaye, J., Freeman, C., Carr, A., and Syme, G. J. (2014). The great experiment with devolved NRM governance: Lessons from community engagement in Australia and New Zealand since the 1980s. *Australasian Journal of Environmental Management*, 21(2), 175–199.

Daniell, K. A. (2011). Enhancing collaborative management in the Basin. In D. Connell, and R. Q. Grafton (Eds.), *Basin Futures: Water Reform in the Murray-Darling Basin*. Canberra, Australia: ANU E-press. Online at http://press.anu.edu.au/apps/bookworm/view/Basin+Futures+Water+reform+in+the+Murray-Darling+Basin/5971/upfront.xhtml

Daniell, K., and Barrateau, O. (2014). Water governance across competing scales: Coupling land and water management. *Journal of Hydrology*, 519, 2367–2380.

Davies, P., Stewardson, M., Hillman, T., Roberts, J., and Thoms, M. (2013). *Sustainable Rivers Audit 2: The Ecological Health of the Rivers in the Murray-Darling Basin at the End of the Millennium Drought (2008–2010)*, Volume 3. Report prepared for the Murray-Darling Basin Authority by the Independent Sustainable Rivers Audit Group. Canberra: MDBA.

Dryzek, J. (1997). *The Politics of the Earth: Environmental Discourses*. New York: Oxford University Press.

Dunlevy, G. (2010). Young men burn guide to Murray-Darling Basin plan, *ABC News*. Online at http://www.abc.net.au/news/2011-11-17/young-men-burn-copies-of-the-guide-to-the-murray/367841

Evans, J. P. (2006). Lost in translation? Exploring the interface between local environmental research and policymaking. *Environment and Planning A*, 38(3), 517–531.

Evans, M., and Pratchett, L. (2013). The localism gap – The CLEAR failings of official consultation in the Murray Darling Basin. *Policy Studies*, 34(5–6), 541–558.

Eyles, R. J. (1977). Changes in drainage networks since 1820, Southern tablelands, NSW. *Australian Geographer*, 13(6), 377–386.

Falkenmark, M. (2003). Freshwater as shared between society and ecosystems: From divided approaches to integrated challenges. *Philosophical Transactions of the Royal Society of London B*, 358(1440), 2037–2049.

Farrier, D. (2002). Fragmented law in fragmented landscapes: The slow evolution of integrated natural resource management legislation in NSW. *Environmental and Planning Law Journal*, 19, 89–108.

Faysee, N., Errahj, M., Imache, A., Kemmoun, H., and Labbaci, T. (2014). Paving the way for social learning when governance is weak: Supporting dialogue between stakeholders to face a groundwater crisis in Morocco. *Society and Natural Resources*, 27(3), 249–264.

Fazey, I., Fazey, J., Salisbury, J., Lindenmayer, D. B., and Dovers, S. (2010). The nature and role of experiential knowledge for environmental conservation. *Environmental Conservation* 33(1), 1–10.

Fischer, F. (2000). *Citizens, Experts and the Environment: Local Knowledge*. Durham: Duke University Press.

Gibson-Graham, J. K. (2003). An ethics of the local. *Rethinking Marxism*, 15(1), 49–74.

Glaser, B., and Strauss, A. (1967). *The Discovery of Grounded Theory*. Chicago: Aldine.

Gleick, P. H. (2002). Soft water paths. *Nature*, 418(6896), 373.

Goldsmith, E., and Hildyard, N. (1984) *The Social and Environmental Effects of Large Dams*. Camelford, Cornwall: Wadebridge Ecological Centre.

Green D., Petrovic J., Moss P., and Burrell M. (2011). *Water Resources and Management Overview: Namoi Catchment*. Sydney: NSW Office of Water.

Greenhill, J., King, D., Lane, A., and MacDougal, C. (2009). Understanding resilience in South Australian farming families. *Rural Society*, 19(4), 318–325.

Griew, R. (2010). Drawing on powerful practitioner-based knowledge to drive policy development, implementation and evaluation. In Productivity Commission, *Strengthening Evidence Based Policy in the Australian Federation*, Volume 1: Proceedings, Roundtable Proceedings (pp. 249–258). Canberra: Productivity Commission.

Gruber, J. S. (2010). Key principles of community-based natural resource management: A synthesis and interpretation of identified effective approaches for managing the commons. *Environmental Management*, 45(1), 52–66.

Guerrero, A. M., McAllister, R. R. J., Corcoran, J., and Wilson, K. A. (2013). Scale mismatches, conservation planning, and the value of social-network analyses. *Conservation Biology*, 27, 35–44.

Gunningham, N. (2009). The new collaborative environmental governance: The localization of regulation. *Journal of Law and Society*, 36(1), 145–166.

Gupta, J., Pahl-Wostl, C., and Zondervan, R. (2013). 'Glocal' water governance: A multilevel challenge in the Anthropocene. *Current Opinion in Environmental Sustainability*, 5(6), 581–589.

Hannam, P. (2015). Battle fields, *The Sydney Morning Herald*, July 11–12, p. 28.

Hart, B. (2004). Environmental risks associated with new irrigation schemes in Northern Australia. *Ecological Management and Restoration*, 5(2), 106–110.

Harvey, D. (2000). *Spaces of Hope*. Edinburgh, UK: Edinburgh University Press.

Hoekstra A. Y., and Mekonnen, M. M. (2012). The water footprint of humanity. *PNAS*, 109(9), 3232–3237.

Holley, C., Gunningham, N., and Shearing, C. (2011). *New Environmental Governance*. Abingdon: Earthscan.

Holling, C. S., and Meffe, G. K. (1996). Command and control and the pathology of natural resource management. *Conservation Biology*, 10(2), 328–337.

Horne, J. (2014). The 2012 Murray-Darling Basin Plan – Issues to watch. *International Journal of Water Resources Development*, 30(1), 152–163.

House Standing Committee on Regional Australia. (2011). *Of Drought and Flooding Rains. Inquiry into the Impact of the Murray-Darling Basin Plan in Regional Australia*. Canberra: Australian Government.

Hulme, M. (2010). Problems with making and governing global kinds of knowledge. *Global Environmental Change*, 20(4), 558–564.

Hussey, K., and Dovers, S. (2006). Trajectories in Australian water policy. *Journal of Contemporary Water Research and Education*, 135(1), 36–50.

Innes, J., and Booher, D. (2004). Reframing public participation: Strategies for the 21st century. *Planning Theory and Practice*, 5(4), 419–436.

IPSOS. (2007). *Namoi catchment management authority stakeholder and community benchmarking study 2007*. Sydney, Australia: IPSOS Consultants.

Irwin, A. (1995). *Citizen Science*. London: Routledge.

Jackson, S. (2017). Enduring and persistent injustices in water access in Australia. In A. Lukasiewicz, S. Dovers, L. Robun, and J. McKay (Eds.), *Natural Resources and Environmental Justice: Australian Perspectives* (pp. 121–132). Melbourne: CSIRO Publishing.

Jackson, W. J., Argent, R. M., Bax, N. J., Clark, G. F., Coleman, S., Cresswell, I. D., Emmerson, K. M., Evans, K., Hibberd, M. F., Johnston, E. L., Keywood, M. D., Klekociuk, A., Mackay, R., Metcalfe, D., Murphy, H., Rankin, A., Smith, D. C., and Wienecke, B. (2017). *Australia State of the Environment 2016: Overview*, Independent report to the Australian Government Minister for the Environment and Energy, Department of the Environment and Energy. Canberra: Australian Government.

Kallis, G., Kiparsky, M., and Norgaard, R. (2009). Collaborative governance and adaptive management: Lessons from California's CALFED Water Program. *Environmental Science and Policy*, 12(6), 631–643.

Kemmis, D. (2002). Science's role in natural resource decisions: Collaborative efforts that rely on local knowledge as well as science are key to resolving difficult land use issues. *Issues in Science and Technology*, 18(4), 31–34.

King, C. (2015). In deep water, *The Saturday Paper*, July 11–17, p. 3.

Kingsford, R., and Thomas, R. (1995). The Macquarie marshes in arid Australia and their waterbirds: A 50 year history of decline. *Environmental Management*, 19(6), 867–878.

Lamaro, E., Stokes, R., and Taylor, M. P. (2007). Riverbanks and the law: The arbitrary nature of river boundaries in New South Wales. *The Environmentalist*, 27(1), 131–142.

Latour, B. (2014). Agency at the time of the Anthropocene. *New Literary History*, 45(1), 1–18.

Linton, J. (2010). *What Is Water? The History of a Modern Abstraction*. UBC Press: Vancouver and Toronto.

Loch, A., Wheeler, S., and Adamson, D. (2014). People versus place in Australia's Murray-Darling Basin: Balancing economic, social ecosystem and community outcomes. In V. R. Squires, H. M. Milner, and K.A. Daniell (Eds.), *River Basin Management in the Twenty-First Century Understanding People and Place* (pp. 275–303). Boca Raton: CRC Press.

Lockwood, M., and Davidson, J. (2010). Environmental governance and the hybrid regime of Australian natural resource management. *Geoforum*, 41(3), 388–398.

Lockwood, M., Davidson, J., Curtis, A., Stratford, E., and Griffith, R. (2009). Multi-level environmental governance: Lessons from Australian natural resource management. *Australian Geographer*, 40(2), 169–186.

Long, W., and Pepper, F. (2015). Murray-Darling community pleads for changes to the Basin plan, *ABC Rural News*. Online at http://www.abc.net.au/news/2015-07-09/murra-darling-basin-meeting-with-senators-social-impacts/6606604

MacDonald, D. H., and Young, M. (2001). *A Case Study of the Murray Darling Basin: Final Report Prepared for the International Water Management Institute*, August 2000 Revised February 2001. Canberra, Australia: CSIRO Land and Water.

Markulev, A., and Long, A. (2013). *On Sustainability: An Economic Approach*, Productivity Commission Staff Research Note, May 2013. Canberra: Productivity Commission.

Marshall, G. (2013). Transaction costs, collective action and adaptation in managing complex social-ecological systems. *Ecological Economics*, 88, 185–194.

Maybeck, M. (2003). Global analysis of river systems: From Earth system controls to Anthropocene syndromes. *Philosophical Transactions of the Royal Society of London*, Series B: Biological Sciences, 358(1440), 1935–1955.

McNamara, K. E., and Westoby, R. (2011). Local knowledge and climate change adaptation on Erub Island in the Torres Strait. *Local Environment: The International Journal of Justice and Sustainability*, 16(9), 887–901.

Murray, B., Zeppel, M., Hose, G., and Eamus, D. (2003). Groundwater dependant ecosystems in Australia: It's more than just water for rivers. *Environmental Management and Restoration*, 4(2), 110–113.

Murray Darling Basin Authority (MDBA). (2010). *Guide to the Proposed Basin Plan*, Namoi community profile. Technical Background, Part III. Canberra: MDBA. pp. 938–963.

Murray Darling Basin Authority (MDBA). (2014). *Annual Report 2013–2014*, Murray Darling Basin Authority. Canberra, Australia: Australian Government.

Namoi Catchment Management Authority (CMA). (2013). *Namoi Catchment Action Plan 2010–2020* (2013 update). Namoic Catchment Management Authority. Sydney, Australia: NSW Government.

Nash, C. (2000). Performativity in practice: Some recent work in cultural geography. *Progress in Human Geography*, 24(4), 653–664.

National Programme for Sustainable Irrigation (NPSI). (n.d.). *Irrigation in Australia, Facts and Figures*, National Programme for Sustainable Irrigation, Canberra, Australia: Australian Government.

National Water Commission (NWC). (2014). *Australia's Water Blueprint: National Reform Assessment 2014*. Online at http://www.nwc.gov.au/publications/topic/assessments/australias-water-blueprint-national-reform-assessment-2014

Natural Resource Commission of NSW (NRC). (2013). *Review of Water Sharing Plans*. Sydney: NSW Government.

Natural Resource Commission of NSW (NRC). (2016). *Review of Water Sharing Plans Due to Expire in 2017 or 2018*. Sydney: NSW Government.

Neales, S. (2015). Farmers see water rights sold down the river, *The Australian*, July 8, p. 7.

Nheu, N. (2002). The continuing challenge of water management reform in NSW. *Environmental and Planning Law Journal*, 19(3), 217–225.

NSW Legislative Council General Purpose Standing Committee (NSWLC). (2012). *Final Report – Coal Seam Gas*, Sydney: NSW Government.

NSWSoC. (2010). *State of the Catchments 2010*. Sydney: NSW Government.

NSWSoE. (2015). *State of the Environment, 2015*, Chapter 19 Groundwater. Sydney: NSW Government.

OECD. (2008). *OECD Environmental Data Compendium 2006–2008, Inland Waters*. Paris: Organization for Economic Co-operation and Development.

O'Gorman, E. (2012a). Local knowledge and the state: The 1990 Floods in Cunnumulla, Qld, Australia. *Environmental History*, 17(3), 512–546.

O'Gorman, E. (2012b). *Flood Country: An Environmental History of the Murray Darling Basin*. Canberra, Australia: CSIRO Publishing.

O'Gorman, E. (2013). Growing rice on the Murrumbidgee River: Cultures, politics and practices of food production and water use, 1900–2012. *Journal of Australian Studies*, 37(1), 96–115.

Ostrom, E. (2000). Crowding out citizenship. *Scandinavian Political Studies*, 23(1), 3–15.

Ostrom, E. (2007). A diagnostic approach for going beyond panaceas. *PNAS*, 104(39), 15181–15187.

Pigram, J. J. (1986). *Issues in the Management of Australia's Water Resources*. Melbourne, Australia: Longman Cheshire.

Quiggin, J. (2012). Why the guide to the proposed basin plan failed, and what can be done to fix it. In J. Quiggin, T. Mallawaarachchi, and S. Chambers (Eds.), *Water Policy Reform* (pp. 49–60). Cheltenham, UK: Edward Elgar Publishing.

Robertson, M., Nichols, P., Horwitz, P., Bradby, K., & MacKintosh, D. (2000). Environmental narratives and the need for multiple perspectives to restore degraded landscapes in Australia. *Ecosystem Health*, 6(2), 119–133.

Rolls, E. (1974). *The River*. Cremorne, Sydney: Angus and Robertson Publishers.

Ross, H., Driml, S., and Zarrezadeh, Z. (2012). Water allocation, social change and resilience. In J. Quiggin, T. Mallawaarachchi, and S. Chambers (Eds.), *Water Policy Reform* (pp. 170–192). Cheltenham, UK: Edward Elgar.

Rural Affairs and Transport References Committee (RATRC). (2011). *Senate Committee Interim Report: The Impact of Mining Coal Seam Gas on the Management of the Murray-Darling Basin*, The Commonwealth of Australia. Canberra: The Australian Government.

Schlumberger Water Services (SWS). (2012). *Namoi Catchment Water Study Independent Expert Final Study Report*, Prepared for Department of Trade and Investment, Regional Infrastructure and Services. Sydney: NSW Government.

Schoeman, J., Allan. C., and Finlayson, M. (2014). A new paradigm for water? A comparative review of integrated, adaptive and ecosystem-based water management in the Anthropocene. *International Journal of Water Resources Development*, 30(3), 377–390.

Scott, D. (1998). *Thinking Like a State*. New Haven: Yale University Press.

Selby, J. (1981). A salty problem for the River Murray. *New Scientist*, 90(1259), 842–843.

Senate Select Committee (SSC). (2016). *Refreshing the Plan*, The Senate Select Committee on the Murray-Darling Basin Plan, Commonwealth Government. Canberra: Australian Government.

Senate Select Committee on Unconventional Gas Mining (SCUGM). (2016). *Interim Report*. Canberra: Australian Government.

Sendzimir, J., Magnuszewski, P., Flachner, Z., Balogh, P., Molnar, G., Sarvari, A., and Nagy, Z. (2007). Assessing the resilience of a river management regime: Informal learning in a shadow network in the Tisza River Basin. *Ecology and Society*, 13(1), 11.

Sharp, E., and Curtis, A. (2012). *Groundwater Management in the Namoi: A Social Perspective*, Report No 67 for the Institute for Land, Water and Society, Albury, NSW: Charles Sturt University.

Smith, P. D., and McDonough, M. H. (2001). Beyond public participation: Fairness in natural resource decision making. *Society and Natural Resources*, 14(3), 239–249.

Strang, V. (2004). *The Meaning of Water*. Oxford and New York: Berg.

Strang, V. (2009). *Gardening the World: Agency, Identity and the Ownership of Water*. New York: Berghahn Books.

Sturgess, G. L., and Wright, M. (1993). *Water Rights in Rural NSW: The Evolution of a Property Rights System*. Sydney: Centre for Independent Studies.

Sydney Morning Herald (SMH). (2013). Coalition readies $30b plan for 100 dams: Report, *Sydney Morning Herald*, 14 February. Online at http://www.smh.com.au/environment/water-issues/coalition-readies-30b-plan-for-100-dams-report-20130213 2edtv.html#ixzz3fkszlgV4

Tan, P. (2002). An historical introduction to water reform in NSW – 1975 to 1994. *Environmental and Planning Law Journal*, 19(6), 445–460.

Tan, P. (2003). Water Law Reform in NSW – 1995 to 1999. *Environmental and Planning Law Journal*, 20(3), 165–194.

Tan, P., Bowmer, K., and Mackenzie, J. (2012). Deliberative tools for meeting the challenges of water planning in Australia. *Journal of Hydrology*, 474, 2–10.

Tarlock, A. D. (2000). Putting rivers back in the landscape: The revival of watershed management in the United States. *Hastings West-Northwest Journal of Environmental Law and Policy*, 6, 167–195.

Tuan, Y-F. (1968). *The Hydrologic Cycle and the Wisdom of God: A Theme in Geoteleology*. Toronto: University of Toronto Press.

Tuck, E., & McKenzie, M. (2015). Relational validity and the 'where' of inquiry. *Qualitative Inquiry*, 21(7), 633–638.

van Tol Smit, E., de Loë E., and Plummer, R. (2015). How knowledge is used in collaborative environmental governance: Water classification in New Brunswick. *Canada Journal of Environmental Planning and Management*, 58(3), 423–444.

Vörösmarty, C. J., and Sahagian, D. (2000). Anthropogenic disturbance of the terrestrial water cycle. *BioScience*, 50(9), 753–765.

Vörösmarty, C. J., Green, P., Salisbury, J., and Lammers, R. B. (2000). Global water resources: Vulnerability from climate change and population growth. *Science*, 289(5477), 284–288.

Vörösmarty, C. J., Hoekstra, A. Y., Bunn, S. E., Conway, D., and Gupta, J. (2015). What scale for water governance? *Science*, 349(6247), 478–479.

Wahlquist, A. (2008). *Thirsty Country*. Sydney: Allen and Unwin.

Wahlqvist, A. (2011). The media and the guide to the Basin Plan. In D. Connell, and R. Q. Grafton (Eds.), *Basin Futures: Water Reform in the Murray-Darling Basin*. Canberra, Australia: ANU E-press. Online at http://press.anu.edu.au/apps/bookworm/view/Basin+Futures+Water+reform+in+the+Murray-Darling+Basin/5971/upfront.xhtml

Wallace, J. S., Acreman, M. C., and Sullivan, C. A. (2003). The sharing of water between society and ecosystems: From conflict to catchment-based co-management. *Philosophical Transactions of the Royal Society, London B*, 358(1440), 2011–2026.

Weir, J. (2009). *Murray River Country: An Ecological Dialogue with Traditional Owners*. Canberra: Aboriginal Studies Press.

Wheeler, S. A., MacDonland, D. H., and Boxall, P. (2017). Water policy debate in Australia: Understanding the tenets of stakeholders' social trust. *Land Use Policy*, 63, 246–254.

Wilcock, D., Brierley, G., and Howitt, R. (2013). Ethnogeomorphology. *Progress in Physical Geography*, 37(5), 573–600.

Williams, D. (2013). Science, practice and place. In W. P. Stewart, D. Williams, and L. Kruger (Eds.), *Place Based Conservation: Perspectives from the Social Sciences* (pp. 32–56). Dordrecht: Springer.

Williams, J. B., and McNeill, J. M. (2005). *The Current Crisis in Neoclassical Economics and the Case for an Economic Analysis based on Sustainable Development*, UNiversitas 21 Global, Working Paper, No 001/05. Online at http://papers.ssrn.com/sol3/papers.cfm?abstract_id=1606342

Woodward, E., and McTaggart, P. M. (2015). Transforming cross-cultural water research through trust, participation and place. *Geographical Research*, 54(2), 129–142.

World Economic Forum (WEF). (2015). *Global Risks*, 10th Edition, Insight Report, Geneva Switzerland.

Wright, S. (2015). More-than-human, emergent belongings: A weak theory approach. *Progress in Human Geography*, 39(4), 391–411.

12 Heterotic water policy futures using place agency, vernacular knowledge, transformative learning and syncretic governance

Robyn Bartel, Louise Noble and Wendy Beck

Introduction

The interrelated existence of humanity and environment is made starkly apparent by the dire consequences forecast for both humans and the globe in the (proposed) Anthropocene epoch (Crutzen & Stoermer, 2000; Crutzen, 2002). Delusions of human 'superiority' amidst human/non-human divisions have contributed to a collapse of Earth systems, and it is perhaps the ultimate irony that this may lead to our own demise at the hands of nature. As Latour (2014) has observed, 'through a complete reversal of Western philosophy's most cherished trope, human societies have resigned themselves to playing the role of dumb object, while nature has unexpectedly taken on that of the active subject!' (pp. 11–12).

Several neologisms have been crafted in an attempt to describe human-nature relationships beyond dualism, including envirosocial (Bartel et al., 2014), hydrosocial (Linton, 2010, 2014; Linton & Budds, 2014; Swyngedouw et al., 2002; Wilson, 2014), social-ecological (Berkes et al., 2003; Folke et al., 2005), earth system science (Gifford et al., 2010), socionatures and naturecultures (see Haraway, 2008; White, 2006), as well as waterscapes (see Karpouzoglou and Vij, 2017). Such terms have arisen from the decentring of the human, particularly in environmental (especially ecocentric) research, and from the increasing recognition in the humanities of non-human agency, as well as the relational and new materialism turns in scholarship more broadly (see Castree & MacMillan, 2001). Rivers have been re-imagined as actor-networks (Kortelainen, 1999) and as legal persons (see Chapter 5, this volume). Comparable ontologies – in which nature and human are not separate (or separable) but co-constituted, and material things, including places and landscapes (and waterscapes), rivers and rocks, are not inanimate but exhibit agency – have long been held and practised by First Nations peoples (see for example Strang, 1997; 2009; Watts, 2013; Weir, 2009; and see also Latour, 2014). Similarly, local and cultural knowledge of place, also known as vernacular knowledge (Bartel, 2014), often demonstrates deeper appreciation for the contingency, complexity, emergence and dynamism of human-environmental interactions (see Chapter 11, this volume). However, such appreciations have been for the most part marginalised – perpetrating further social injustice – as well as producing sub-optimal water management and poor environmental outcomes (see for example Jackson, 2017; Gibbs, 2009; McLean, 2014; Linton, 2010, p. 59; Strang, 2009; Tipa & Nelson, 2017). Irrespective of their level of sophistication, alternative knowledges and ways of understanding our world have often been ignored or considered deficient due to the preference in policy circles for 'modern' scientific and technocratic approaches (see for example Bartel, 2014; Delvaux & Schoenaers, 2012; Gregory, 2001, p. 96; Taylor & De Löe, 2012; Watts, 2013, among others).

This chapter will use the views of key stakeholders in the Australian case study described in Chapter 11 (this volume) to further demonstrate the value of incorporating vernacular (place-based) knowledges in water policy. There is a significant opportunity to realize more inclusive governance, and to enable and support transformative learning, in order to derive more diverse and dynamic policy outcomes than are presently being generated.

Background: The voices of the Namoi catchment, NSW, Australia

This chapter draws on a case study of the Namoi catchment of New South Wales (NSW), Australia (see Figure 11.1, Chapter 11, this volume), including interviews conducted with a purposive sample of key stakeholders, irrigators and landholders. A grounded theory approach disclosed major themes in the data, several of which, in relation to water policy trajectories, and some of the major social and environmental consequences, as well as pivotal regulatory failures, have already been discussed in Chapter 11 (this volume). In this chapter, further themes are reported that concentrate on the need for increased recognition of the value of place-based voices and how best these may be incorporated into policy and governance, particularly through transformative learning. One of the chief features of such voices are their co-creation by and in place – the agency of place (particularly water) – in generating the vernacular knowledge/s that the voices communicate (see Bartel, 2014). Selected quotes are used here to highlight and illustrate areas critical to this examination. Generic appellations are used to preserve the anonymity of the ten interviewees who participated in semi-structured interviews of one to two hours in length (see further Chapter 11). The participants were interviewed individually but together comprise a mixed age and gender cross-section of long-term residents of the catchment, either living on working properties (productive agriculture, including irrigation) and/or engaged in off-farm occupations involving natural resource management, including in government positions and the private sector.

Discussion

Transformation beckons

Given the culpability of current water policies and approaches in the creation of several impending water crises, it is imperative that we develop new ways of relating to water (see for example Cummins & Watson, 2012; Moore et al., 2014; Marshall & Alexandra, 2016). In announcing the Murray-Darling Basin (MDB) Plan in 2007, the then Australian Prime Minister John Howard declared that 'the old ways of managing the Murray-Darling Basin has reached its use-by-date'. However, the approach subsequently taken in the MDB (see Table 11.1 in Chapter 11, this volume), was appraised by local stakeholders as destined for failure:

> *It's our approach that's flawed, you know, like how to fix rivers is flawed... the conventional way will never deal with [the problems].* (Namoi 5)

Yet failure is not always a bad thing – as learning from failure is a significant opportunity to craft improvements (Howlett, 2012; May, 1992). While there are no guarantees that any learning is effective (Armitage et al., 2008), and there are political risks associated with acknowledging policy failure (see Bartel, 2016), it is much harder to make any gains without some kind of conscious reflective practice (see for example Mezirow, 1990; 1996; Sanderson, 2002). Learning from experience is considered essential (see for example Tan, 2010; Vella et al., 2015) and identifying what exactly needs to be learnt (and by whom) is also vitally important.

A wide range of evaluative approaches has been described in public policy circles. Within the environmental arena, 'adaptive management' is one of the most popular (Holling, 1978; Williams, 2011). It is also sometimes referred to as 'adaptive governance' (see for example Cosens et al., 2014; Folke et al., 2005; Nelson et al., 2008), and in Australia by the acronym 'MERI', which stands for: Monitoring, Evaluation, Reporting, Improvement. Adaptive management and similar approaches describe a mode of learning in which evidence of the consequences of an intervention is fed back into an iterative 'Plan-Do-Check-Act' improvement cycle. Also known variously as 'learning-by-doing', 'experiential' and 'action learning', solutions are incrementally modified after repeated application and testing (see Argyris & Schön, 1974).

Simple learning cycles have limitations. The approaches may not be taken at the requisite government level, and even when performed, may only capture the most immediate, apparent and measurable features, including those related to policy failures, and overlook, and therefore leave untreated, more fundamental structural, systemic and philosophical issues (Pahl-Wostl, 2009; Waddell, 2002). They may also suffer from being exclusionary rather than pluralistic and participatory (Davidson-Hunt, 2006). For initiating new and innovate insights and approaches, transformative learning, also known as triple-loop learning (and also epistemic learning, see Sterling, 2011), has been recommended. Transformative learning was originally described in adult education (see for example Taylor, 2008; also Mezirow, 1996), and its effects on the individual characterised as learning that is 'about change – dramatic and fundamental change in the way we see ourselves and the world in which we live' (Brown & Posner, 2001 p. 2).

Transformative learning is designed to question assumptions, worldviews, values and paradigms that may constitute the ultimate (as opposed to proximate) causes of problems, including environmental crises, and therefore also promote more sustainable, holistic, lasting and effective solutions (Pahl-Wostl, 2009; Sterling, 2011; Tàbara & Chabay, 2013; Wilner et al., 2012; and see also Bateson, 1972; Flood & Romm, 1996; Freire, 1970). This may include the triggering of 'regime shifts' into different paradigms (see Chaffin et al., 2016). As Pahl-Wostl (2009) has observed, 'there would be no innovation or evolution to higher adaptive capacity if individuals or organizations never revisited basic values and beliefs' (p. 359).

The Namoi Catchment Management Authority (CMA), the watershed-based administrative unit responsible for water from 2004 until 2014, described and committed to a process of adaptive management based on single, double and triple-loop learning, although their description of triple-loop learning was focused on people and processes rather than values and paradigms (Namoi CMA, 2013, p. 53). Their successor, the Local Land Services, presented a similar conceptualisation in their Strategic Plan (NWLLS, 2016), which also appears to conflate value systems largely with processes, and somewhat inaccurately describes learning for transformation as entailing 'more specific levels of questioning' (NWLLS, 2016, p. 33). This approach also appears more focused on governance structures and local participation in decision-making than the interrogation of deeply-held assumptions. Although processes are an essential component of transformative learning, much more is required. However, there may be a useful bridge built between processes and values: by using more inclusive processes to facilitate the incorporation of views that challenge the *status quo*. This would mirror the aims of some social and collaborative learning models in environmental management (see for example Collins & Ison, 2009; Faysee et al., 2014; Woodhill & Röling; 1998), as well as sustainability learning, which, according to Tàbara and Pahl-Wostl (2007):

> relates to learning to develop the capacity to manage options for the adaptation of human societies to the limits and changing conditions that are imposed by their own social-ecological systems… [which entails] … A more hybrid, relational, and co-evolutionary holistic understanding of human-natural interactions. (p. 3)

As one of our interviewees also asked:

> *[can we have] a symbiotic relationship [with nature?]* (Namoi 2)

Symbiosis means mutually beneficial, and a mutually constituted, de-binarised view of human and nature has been championed elsewhere as essential to address the ultimate causes of many of our environmental failings (see for example Bartel et al., 2014; Weir, 2009). The Senate Committee charged with 'refreshing' the Basin Plan (see Chapter 11, this volume) recommended that the governing legislation be amended 'to make clear the *equal* standing of economic, social and environmental needs and outcomes' [emphasis added] (SSC, 2016). This was a response to a political perception that the environment was outweighing, or, in other words, 'trouncing' other objectives, and, if pursued, would demote environmental sustainability from its original priority status (Quiggin, 2012, p. 53). Such a recommendation assumes that there is a competition between environmental, social and economic goals, rather than a recognition that the economy is embedded in society, which is in turn dependent upon the environment (Bartel et al., 2014). To go beyond this, and to articulate and achieve de-binarised envirosocial outcomes requires that environmental objectives be framed as more than mere trade-offs that need to be 'balanced' so that they do not 'outweigh' economic goals (Bartel et al., 2014).

Current frames persist in fragmenting social from environmental concerns, and, further, economic from social and environmental aims. This may be another example of path dependence (see Chapters 1 and 9, this volume), as the goals, processes and systems of previous approaches are never completely overtaken by the new, and to which even more radical ideas may need to seek some conformity to obtain any purchase. Cummins and Watson (2012, p. 9) observe that current policy makers are heavily constrained by the decisions made in the past: policies, as intended, leave consequences. Industries are born and encouraged; towns grow and become dependent on their industries, which exert self-interested influence for their own survival; and of course, the worldviews that generated the policies also remain (see Schmidt, 2014). Certain ideological threads have proven remarkably resilient to changing circumstances (adopting the narrower definition of resilience as 'bouncing back', rather than adapting to new circumstances or changing state, see discussion in Brown, 2014; see also Folke et al., 2010). As Marshall and Alexandra (2016) contend, there is 'a strong tendency to reinforce the *status quo* and limit innovation' (p. 379).

This is true also of the word 'management' itself, and the ontologies inherent in the word, which privilege human agency and perpetuate the arrogance of humanism (Hamilton, 2010; Linton, 2010, p. 58; Weir, 2009). It is anthropocentric (and anthroparchic), especially when also considering the dominant frame of water as a resource for human use (see for example Linton, 2010; Weir, 2009), which survives, even when using the apparently more reflective approach of adaptive management. The Namoi Strategic Plan (NWLLS, 2016), although strong on participation, is also heavily aimed at being 'commercially-focused', which is a marked regression from the regulatory role undertaken by their forebears. This move reflects an economic and resourcist agenda driving commodification and commensuration, including of the natural environment (see for example Bartel et al., 2014; Havemann, 2005; Merchant, 1980; 1989; Tarlock, 2000), through ecosystem services (see for example Wallace et al., 2003) and now also cultural ecosystem services (Bark et al., 2015). There is a strong historic lineage here with popular cultural notions of a natural relation between private property (Graham, 2011) and productivism (see for example Argent, 2002; Cocklin et al., 2006), in which ownership, labour and productivity are valorised (see for example Strang, 2009, p. 123; see also Chapter 3, this volume). As one of our participants observed:

> *there's this paradigm of land use… that if you're not farming land in Australia, then it's wasted.* (Namoi 5)

Thus, desirable social and cultural outcomes may continue to be conflated with economic growth (see for example Axelrod, 2011; Goss, 2003). However, as one interviewee identified, social interests may in reality be ill-served by the growth agenda:

> *The social and economic are often diametrically opposed.* (Namoi 5)

Anthropocentric policies, however, still persist, although it is (as always) a select group of humans, and select group of interests and worldviews, that are most directly served, and other humans, non-humans and other worldviews are often excluded (see Bartel, forthcoming; see also Ballantine, 2017; Jackson & Barber, 2016; Hidalgo et al., 2017). The value of a more inclusive approach to encompass these diverse views is increasingly being recognized, through greater collaboration, as within the governance agenda discussed in Chapter 11 (this volume); however, overall these proposals have failed to engage with a diversity of voices, or interrogate what such a process might mean for policy content and implementation.

Vernacular knowledge and verisimilitude

Participants in the Namoi case study (see also Chapter 11, this volume) identified that a critical first step towards change was local salience and verisimilitude gained through local knowledge of place:

> *What would be a better method, would be scale-related… It would be the catchment level… That's actually mirroring the resource.* (Namoi 1)

It is increasingly recognized that policies, once intended to be universal, cannot be place-transcendent or 'one size fits all' (NZ Productivity Commission, 2012, p. 29). The NSW Natural Resource Commission, in its review of the 2004 Water Sharing Plans (see Chapter 11, this volume), devoted much attention to the importance of the local scale in its recommendations – specifically that water management should focus on addressing local issues, recognising social and environmental connections within the landscape, encouraging collaboration and supporting meaningful community engagement, including the sharing of information and incorporation of new knowledge (NRC, 2013). The Independent Review of the Water Act (see Chapter 11, this volume) was highly complimentary of the localism pursued by the Commonwealth Environmental Water Holder (Australian Government, 2014, p. 74ff). The environment is heterogeneous and policies must be also – which to a certain degree is to return to the almost infinite variation in locale-based responses to water that existed in pre-modern Europe, and that continue elsewhere in the world today (Linton, 2010, pp. 75, 81, 87).

Policies are also rarely future-proof or time-transcendent – as both biophysical and socioeconomic environments are dynamic and undergo continual change. Or, in other words, what worked in the environment and context of yesterday might not work today (see for example Milly et al., 2008). The NSW Natural Resource Commission (NRC, 2016) in its review of the Groundwater Sharing Plans (see Chapter 11, this volume) recommended that: 'adaptive management provisions should be retained in plan replacements so that amendments can be made as new information becomes available' (p. 33).

Furthermore, local knowledge is not static – it is dynamic, in order to reflect changing conditions, and ways of knowing change over time. Nor is science static – the best available science yesterday is not the best available science today, continual refinement and testing may refine a fact or debunk a hypothesis and as the environment is changing, observations and hypotheses will also need to change to reflect current conditions more accurately (see Chapter 4, this volume). Nor are political environments static. Therefore, policies, to be effective, also cannot be static; and rather than reactive and responsive to politically expedient cycles, they must undergo continual adjustments to reflect the reality of dynamic, heterogeneous complexity in the natural world. In the Namoi it was agreed that dynamism was essential as management is not only contingent but also emergent. As one of our research participants expressed it:

> *it's gotta be a dynamic process.* (Namoi 5)

While simple loop learning may be appropriate where only minor modifications are required, transformative learning is needed where circumstances demand it, and such changes must be supported and facilitated. Examples and comparison of the application of the concepts of multiple loop learning is provided in Table 12.1.

Multi-vocal, multi-scalar solutions

Localism, while possessing advantages, has also been critiqued for promoting insularity and slipping towards environmental determinism (see discussion in Arnold, 2005a; Tomaney, 2013; and see also Bartel, 2014). And any experience, whether anecdotal or empirical, if limited or biased, may be less useful for policy (see Chapter 4, this volume), although may still be essential in consensus building (see Baldwin & Ross, 2012). The potential downsides of localism were explicitly acknowledged by our case study participants:

> *I guess we're all selfish to a pretty big extent, but we think, well, you can't have our water.* (Namoi 3)
> *…to have community consultation, you are inevitably talking with people who've got a heavy vested interest, economically, in the short term.* (Namoi 5)

Table 12.1 Learning cycles, policy learning and water appreciation

Loops	Learning Description	Application in Policy	Water Application
Single	Single loop learning is a feedback cycle in which the knowledge and lessons learned from repeated iterations of policy or other intervention are 'fed back' in to improve the action and therefore also, hopefully, the outcomes. It is repeated to ensure continuous learning and improvement. According to Pahl-Wostl (2009): 'single-loop learning refers to an incremental improvement of action strategies without questioning the underlying assumptions' (p. 359).	According to May (1992): 'instrumental policy learning; entails lessons about the viability of policy instruments or implementation designs. . . . [It] entails new understandings about the viability of policy interventions or implementation designs' (pp. 332, 335). It may result in new 'policy instruments or policy goals' (p. 336).	Expanding water supply by the construction of large dams and impoundments in order to meet growing demand. Huntjens et al. (2011, p. 150) provide the example of increasing dykes to improve flood protection.
Double	According to Pahl-Wostl (2009) '[d]ouble-loop learning refers to a revisiting of assumptions (e.g. about cause-effect relationships) within a value-normative framework' (p. 359).	According to May (1992) 'social policy learning: entails lessons about the social construction of policy problems, the scope of policy, or policy goals. . . [It] entails a new or reaffirmed social construction of a policy by the policy elites of a given policy domain. . . . It entails reaffirmation or revision of the dominant causal reasoning about policy problems, interventions, or objectives.' (pp. 332, 337). It may result in 'a change in policy aims or goals' (p. 336).	Huntjens et al. (2011, pp. 150, 151) provide examples of reducing demand to match supply, increasing boundaries for flood management and including some public participation (see for example Huntjens et al., 2011, p. 151).

(continued)

Table 12.1 (continued)

Loops	Learning Description	Application in Policy	Water Application
Triple/ Transformative	According to Pahl-Wostl (2009), in 'triple-loop learning one starts to reconsider underlying values and beliefs, world views, if assumptions within a world view do not hold anymore' (p. 359).	Transformative regulation entails lessons about the normative elements and epistemological and ontological basis and biases of public policy and institutions, includes questioning how a public policy problem is and has been constructed and perceived as a problem, how current systems may bias the construction of problems and solutions and obscure alternative directions. It entails a new or reaffirmed appreciation and understanding of problems. This may result in the reconstruction of policy problems as well as policy responses, institutions and systems.	An example would be to look at how the environment is understood and constructed, and how harms are appreciated and perceived. Solutions here are radical and entail paradigm shifts in thinking and behaviour, for example entirely new management measures or interventions (see Huntjens et al., 2011, p. 151; and see also Watts, 2013).

However, transformative learning is about questioning values, and it was also identified that much short-term thinking was due to the external influence of the market:

> *The challenge is to get our ... economies working within the constraints of the natural environment and within the patterns of variability and unpredictability of the natural environment ... and that's particularly the case with our rivers. We've got highly unpredictable river systems and yet our bank manager wants a payment every year.* (Namoi 5)

Landholders recognized the constraints of not being able to be green when they were in the 'red' (see for example Cocklin et al., 2006; Lawrence et al., 2004; Vanclay, 2004):

> *[before] a lot of the crops we grew were luxury crops for the soil... we'd turn it back into soil... [now] we have to go for whatever crop's going to make the most dollars, in order to survive. And it's the old problem, that when you're in the red, you can't be green.* (Namoi 2)

To counter the influence of the market, and the tensions associated with competing endogeneous and exogeneous interests (Strang, 2009, p. 51), it was acknowledged that there needed to be many local voices, and that the process needed to take into account multiple, place-based meanings:

> *[Water] has completely different meanings and it's in [the] context of where you are in the landscape.* (Namoi 6)

Local knowledge is heterogeneous and a process to encompass and support poly-vocality was seen as essential:

> *...doing it as a cumulative approach and everybody putting in their knowledge... And everybody coming up with that local solution to a local problem... They're also going to be resilient in being able to respond to what's required for their environment at the local level.* (Namoi 1)

It was furthermore acknowledged that this would and should encompass multiple disciplines and ontologies, including Indigenous knowledge, which has so often been excluded.[1] In the Namoi, both 'mainstream' science and 'Aboriginal science' were recognized as desirable:

> *I reckon it'd be quite simple to marry the (sic) science up with Aboriginal science.* (Namoi 6)

Such a plural ontological approach may also be applied to other knowledges, and becoming inter-, multi-, trans-, pan- and post-disciplinary has been identified as necessary (see Norgaard & Kallis, 2011; Stember, 1991; Weisz & Clark, 2011). Tàbara and Chabay (2013) suggest the coupling of diverse human knowledge with social-ecological systems, each dynamic, which would require a paradigm shift in our traditional approaches to policy – in all areas. Drawing on transformative learning theory and principles, such approaches to policy would require the utilisation of a wider range of knowledges and multiple views and experiences, at a range of scales and perspectives, rather than limiting ourselves to the current range of epistemologically narrower sources and areas (Kahane, 2004; Robertson et al., 2000; see also Chapters 8 and 9, this volume). Our interviewees appreciated that a range of scales, such as 'nested'-ness and poly-centricity, was required:

We need [citizen] juries nested within juries. (Namoi 5)

Such a system (see for example Irvin & Stansbury, 2004, also for critique) can be used to effect deliberative democracy (Smith & Wales, 2000) and is one route by which vested interests may be addressed (since juries are not selected only from stakeholder groups but the general public). Similarly, multi-scalar 'nested'-ness may be able to address bias-related issues and be aligned according to subsidiarity principles (see for example Curtis et al., 2014; Linton, 2010, p. 56; Marshall et al., 2013).

Syncretic governance and heterotic futures

The reforms recommended here are designed to enable transformative learning through multi-vocality, multi-scalarity and multi-temporal[2] integrated engagement in water appreciation and decision-making, with the intention of modifying our current paradigm, and present circumstances. This process is labelled here as syncretic governance, a neologism drawing on the idea of syncretism from anthropology and sociology (Scott, 1977a and b; 1998; 2013; Shaw & Stewart, 1994; Stewart, 1999), as well as geography (see for example Curti & Johnson, 2017), and which is used here to refer to the combination, in water policy, processes, management and institutions, of different ontologies.

Syncretism and similar concepts, such as creolisation (a term borrowed from linguistics, see Palmié, 2006; Stewart, 1999), have been criticised for maintaining and reifying, rather than problematising and addressing, essentialism and prejudice. Resulting 'mixes' of supposedly bounded and homogeneous component parts may be considered inferior to the superior 'unleavened' product or standard. This is especially the case in colonial and imperial contexts, where histories of violence and subjugation may continue as well as permeate contemporary acculturation and appropriation (including of academic research, see Ballantine, 2017, and also Stewart & Ernst, 2003). This is particularly relevant for the current situation, as traditional and other subaltern knowledges may be ignored, destroyed or subsumed and assimilated within dominant and more mainstream views (see for example Stewart, 1999). However, reappraisals of syncretism from

more critical perspectives have identified that there may be alternative consequences of the agency of subaltern peoples, knowledge and systems, including
resilience, 'survivance' (Vizenor, 2008) and the potential creativity of generative
conflict and resistance (see Stewart, 1999; 2007). Bhabha (1994) has deployed
the term hybridity to describe a truly generative and often also subversive and
transgressive re-imagining in a 'third' space between and beyond any imaginary
discrete antecedents (see also Chapter 5, this volume). Hybridity has also been
used by Latour (1993), Swyngedouw (1996), Whatmore (2002), Castree (2003),
Demeritt (2005) and Linton (2010, p. 68). It is a label borrowed from biology,
where its meaning has also included sterility, and a more appropriate biological
term may therefore be symbiosis (already referred to above). However, the latter has had little purchase in the literature (although it has been used by Scott,
1977a; 2013). Other terms include synthesis (Stewart, 1999; Stewart & Ernst,
2003), pluralism (Berman, 2012), bricolage (Cleaver, 2002), cosmopolitanism
(Berman, 2012), co-production (Armitage et al., 2011; Carolan, 2006; Jasanoff,
2004), co-evolution (Burton, 2002; Kallis, 2007; Norgaard & Kallis, 2011; Weisz
& Clark, 2011) and co-becoming (Bawaka et al., 2013; 2016; and see also Jones &
Instone, 2016).

Co-becoming – the most resonant to the meaning intended here – involves
a plurality of perspectives and multiplicity of voices that encompasses humans,
non-humans (including place and place agents such as water and groundwater)
and more-than-humans (including technology but also material-plus-human
agents such as landscapes) (see also Anderson et al., 2010; Tuck & McKenzie,
2015). Such an approach may be critiqued on the basis that it could amount to
cultural relativism. However, we adopt Wilcock, Brierley and Howitt's (2013)
view that recognising multiple ways of connecting with the environment is
not synonymous with recognising *any* way, and that pluralistic and multivalent
conceptualisations are to be lauded rather than avoided (see also Broderick,
2005; Braun, 2006; Brooks, 2002; Gonzalez et al., 2009), particularly perhaps
as unitary or monadic approaches have hardly proven themselves superior
(Linton, 2010, p. 50).

The synthesis of knowledges, however, is pertinent in the context of inter-
disciplinarity, which similarly describes the integration and synthesis of multiple
knowledges and approaches and generative processes of collaboration (Stember,
1991). As Wilkinson (1992) has observed:

> Western water has long been the province of 'experts', mostly engineers and
> lawyers. Professionals from many other disciplines – economists, historians,
> biologists, sociologists, political scientists, and ecologists are just a few – have
> much to offer to water policy. So, too, does the generalist, the conscientious
> citizen, have much to offer this field, where a fresh look is so critical . . . The
> engineering mentality has been one factor in making water policy one-sided
> in favor of building and extractive uses. It is now clear that there is much
> more in our rivers than we are allowed to see through the lens with which
> our policies view them. (p. 291)

Heterotic is used here to describe the outcomes of syncretic governance. Although in physics and biology it may refer to the combination from two parents, like heterodoxy, it can also mean simply *more* alternatives and diversity (i.e. from the roots 'hetero' + 'otic' = 'other process').[3] As Lucashenko (2005) describes it, we need to replace our 'big' narratives of abundance with 'earthspeaking', that is, multiple, smaller, place-based stories:

> You will think of it as a big story, a national story . . . Turn the rivers backwards, or find that inland sea. It's Burke and Wills, it's the Bushtucker Man, its drought and flooding rains, but no. Stop. Pause for breath, since people in a hurry cannot feel. You say: It's a big country. I say: there isn't much room for everyone's big stories.... . I am earthspeaking, talking about this place, my home and it is first, a very small story . . . One valley.

There is a need to move away from 'broad abstractions' towards 'local realities' (Strang, 2009, p. 287) and towards tailoring bespoke and unique solutions (Bartel, 2014; Brooks, 2002; Linton, 2010, p. 56). Re-locating water, recognising its *genius loci*, would counter the historic and current deterritorialisation and dematerialisation that occurs through conceptual abstraction and technical control (Linton, 2010, pp. 18, 82). Currently there is little to 'no place' for '"local" knowledge' (Gregory, 2001, p. 97), and engagement with water is 'hardly on the terms of the people who live in these places' (Linton, 2010, p. 227). Nor is there place for water – or place itself.

Places and water have their own agency and may also exercise their agency through people – through local stories and appreciations. There is also a need for appreciation of the Earth as a place – the only place where surface water has been found to flow, and at the global scale the magnitude of our water security and environmental health challenges become evermore apparent – the global water story is one of the chief features of the Anthropocene (Bogardi et al., 2013; Steffen et al., 2015). Appreciating water issues at a global scale should not imply universal or universalist conceptualisations, rather a multi-scalar, and poly-centric, understanding; for such appreciations are adapted and adopted locally as well – the local and the cosmopolitan are imbricated rather than antithetical (see Palmié, 2006; and see also Arnold, 2005b; Gupta et al., 2013).

According to Scott (1977a; 1977b), syncretic folk traditions are reflexive: there are always etic as well as emic dimensions to place, and there are many external aspects that may be adopted and adapted, emerging and evolving over time into localised 'little traditions'. Forces that are reductive and coercive may become part of the 'great traditions' of the elite, with which 'little traditions' may conflict (Scott, 1977a; 1977b). Such resistance should not be dismissed as narrowly parochial or reactionary – as here discussed between local heterogeneous understandings of place and water compared to centralised modernist, anthropocentric and managerialist conceptions. Scott (1977a; see also 2013) refers to such antithetical divergences in the religious context as 'profanations' that offer a 'critique of the existing order' (Scott, 1977a, p. 31) and 'an alternative symbolic universe' (Scott, 1977b, p. 224). They may underpin various forms of creative and violent rebellion – as well as traditional court action that may also be used to generate better outcomes – even if only as a means to motivate negotiation (Macdonald & Young, 2001, p. 83), and more imaginative consensus and co-construction (Linton, 2010, p. 246). Thus may alternative ontologies of water as agent have a transformative effect on policy, ideally peaceably, and with long-term continuous improvement, through collaborative, syncretic govern-ance and triple-loop learning to generate heterotic governance for heterogeneous social and biophysical environments.

Past practices have created legacy issues and political and social systems path dependencies (see Chapter 1, this volume) that will be difficult to address and divert. This includes patterns of thinking. However, our historical practices have not just been destructive, and at least one consequence of poor water man-agement may be creative: the recognition of environmental consequences, or in other words, the agency of place in responding and communicating change, may assist us in challenging our reliance on dominant techniques and ontologies regarding water in Australia.

Conclusion

> … *a lot of natural resource management is based on ignorance – we don't know how the system works over hundreds of years.* (Namoi 5)

Much about the environment remains unknown, particularly when factoring in the Anthropocene and a 'climate changed future' (see for example Horne, 2014; Murray et al., 2003). There is much to be said for embracing humility and acknowledging our deficiencies when it comes to knowledge and knowing: for humanist arrogance to know, and control, via management, is a large part of the problem we face in our relationships with water. It masks and prevents a deeper appreciation of and enquiry into how our policies and practices mar rather than mirror natural landscapes, including increasingly critical questions of water health, quality and scarcity. It includes the ontology, and sympathetic ideologies (see Bartel, forthcoming), of a positivist world 'out there' with resources to be exploited in the service of humanity, and more specifically the interests of a few, and of a narrow growth agenda. For both equity and efficacy reasons it is there-fore critical to ask whose and what types of knowledge are sought, and whose are overlooked in the current management frame (see for example Howitt & Suchet-Pearson, 2006; Tadaki & Sinner, 2014; Taylor & De Löe, 2012).

Diverse ontologies may be able to challenge the hegemony of current appre-ciations and, through respectful and generative engagement, creative and innovative reforms. We need to acknowledge the tensions between 'control and management' compared to 'humility and reflexiveness' (Lövbrand et al., 2009, p. 12). Vernacular knowledge derived from living on and with the land (and water) is often already reflexive to place agency, rather than seeing nature as sep-arate and divisible. Tellingly, however, most formal policies still attempt to 'deal' with water as an isolated issue rather than integrated within the environments and cultures of peoples. Thus vernacular knowledge has vital lessons and transformative potential for developing understandings that may better support our natural environments and water, as well as ourselves.

Acknowledgements

Thanks to Arts New England, our interview participants and the Namoi catchment for supporting our fieldwork. The research was conducted following Australian Human Research Ethics protocols, University of New England approval no HE10/193.

Notes

1 See for example Cranney and Tan, 2011; House Standing Committee on Regional Australia, 2011; Howitt and Suchet-Pearson, 2006; Jackson, 2005; Jackson and Barber, 2013; 2016; McNamara and Westoby, 2011; Tadaki and Sinner, 2014; Tipa and Nelson, 2017; Weir, 2009; Woodward and McTaggart, 2015, among others.
2 The prefixes multi- and poly- have been used somewhat interchangeably.
3 Any apparent relationship to heterotica or heterotopia is unintended. One may wish to question, if not also decry the formulation of yet more 'buzz' words, but language is important as it shapes, constrains and maintains ways of thinking. If change is required then new words describing new ways may assist us (see for example Boroditsky, 2011).

References

Anderson, J., Adey, P., & Bevan, P. (2010). Positioning place: Polylogic approaches to research methodology. *Qualitative Research*, 10(5), 589–604.

Argent, N. (2002). From pillar to post? In search of the post-productivist countryside in Australia. *Australian Geographer*, 33(1), 97–114.

Argyris, C., & Schön, D. (1974). *Theory in Practice. Increasing Professional Effectiveness.* San Francisco: Jossey-Bass.

Armitage, D., Berkes, F., Dale, A., Kocho-Schellenberg, E., & Patton, E. (2011). Co-management and the co-production of knowledge: Learning to adapt in Canada's Arctic. *Global Environmental Change*, 21(3), 995–1004.

Armitage, D., Marschkeb, M., & Plummerc, R. (2008). Adaptive co-management and the paradox of learning. *Global Environmental Change*, 18(1), 86–98.

Arnold, C. A. (2005a). Introduction. Integrating water controls and land use controls: New ideas and old obstacles. In C. A. Arnold (Ed.), *Wet Growth: Should Water Law Control Land Use* (pp. 1–56). Washington, D. C.: Environmental Law Institute.

Arnold, C. A. (2005b). Polycentric wet growth: Policy diversity and local land use regulation in integrating land and water. In C. A. Arnold (Ed.), *Wet Growth: Should Water Law Control Land Use* (pp. 393–433). Washington, D. C.: Environmental Law Institute.

Australian Government. (2014). Report of the Independent Review of the Water Act 2007. Canberra: Australian Government.

Axelrod, J. (2011). Water crisis in the Murray-Darling Basin: Australia attempts to balance agricultural need with environmental reality. *Sustainable Development Law & Policy*, 12(1), 51–53.

Baldwin, C., & Ross, H. (2012) Bridging troubled waters applying consensus-building techniques to water planning. *Society and Natural Resources*, 25(3), 217–234.

Ballantine, A. (2017). The river mouth speaks: Water quality as storyteller in decolonisation of the Port of Tacoma. *Water History*, 9(1), 45–66. doi:10.1007/s12685-016-0179-5.

Bark, R. H., Barber, M., Jackson, S., Maclean, K., Pollino, C., & Moggridge, B. (2015). Operationalising the ecosystem services approach in water planning: A case study of indigenous cultural values from the Murray-Darling Basin, Australia. *International Journal of Biodiversity Science, Ecosystem Services & Management*, 11(3), 239–249.

Bartel, R. (2014). Vernacular knowledge and environmental law: Cause and cure for regulatory failure. *Local Environment: The International Journal of Justice and Sustainability*, 19(8), 891–914.

Bartel, R. (2016). Legal geography, geography and the research-policy nexus. *Geographical Research*, 54(3), 233–244.

Bartel, R. (Forthcoming). Place-thinking: The hidden geography of environmental law. In A. Philippopoulos-Mihalopoulos, & V. Brooks (Eds.), *Handbook of Research Methods in Environmental Law*. Cheltenham, UK: Edward Elgar Publishing.

Bartel, R., McFarland, P., & Hearfield, C. (2014). Taking a de-binarized envirosocial approach to reconciling the environment vs economy debate: Lessons from climate change litigation for planning in NSW, Australia. *Town Planning Review*, 85(1), 67–96.

Bateson, G. (1972). *Steps to an Ecology of the Mind*. New York: Ballantine.

Bawaka Country including Suchet-Pearson, S., Wright, S., Lloyd, K., & Burarrwanga, L. (2013). Caring *as* country: Towards an ontology of co-becoming in natural resource management. *Asia Pacific Viewpoint*, 54(2), 185–197.

Bawaka Country, Wright, S., Suchet-Pearson, S., Lloyd, K., Burarrwanga, L., Ganambarr, R., Ganambarr-Stubbs, M., Ganambarr, B., Maymuru, D., & Sweeney, J. (2016). Co-becoming Bawaka: Towards a relational understanding of place/space (2015). *Progress in Human Geography*, 40(4), 455–475.

Berkes, F., Colding, J., & Folke, C. (2003). *Navigating Social-Ecological Systems: Building Resilience for Complexity and Change*. Cambridge: Cambridge University Press.

Berman, P. S. (2012). *Global Legal Pluralism: A Jurisprudence of Law Beyond Borders*. New York: Cambridge University Press.

Bhabha, H. K. (1994). *The Location of Culture*. New York: Routledge.

Bogardi, J. J., Fekete, B. M., & Vörösmarty, C. J. (2013). Planetary boundaries revisited: A view through the water lens. *Current Opinion in Environmental Sustainability*, 5(6), 581–589.

Boroditsky, L. (2011). How language shapes thought. *Scientific American*, 304(2), 62–65.

Braun, B. (2006). Towards a new earth and a new humanity: Nature, ontology, politics. In Castree, N., & Gregory, D. (Eds.), *David Harvey: A Critical Reader* (pp. 191–222). Malden, US: Blackwell.

Broderick, K. (2005). Communities in catchments: Implications for natural resource management. *Geographical Research*, 43(3), 286–296.

Brooks, D. B. (2002). *Water: Local Level Management*. Ottawa: International Development Research Institute.

Brown, K. (2014). Global environmental change I: A social turn for resilience? *Progress in Human Geography*, 38(1), 107–117.

Brown, L. M., & Posner, B. Z. (2001). Exploring the relationship between learning and leadership. *Leadership & Organization Development Journal*, 22(6), 274–280.

Burton, L. (2002). *Worship and Wilderness: Culture, Religion and Law in Public Lands Management.* University of Wisconsin Press.

Carolan, M. S. (2006). Sustainable agriculture, science and the coproduction of 'expert' knowledge: The value of interactional expertise. *Local Environment: The International Journal of Justice and Sustainability*, 11(4), 421–431.

Castree, N. (2003). Environmental issues: Relational ontologies and hybrid politics. *Progress in Human Geography*, 27(2), 203–211.

Castree, N., & MacMillan, T. (2001). Dissolving dualisms: Actor-networks and the reimagination of nature. In N. Castree, & B. Braun (Eds.), *Social Nature: Theory, Practice and Politics* (pp. 208–224). Malden, MA: Blackwell.

Chaffin, B. C., Garmestani, A. S., Gunderson, L., Benson, M. H., Angeler, D. G., Arnold, C. T., Cosens, B. A., Craig, R. K., Ruhl, J. B., & Allen, C. R. (2016). Transformative environmental governance. *Annual Review of Environment and Resources*, 41, 399–423.

Cleaver, F. (2002). Reinventing institutions: Bricolage and the social embeddedness of natural resource management. *The European Journal of Development Research*, 14(2), 11–30.Cocklin, C., Dibden, J., & Mautner, N. (2006). From market to multifunctionality? Land stewardship in Australia. *The Geographical Journal*, 172(3), 197–205.

Collins, K., & Ison, R. (2009). Jumping off Arnstein's ladder: Social learning as a new policy paradigm for climate change adaptation. *Environmental Policy and Governance*, 19(6), 358–373.

Cosens, B. A., Gunderson, L., & Chaffin, B. C. (2014). The adaptive water governance project: Assessing law, resilience and governance in regional socio-ecological water systems facing a changing climate. *Idaho Law Review, Natural Resources and Environmental Law Edition*, 51(1), 1–27.

Cranney, K., & Tan, P. (2011). Old knowledge in freshwater: Traditional ecological knowledge in the Richmond catchment. *Australasian Journal of Natural Resources Law and Policy*, 14(2), 71–113.

Crutzen, P. J. (2002) The geology of mankind. *Nature*, 415(6867), 23.

Crutzen, P. J., & Stoermer, E.F. (2000). The Anthropocene. *IGBP Newsletter*, 41(17), 17–18.

Cummins, T., & Watson, A. (2012). A hundred year policy experiment: The Murray-Darling Basin in Australia. In J. Quiggan, T. Mallawaarachchi, & S. Chambers (Eds.), *Water Policy Reform* (pp. 9–36). Cheltenham, UK: Edward Elgar Publishing.

Curti, G. H., & Johnson, T. M. (2017). Chapter 12: Syncretic (S)p[l]aces. In P. C. Adams, J. Crane, & J. Dittmer (Eds.), *The Ashgate Research Companion to Media Geography*. Farnham, England: Ashgate Publishing.

Curtis, A., Ross, H., Marshall, G. R., Baldwin, C., Cavaye, J., Freeman, C., Carr, A., & Syme, G. J. (2014). The great experiment with devolved NRM governance: Lessons from community engagement in Australia and New Zealand since the 1980s. *Australasian Journal of Environmental Management*, 21(2), 175–199.

Davidson-Hunt, I. (2006). Adaptive learning networks: Developing resource management knowledge through social learning forums. *Human Ecology*, 34(4), 593–614.

Delvaux, B., & Schoenaers, F. (2012). Knowledge, local actors and public action. *Policy and Society*, 31(2), 105–117.

Demeritt, D. (2005). Hybrid geographies, related ontologies and situated knowledges. *Antipode*, 37(4), 818–823.

Faysee, N., Errahj, M., Imache, A., Kemmoun, H., & Labbaci, T. (2014). Paving the way for social learning when governance is weak: Supporting dialogue between stakeholders to face a groundwater crisis in Morocco. *Society and Natural Resources*, 27(3), 249–264.

Flood, R. L., & Romm, N. R. A. (1996). *Diversity Management: Triple Loop Learning.* Chichester, UK: Wiley.

Folke, C., Carpenter, S. R., Walker, B., Scheffer, M., Chapin, T., & Rockström, J. (2010). Resilience thinking: Integrating resilience, adaptability and transformability. *Ecology and Society*, 15(4), 20. Online at http://www.ecologyandsociety.org/vol15/iss4/art20/

Folke, C., Hahn, T., Olsson, P., & Norberg, J. (2005). Adaptive governance of social-ecological Systems. *Annual Review of Environment and Resources*, 30(1), 441–473.

Freire, P. (1970). *Pedagogy of the Oppressed.* New York: Continuum.

Gibbs, L. (2009). Just add water: Colonisation, water governance, and the Australian inland. *Environment and Planning*, 41(12), 2964–2983.

Gifford, R., Steffen, W., Finnigan, J. J., & fellow members of the National Committee for Earth System Science. (2010). *To Live Within Earth's Limits: An Australian Plan to Develop a Science of the Whole Earth System.* Canberra: Australian Academy of Science.

Gonzalez, C., Clemente, A., Nielsen, K. A., Branquinho, C., & Ferreira dos Santos, R. (2009). Human-nature relationship in Mediterranean streams: Integrating different types of knowledge to improve water management. *Ecology and Society*, 14(2), 35. Online at http://www.ecologyandsociety.org/vol14/iss2/art35/

Goss, K. F. (2003). Environmental flows, river salinity and biodiversity conservation: Managing trade-offs in the Murray Darling Basin. *Australian Journal of Botany*, 51(6), 619–625.

Graham, N. (2011). *Lawscape: Property, Environment, Law.* London: Routledge.

Greenhill, J., King, D., Lane, A., & MacDougal, C. (2009). Understanding resilience in South Australian farming families. *Rural Society*, 19(4), 318–325.

Gregory, D. (2001). (Post)colonialism and the construction of Nature. In N. Castree, & B. Braun (Eds.), *Social Nature: Theory, Practice and Politics* (pp. 84–111). Malden, MA: Blackwell.

Gupta, J., Pahl-Wostl, C., & Zondervan, R. (2013). 'Glocal' water governance: A multi-level challenge in the Anthropocene. *Current Opinion in Environmental Sustainability*, 5(6), 573–580.

Hamilton, C. (2010). *Requiem for a Species.* Sydney: Allen & Unwin.

Haraway, D. (2008). *When Species Meet.* Minneapolis, US: University of Minnesota Press.

Havemann, P. (2005). Denial, modernity and exclusion: Indigenous placelessness in Australia. *Macquarie Law Journal*, 5, 57.

Hidalgo, J. P., Boelens, R., & Vos, J. (2017). De-colonizing water. Dispossession, water insecurity, and Indigenous claims for resources, authority and territory. *Water History*, 9(1), 67–85. doi:10.1007/s12685-016-0186-6

Holling, C. S. (1978). *Adaptive Environmental Assessment and Management.* New York: Wiley.

Horne, J. (2014). The 2012 Murray-Darling Basin Plan – Issues to watch. *International Journal of Water Resources Development*, 30(1), 152–163.

House Standing Committee on Regional Australia. (2011). *Of Drought and Flooding Rains.* Canberra: Australian Government.

Howitt, R., & Suchet-Pearson, S. (2006). Rethinking the building blocks: Ontological pluralism and the idea of 'management'. *Geografiska Annaler B*, 88(3), 323–335.

Howlett, M. (2012). The lessons of failure: Learning and blame avoidance in public policy-making. *International Political Science Review*, 33(5), 539–555.

Huntjens, P., Pahl-Wostl, C., Rihoux, B., Schlueter, M., Flachner, Z., Neto, S., Koskova, R., Dickens, C., & Nabide Kiti, I. (2011). Adaptive water management in a changing climate: A formal comparative analysis of eight water management regimes in Europe, Africa, and Asia. *Environmental Policy and Governance*, 21(3), 145–163.

Irvin, R. A., & Stansbury, J. (2004). Is citizen participation worth the effort? *Public Administration Review*, 64(1), 55–65.

Jackson, S. (2005). Indigenous values and water resource management: A case study from the Northern Territory. *Australasian Journal of Environmental Management*, 12(3), 136–146.

Jackson, S. (2017). Enduring and persistent injustices in water access in Australia. In A. Lukasiewicz, S. Dovers, L. Robun & J. McKay (Eds.), *Natural Resources and Environmental Justice: Australian Perspectives* (pp. 121–132). Melbourne: CSIRO Publishing.

Jackson, S., & Barber, M. (2013). Recognition of indigenous water values in Australia's Northern Territory: Current progress and ongoing challenges for social justice in water planning. *Journal of Planning Theory & Practice*, 14(4), 435–454.

Jackson, S., & Barber, M. (2016). Historical and contemporary waterscapes of North Australia: Indigenous attitudes to dams and water diversions. *Water History*, 8(4), 385–404.

Jasanoff, S. (2004). The idiom of co-production. In S. Jasanoff (Ed.), *States of Knowledge: The Co-Production of Science and Social Order* (pp. 1–12). London and New York: Routledge.

Jones, R., & Instone, L. (2016) Becoming-urban, becoming-forest: A historical geography of urban forest projects in Australia. *Geographical Research*, 54(4), 433–445.

Kahane A. (2004). *Solving Tough Problems: An Open Way of Talking, Listening, and Creating New Realities*. San Francisco: Berrett-Koehler.

Kallis, G. (2007). When is it coevolution? *Ecological Economics*, 62(1), 1–6.

Karpouzoglou, T., & Vij, S. (2017). Waterscape: A perspective for understanding the contested geography of water. *WIREs Water*, 4(3). doi:10.1002/wat2.1210.

Kortelainen, J. (1999). The river as an actor-network: The Finnish forest industry utilization of lake and river systems. *Geoforum*, 30(3), 235–247.

Latour, B. (1993). *We Have Never Been Modern*. London: Harvester Wheatsheaf.

Latour, B. (2014). Agency at the time of the Anthropocene. *New Literary History*, 45(1), 1–18.

Lawrence, G., Richards, C. A., & Cheshire, L. (2004) The environmental enigma: Why do producers professing stewardship continue to practice poor natural resource management? *Journal of Environmental Policy & Planning*, 6(3–4), 251–270.

Linton, J. (2010). *What Is Water? The History of a Modern Abstraction*. UBC Press: Vancouver and Toronto.

Linton, J. (2014). Modern water and its discontents: A history of hydrosocial renewal. *WIREs Water*, 1, 111–120.

Linton, J., & Budds, J. (2014). The hydrosocial cycle: Defining and mobilising a relational-dialectic approach to water. *Geoforum*, 57, 170–180.

Lövbrand, E., Stripple, J., & Wiman, B. (2009). Earth System governmentality: Reflections on science in the Anthropocene. *Global Environmental Change*, 19(1), 7–13.

Lucashenko, M. (2005). Not quite white in the head. *Griffith Review*, Edition 2, Dreams of Land. Online at https://griffithreview.com/articles/notquite-white-in-the-head/

Macdonald, D. H., & Young, M. (2001). *A Case Study of the Murray Darling Basin: Final Report Prepared for the International Water Management Institute*, August 2000 Revised February 2001. Canberra, Australia: CSIRO Land and Water.

Marshall, G. R., & Alexandra, J. (2016). Institutional path dependence and environmental water recovery in Australia's Murray-Darling Basin. *Water Alternatives*, 9(3), 679–703.

Marshall, G. R., Connell, D., & Taylor, B. M. (2013). Australia's Murray-Darling Basin: A century of polycentric experiments in cross-border integration of water resources management. *International Journal of Water Governance*, 1, 197–218.

May, P. J. (1992). Policy learning and failure. *Journal of Public Policy*, 12(4), 331–354.

McLean, J. (2014). Still colonising the Ord River, northern Australia: A postcolonial geography of the spaces between Indigenous people's and settlers' interests. *Geographical Journal*, 180(3), 198–210.

McNamara, K. E., & Westoby, R. (2011). Local knowledge and climate change adaptation on Erub Island in the Torres Strait. *Local Environment: The International Journal of Justice and Sustainability*, 16(9), 887–901.

Merchant, C. (1980). *The Death of Nature: Women, Ecology, and the Scientific Revolution*. New York: HarperCollins.

Merchant, C. (1989). *Ecological Revolutions: Nature, Gender, and Science in New England*. Chapel Hill: University of North Carolina Press.

Mezirow, J. (1990). How critical reflection triggers transformative learning. In J. Mezirow, & Associates (Eds.), *Fostering Critical Reflection in Adulthood: A Guide to Transformative and Emancipatory Learning* (pp. 1–20). San Francisco: Jossey-Bass Publisher.

Mezirow, J. (1996). Contemporary paradigms of learning. *Adult Education Quarterly*, 46(3), 158–172.

Milly, P. C. D., Betancourt, J., Falkenmark, M., Hirsch, R. M., Kundzewicz, Z. W., Lettenmaier, D. P., & Stouffer, R. J. (2008). Stationarity is dead: Whither water management. *Science*, 319(5863), 573–574.

Moore, M-L., von der Portern, S., Plummer, R., Brandes, O., & Baird, J. (2014). Water policy reform and innovation: A systematic review. *Environmental Science and Policy*, 38, 263–271.

Murray, B., Zeppel, M., Hose, G., & Eamus, D. (2003). Groundwater dependant ecosystems in Australia: It's more than just water for rivers. *Environmental Management & Restoration*, 4(2), 110–113.

Namoi Catchment Management Authority (CMA). (2013). *Namoi Catchment Action Plan 2010–2020* (2013 update). Namoi Catchment Management Authority. Sydney, Australia: NSW Government.

Nelson, R., Howden, M., & Smith, M. S. (2008). Using adaptive governance to rethink the way science supports Australian drought policy. *Environmental Science & Policy*, 11(7), 588–601.

New Zealand (NZ) Productivity Commission. (2012). *Local Government Regulatory Performance*. Online at http://www.productivity.govt.nz/sites/default/files/FINAL%20 Local%20government%20issues%20paper.pdf.

Norgaard, R. B., & Kallis, G. (2011). Coevolutionary contradictions: Prospects for a research programme on social and environmental change. *Geografiska Annaler B*, 93(4), 289–300.

North West Local Land Service (NWLLS). (2016). *North West Local Strategic Plan 2016–2021*. Sydney, Australia: NSW Government.

Natural Resource Commission of NSW (NRC). (2013). *Review of Water Sharing Plans*. Sydney: NSW Government.

Natural Resource Commission of NSW (NRC). (2016). *Review of Water Sharing Plans due to expire in 2017 or 2018*. Sydney: NSW Government.

Pahl-Wostl, C. (2009). A conceptual framework for analysing adaptive capacity and multi-level learning processes in resource governance regimes. *Global Environmental Change*, 19(3), 354–365.

Palmié, S. (2006). Creolization and its discontents. *Annual Review of Anthropology*, 35, 433–456.

Quiggin, J. (2012). Why the guide to the proposed basin plan failed, and what can be done to fix it. In J. Quiggin, T. Mallawaarachchi, & S. Chambers (Eds.), *Water Policy Reform* (pp. 49–60). Cheltenham, UK: Edward Elgar Publishing.

Robertson, M., Nichols, P., Horwitz, P., Bradby, K., & MacKintosh, D. (2000). Environmental narratives and the need for multiple perspectives to restore degraded landscapes in Australia. *Ecosystem Health*, 6(2), 119–133.

Sanderson, I. (2002). Evaluation, policy learning and evidence based policy. *Public Administration*, 80(1), 1–22.

Schmidt, J. J. (2014). Historicising the hydrosocial cycle. *Water Alternatives*, 7(1), 220–234.

Scott, J. C. (1977a). Protest and profanation: Agrarian revolt and the little tradition, part I. *Theory and Society*, 4(1), 1–38.

Scott, J. C. (1977b). Protest and profanation: Agrarian revolt and the little tradition, part II. *Theory and Society*, 4(2), 211–246.

Scott, J. C. (1998). *Seeing Like a State*. New Haven, US: Yale University Press.

Scott, J. C. (2013). *Decoding Subaltern Politics: Ideology, Disguise, and Resistance in Agrarian Politics*. London and New York: Routledge.

Senate Select Committee (SSC). (2016). *Refreshing the Plan*, The Senate Select Committee on the Murray-Darling Basin Plan, Commonwealth Government. Canberra: Australian Government.

Shaw, R., & Stewart, C. (1994). Introduction: Problematizing syncretism. In C. Stewart, & R. Shaw (Eds.), *Syncretism/anti-syncretism: The Politics of Religious Synthesis* (pp. 1–26). London and New York: Routledge.

Smith, G., & Wales, C. (2000). Citizens' juries and deliberative democracy. *Political Studies*, 48(1), 51–65.

Steffen, W., Richardson, K., Rockström, J., Cornell, S. E., Fetzer, I., Bennett, E. M., Biggs, R., Carpenter, S. R., de Vries, W., de Wit, C. A., Folke, C., Gerten, D., Heinke, J., Mace, G. M., Persson, L. M., Ramanathan, V., Reyers, J., & Sörlin, S. (2015). Sustainability. Planetary boundaries: Guiding human development on a changing planet. *Science*, 347(6223), 1259855.

Stember, M. (1991). Presidential address: Advancing the social sciences through the interdisciplinary enterprise. *The Social Science Journal*, 28(1), 1–14.

Sterling, S. (2011). Transformative learning and sustainability: Sketching the conceptual ground. *Learning and Teaching in Higher Education*, 5, 17–33.

Stewart, C. (1999). Syncretism and its synonyms: Reflections on cultural mixture. *Diacritics*, 29(3), 40–62.

Stewart, C. (2007). *Creolisation: History, Ethnography, Theory*. London and New York: Routledge.

Stewart, T. K., & Ernst, C. W. (2003). Syncretism. In M. A. Mills, P. J. Claus, & S. Diamond (Eds.), *South Asian Folklore: An Encyclopedia*. London and New York: Routledge.

Strang, V. (1997). *Uncommon Ground: Cultural Landscapes and Environmental Values*. Oxford: Berg.

Strang, V. (2009). *Gardening the World: Agency, Identity and the Ownership of Water*. New York: Berghahn Books.

Swyngedouw, E. (1996). The city as a hybrid: On nature, society and cyborg urbanization. *Capitalism, Nature, Socialism*, 7(2), 65–80.

Swyngedouw, E., Kaika M., & Castro, E. (2002). Urban water: A political-ecology perspective. *Built Environment*, 28(2), 124–137.

Tàbara, J. D., & Chabay, I. (2013). Coupling human information and knowledge systems with social-ecological systems change: Reframing research, education, and policy for sustainability. *Environmental Science and Policy*, 28, 71–81.

Tàbara, J. D., & Pahl-Wostl, C. (2007). Sustainability learning in natural resource use and management. *Ecology and Society*, 12(2), 3. Online at http://www.ecologyandsociety.org/viewissue.php?sf=28

Tadaki, M., & Sinner, J. (2014). Measure, model, optimise: Understanding reductionist concepts of value in freshwater governance. *Geoforum*, 51,140–151.

Tan, P. (2010). Learning from water law reform in Australia. In P. Cullet, A. Gowlland-Gualtieri, R. Madhav, & U. Ramanathan (Eds.), *Water Governance in Motion – Towards Socially and Environmentally Sustainable Water Laws* (pp. 447–476). New Delhi: Cambridge University Press.

Tarlock, A. D. (2000). Putting rivers back in the landscape: The revival of watershed management in the United States. *Hastings West-Northwest Journal of Environmental Law and Policy*, 6, 167–195.

Taylor, E. (2008). Transformative learning theory. *New Directions for Adult and Continuing Education*, 119, 5–15.

Taylor, B., & De Löe, R. C. (2012). Conceptualizations of local knowledge in collaborative environmental governance. *Geoforum*, 43(6), 1207–1217.

Tipa, G., & Nelson, K. (2017). Eco-cultural restoration across multiple scales: A New Zealand case study. *Water History*, 9(1), 87–106. doi:10.1007/s12685-016-0175-9

Tomaney, J. (2013). Parochialism – A defence. *Progress in Human Geography*, 37(5), 658–672.

Tuck, E., & McKenzie, M. (2015). Relational validity and the 'where' of inquiry. *Qualitative Inquiry*, 21(7), 633–638.

Vanclay, F. (2004). Social principles for agricultural extension to assist in the promotion of natural resource management. *Australian Journal of Experimental Agriculture*, 44(3), 213–222.

Vella, K., Sipe, N., Dale, A., & Taylor, B. (2015). Not learning from the past: Adaptive governance challenges for Australian natural resource management. *Geographical Research*, 53(4), 379–392.

Vizenor, G. (2008). *Survivance: Narratives of Native Presence*. Lincoln, US: University of Nebraska Press.

Waddell, S. (2002). Six societal learning concepts for a new era of engagement. *REFLECTIONS*, 3(4), 18–26.

Wallace, J. S., Acreman, M. C., & Sullivan, C. A. (2003). The sharing of water between society and ecosystems: From conflict to catchment-based co-management. *Philosophical Transactions of the Royal Society, London B*, 358(1440), 2011–2026.

Watts, V. (2013). Indigenous place-thought and agency amongst humans and non-humans (First Woman and Sky Woman go on a European world tour!). *Decolonization: Indigeneity, Education & Society*, 2(1), 20–34.

Weir, J. (2009). *Murray River Country: An Ecological Dialogue with Traditional Owners*. Canberra: Aboriginal Studies Press.

Weisz, H., & Clark, E. (2011). Society-nature coevolution: Interdisciplinary concept for sustainability. *Geografiska Annaler B*, 93(4), 281–287.

Whatmore, S. (2002). *Hybrid Geographies*. London: Sage.

White, D. (2006). A political sociology of socionatures: Revisionist manoeuvres in environmental sociology. *Environmental Politics*, 15(1), 59–77.

Wilcock, D., Brierley, G., & Howitt, R. (2013). Ethnogeomorphology. *Progress in Physical Geography*, 37(5), 573–600.

Wilkinson, C. F. (1992). *Crossing the Next Meridian: Land, Water, and the Future of the West*. Washington, D.C.: Island Press.

Williams, B. K. (2011). Adaptive management of natural resources and framework and issues. *Journal of Environmental Management*, 92(5), 1346–1353.

Wilner, K. B., Wiber, M., Charles, A., Kearney, J., Landry, M., Wilson, L., & on behalf of the Coastal CURA Team. (2012). Transformative learning for better resource management: The role of critical reflection. *Journal of Environmental Planning and Management*, 55(10), 1331–1347.

Wilson, N. J. (2014). Indigenous water governance: Insights from the hydrosocial relations of the Koyukon Athabascan village of Ruby, Alaska. *Geoforum*, 57, 1–11.

Woodhill, H., & Röling, N. G. (1998). The second wing of the eagle: The human dimension in learning our way to more sustainable futures. In N. G. Röling, & M. A. E. Wagemakers (Eds.), *Facilitating Sustainable Agriculture: Participatory Learning and Adaptive Management in Times of Environmental Uncertainty* (pp. 46–71). Cambridge: Cambridge University Press.

Woodward, E., & McTaggart, P. M. (2015). Transforming cross-cultural water research through trust, participation and place. *Geographical Research*, 54(2), 129–142.

Coda

Jacqueline Williams, Robyn Bartel, Stephen Harris and Louise Noble

The contributions from each of the authors in this volume mark part of a long collaborative journey that began in 2008 when the Water Resources and Innovation Network (WRaIN) was established. Involving colleagues with a passion for innovations in water policy from many disciplines across the University of New England, Australia, WRaIN was interested in developing alternative approaches for water governance based on an interdisciplinary foundation. Increasingly researchers are being motivated to work together to 'solve the grand challenges facing society – energy, water, climate, food, health' (Nature, 2015, p. 305). However, industry imperatives, including those of the higher education sector, as well as the expectations of government, and even non-government institutions, are often barriers to undertaking interdisciplinary research. Siloed approaches are still often the norm, which can make the journey frustratingly difficult at times (see for example Nature, 2016).

Interdisciplinary approaches require a foundation of collegiality and friendship, and it is essential that everyone feels safe to explore differences. Even apparently minor issues such as writing styles, referencing and expression can become points of contention. And there are bigger questions also, as WRaIN member and author of Chapter 3, Adrian Walsh (pers. comm.), observes:

> Methodological challenges… were part and parcel of working within this group, given the remarkable diversity of disciplines represented. … water scientists, economists, geographers, ecologists and literary theorists (to name but some of the disciplines represented) [meant that] much time was spent searching for a common language for discussing issues. There was a genuine desire on the part of all involved to learn from each other despite the different methodological assumptions.

Due to these generative experiences, the collaborative approach represented here is, in our view, the most appropriate means of effecting change in water policy – as it reflects both the intrinsically complex relationship humans have with water, and the political and normative complications attending the challenge of achieving improvements. In this respect, *Water Policy, Imagination and Innovation: Interdisciplinary Approaches* extends on earlier studies that have proceeded from the same diverse methodological premise, and with a similar view to addressing a comparable set of concerns. Of many important examples on which we might draw, *Water: History, Cultures, Ecologies* (2006) is notable in the Australian context. The editors, Marnie Leybourne and Andrea Gaynor, present an impressive selection of essays that entertain diverse views on the meaning and significance of water, from both national and international perspectives. In doing so, they invite the corroborative response offered in the present volume when acknowledging that 'there is clearly room for further ecological conversations and collaborative, integrated research to take place [in which] a series of epistemological and ethical issues will undoubtedly arise and require rigorous attention' (xviii) – precisely the concerns addressed in this volume. In doing so, *Water Policy, Imagination and Innovation: Interdisciplinary Approaches* contributes directly to the political project of inter-disciplinarity – to what Krueger et al. (2016) refer to as the need 'to challenge both certified and noncertified knowledge productively' in the conviction that it is imperative to examine and understand 'how water research itself embeds and is embedded in social context and performs political work' (p. 369). We hope that our concerted interdisciplinary efforts toward enhancing imagination and innovation in water policy in this volume further contributes to collective endeavours to address the global water issues of the Anthropocene.

References

Krueger, T., Maynard, C., Carr, G., Bruns, A., Mueller, E. N., & Lane, S. (2016). A transdisciplinary account of water research. *WIREs Water*, 3, 369–389.

Leybourne, M., & Gaynor, A. (2006). *Water: History, Cultures, Ecologies*. University of Western Australia Press.

Nature (2015). Interdisciplinarity: Scientists must work together to save the world. A special issue asks how they can scale disciplinary walls. *Nature*, 525, 305.

Nature (2016). Editorial: The big picture: Interdisciplinary research is vital if we are to meet the diverse needs of modern society. *Nature*, 534, 589–590.

Index

Printed and bound by CPI Group (UK) Ltd, Croydon, CR0 4YY

23/10/2024

01778242-0013